Appetite and Its Discontents

Appetite and Its Discontents

Science, Medicine, and the Urge to Eat, 1750–1950

ELIZABETH A. WILLIAMS

The University of Chicago Press   Chicago and London

The University of Chicago Press, Chicago 60637
The University of Chicago Press, Ltd., London

Published 2020
Printed in the United States of America

29 28 27 26 25 24 23 22 21 20   1 2 3 4 5

ISBN-13: 978-0-226-69299-9 (cloth)
ISBN-13: 978-0-226-69304-0 (paper)
ISBN-13: 978-0-226-69318-7 (e-book)
DOI: https://doi.org/10.7208/chicago/9780226693187.001.0001

Published with the support of the Susan E. Abrams Fund

Library of Congress Cataloging-in-Publication Data

Names: Williams, Elizabeth A. (Elizabeth Ann), 1950– author.
Title: Appetite and its discontents: science, medicine, and the urge to eat, 1750–1950 / Elizabeth A. Williams.
Description: Chicago ; London : University of Chicago Press, 2020. | Includes bibliographical references and index.
Identifiers: LCCN 2019035093 | ISBN 9780226692999 (cloth) | ISBN 9780226693040 (paperback) | ISBN 9780226693187 (ebook)
Subjects: LCSH: Appetite—Research—History. | Appetite—Research—History—19th century. | Appetite—Research—History—18th century. | Appetite—Research—History—20th century. | Appetite disorders—Research—History. | Science—History. | Medicine—History.
Classification: LCC QP136.W55 2020 | DDC 612.3—dc23
LC record available at https://lccn.loc.gov/2019035093

♾ This paper meets the requirements of ANSI/NISO Z39.48-1992 (Permanence of Paper).

To Robert, for his enduring love and the joys of our shared life

Contents

Illustrations

Introduction

"We all have an eating disorder now"—not long ago I heard this remark by a young man at the gym that I frequent. At the time he was training for a bodybuilding competition, and he talked to anyone who would listen about the strange and difficult diet his training entailed. His declaration was perhaps a bit self-aggrandizing, as it put him at the center of major cultural trends. Nonetheless, it struck me as containing an element of truth. Certainly, there can be little doubt that anxieties over eating—how to choose foods, how to eat to be healthy or thin or attractive, how to cope with the conditions now termed "eating disorders"—are widespread in our culture, as are claims by diverse authorities that such problems are rapidly on the rise. Why and how eating has become problematic for more and more people are questions that historians have come to only recently. This new historical literature has focused on economic, social, and cultural issues linked to food and eating, but so far little attention has been given to the problem of appetite—the changing ways that the appetite for food is formed or how the views of scientific and medical experts on the subject have developed through time.[1] My own study rests on the paired assumptions that our current eating predicament involves troubles of appetite and that, since the eighteenth century, thinking about appetite has been shaped in important ways by scientists and physicians who have striven to establish authority over its development and exercise.

This book traces, then, the history of scientific and medical inquiry into the nature and functioning of the appetite for food in the two centuries between 1750 and 1950, from the mid-Enlightenment to the years just after the Second World War and the dawn of "big science."[2] In geographical and linguistic coverage the book is simultaneously expansive and selective, following major lines of argument within the dominant traditions of modern science and medicine. Before 1900 I focus chiefly on work undertaken in France, Germany, and Britain (with some Italian and Russian entries as well). After 1900, in accord with the rise of the vast scientific and medical establishment of the United States and the increasingly powerful sway of American culture in general, I shift emphasis to American investigators while also exploring transatlantic interactions.

In focusing this book chiefly on the story of appetite as written by scientists and physicians, I move among different levels and types of discourse. Disputes about appetite were sometimes largely semantic, and to get at such verbal sparring, I make use of a wealth of definitions offered in dictionaries, encyclopedias, and textbooks. Other developments entailed more elaborate conceptualization and theory building, and to explore these I rely heavily on formal writings—essays, treatises, journal articles—produced by such broadly influential figures as Georges-Louis Leclerc de Buffon, Erasmus Darwin, François Magendie, Carl Ludwig, Ivan Pavlov, Sigmund Freud, and Walter B. Cannon. Where appetite appears as part of highly elaborated systems or theories proposed by such investigators, I have attempted to explicate the larger contours of their thinking before turning to the place of appetite within it.

The book is divided into four parts covering developments in the four half-century periods between 1750 and 1950. Although I have used these divisions for convenience, they also involve major shifts in the problematics and methodologies dominant in the study of appetite. The second half of the eighteenth century (explored in part 1) saw enhanced interest in a medical dietetics informed by ancient doctrines, especially Hippocratic ones, that upheld the individuality of appetite and its dependability as a guide to eating. Yet it also witnessed the first sustained efforts to challenge these precepts by investigators who, often using experimental methods, sought to identify uniformities in ingestive and digestive processes. The decades between 1800 and 1850 (part 2) confirmed the dominance of the experimental approach to bodily functions but also encompassed redoubled efforts by physicians, especially in emergent "mental medicine," to document and comprehend extremes of individual appetite. Between 1850 and 1900 (part 3), experimental physiology, which rested increasingly on the use of precision instruments and forms of measurement, sought a clear somatic site of or cause for the urge to eat while physicians seeking the source of ills of appetite separated along a "cerebralist-visceralist" divide. This era also registered the impact of Darwinism, which, among many powerful effects, encouraged the emergence within the previously mind-centered discipline of psychology of new subfields devoted to "physiological" and "animal" psychology, arenas in which debate centered on the governance of eating by "instinct" or by "intelligence." Finally, the era between 1900 and 1950 (part 4) saw the rise to dominance of "scientific medicine," whose practitioners searched, often in alliance with physiologists, for effective somatic remedies for troubled appetite and eating. At the same time, the increasingly influential fields of psychology, psychiatry, and ethology offered psychodynamic and behavioral approaches to appetite that challenged or complicated a straightforward somaticism. By 1950 the study of appetite continued to be marked by persistent dichotomies—somatic-psychic, innate-acquired, objective-subjective; the few "holistic" interpretations attempted did not prevail. In the epilogue I briefly survey developments after 1950, suggesting how, in the era of "big science," research on appetite dispersed into a complicated range of specialties and subspecialties that largely spoke their own language, employed their own methodologies, and disregarded developments beyond their own borders. By around 1980, assorted efforts were launched—a new academic journal, a book series sponsored by a major European publisher, international conferences—to "integrate" research on appetite, but even the organizers of these projects lamented the apparently unstoppable fragmentation of research. All the while, despite persistent uncertainties surrounding appetite, technical and utilitarian instruments for its control ceaselessly proliferated.

My goals in this work are several. I seek, on the one hand, to illuminate how, in the two centuries I cover, scientists and physicians viewed the ontology and epistemology of appetite—what kind of phenomenon it is and by which methods it can be known in both its "normal" and its "pathological" functioning.[3] Second, I explore what I see as the long-term thrust of biomedical work on appetite, that of promoting its management by dietetic authorities, an endeavor that is one of two processes aimed at homogenizing appetite and eating in our time. One of these processes (in recent years explicated by food historians) is the work of the modern food industry, which has created a standardized food supply consisting in good part of scientifically designed "food products" meant to ensure profitability.[4] The other, which has drawn much less attention from historians, is the long-term movement within science and medicine to promote health and longevity by encouraging eaters to distrust their own appetite and to heed expert direction. The tension between these two homogenizing processes lies at the center, I argue, of what one observer has referred to as our era of "nutritional disarray."[5]

In the two centuries covered by this book, writers of many different sorts—from theologians and philosophers to essayists, poets, and satirists, along with scientists and physicians—commented on the appetite for food. Given that eating is both an inescapable necessity and, for most people, one of the great pleasures of human life, the fact that so many types of observers commented on appetite is unsurprising. No one book could cover this diverse history, and I do not try to do so. Even in respect to science and medicine my coverage is selective. I only glance, for example, at the writings of ethnologists and anthropologists, whose accounts of the eating choices of "savage" or far-flung peoples gradually replaced the reports of travelers and other untrained observers of earlier years. Instead, I emphasize what I see as the core disciplines in investigations of appetite and eating, including physiology, natural history, medicine, and, from the late nineteenth century, psychology and ethology. Already in the eighteenth century, when the learned world had few fixed borders, these domains of learning were coming to be differentiated, as physiologists took on the task of describing "normal" ingestion and digestion, natural historians explored differences and similarities between human and animal appetite, and physicians worked to define, classify, and treat ills marked by disturbed appetite. As I move into the nineteenth century, my focus on physiology and medicine continues while natural history drops out for a time, as its successor sciences focused chiefly on animal morphology and anatomy rather than function. In turning to the years after 1850, I take up the story of how psychology, long a discipline limited to mental phenomena, came to embrace psychophysiological problems such as appetite, first within "physiological" and "animal" psychology and later within the emergent field of ethology (in the United States more commonly called animal behavior studies). Similarly, I trace the role of appetite in developing specialties within general medicine. From the point in the early nineteenth century when "mental medicine" first emerged, and continuing into the twentieth, I explore how different forms of psychiatry (alienism, eclectic forms of psychotherapy, psychoanalysis, psychosomatic medicine) sought to take charge of troubled eating and to push to the sidelines somaticist specialties (neurology, gastroenterology, endocrinology) that simultaneously staked a claim to appetite.

As these remarks suggest, this study is organized in accordance with disciplinary boundaries. In respect to the study of appetite in the eighteenth century, this approach may seem artificial, as such boundaries were only beginning to emerge. And certainly shifts in the borders between disciplines and fields continued long thereafter. Nonetheless, maneuvering toward disciplinary distinctiveness—on methodological, conceptual, and institutional terrain—was, even in the eighteenth century, an important element of this history, because appetite was never an easily classifiable phenomenon. In physiology in particular the question of whether the emergent discipline should be limited to physical phenomena or should encompass what the French called "relations of the physical and the moral" was crucial to thinking about appetite.[6] I argue that it was precisely the late eighteenth- and early nineteenth-century transformation of physiology, its gradual shift away from an expansive physico-moral framework toward its constitution as an experimental science, that accounted for the uncertain status of appetite in that discipline for decades. Equally crucial to the history of appetite was the long-term hostility toward experimental, laboratory-based physiology that caused defenders of bedside medicine as late as 1918 to declare that "'laboratory training *unfits* a man for his work as a physician.'"[7] Among other issues, this contest between physiologists and physicians revealed deep fissures in attitudes toward the utility of knowledge gained from animals for understanding human beings in health and sickness. While all physiologists eventually affirmed that knowledge of animals was essential to scientific and medical progress, they were divided among themselves (as psychologists would be, in time, as well) over questions about how such knowledge should be obtained, which kinds of animals should be observed, and where lines (if any) should be drawn between the nature and experience of animals and humans. Given contestations of this sort, attention to disciplinary boundaries is not an option but an essential feature of the history of appetite.[8]

Some readers may find my limitation of this study to the science and medicine of appetite frustrating. It might seem that few topics more urgently call on the historian systematically to explore interconnections between science and medicine and larger political, social, and cultural imperatives. Having spent many years with this material, no one is more cognizant of the need to establish such connections than I am, and I have discussed at important moments what I see as key elements of the political and social settings in which thinking about appetite evolved. All the same, I do focus closely on scientific and medical developments, and here is why: my goal is to explain the historical process by which appetite—once a simple reality of daily life—became an "object" to be defined and, ultimately, managed by scientists and physicians.[9] How this happened is a long and complex story. It is also one that has scarcely begun to be written; on the theme of appetite historians have been largely silent.[10] Given the near absence of historical literature on changing forms and views of appetite, I believe the essential first step toward a comprehensive history must be to relate how appetite came to be an object of biological and medical science.

My focus is on science and medicine also because—as I argue throughout the book—the investigators who studied appetite largely did so within the confines of remarkably restricted disciplinary perspectives. The connectedness of investigators to broader currents, whether political, socioeconomic, or cultural, varied of course from one historical setting to another. I begin with the later Enlightenment, an era when natural philosophers and physicians were not separated from the broader social world and often eagerly framed their investigations in reference to such themes as the demise of "natural" appetite amid advancing culinary luxury. And this history ends with the work of experts in proliferating specialties who actively seek to intervene in the quotidian experience of eating. Nonetheless, through good stretches of this history the investigators whose work I explore self-consciously and purposely approached appetite as an object isolated from "real-world" contexts of food and eating. This means that the kinds of arguments they engaged in—over definitions, concepts, methods, varieties of evidence—developed largely in response to internal dynamics rather than external demands or pressures. Indeed, the restricted vision of scientific and medical researchers may be judged one of the paradoxes of the history of appetite if we agree with observers as diverse as Aristotle and the American poet Wendell Berry that eating is "the profoundest enactment of our connection with the world."[11] Although this may be the existential reality of appetite and eating, it is an essential fact in the history of appetite that those who claimed special knowledge of it mostly did so by observing the eating choices of animal and human "subjects" only in the preferred venues of hospitals, laboratories, and contrived outdoor environments. The implications of this sequestering of appetite were, and are, profound. I have written this book in part to argue that few human experiences demand more forcefully a break with inward-looking science and the pursuit of forms of knowledge in which the fateful divide between the expert and the ordinary human actor gives way.

In the meantime, the sequestering of appetite marks not only ongoing scientific and medical research but also the historiography of the subject. Indeed, the only histories that cover general developments in this area over any span of time are works written by researchers themselves, who focus chiefly on the successes and failures of particular methods and approaches.[12] I could not have undertaken this study without the guidance these investigations provide, both to the historical corpus of work on appetite and to the technical issues involved. Nonetheless, my goals as a historian obviously differ from those of scientists who probe the history of research problems that they see as unsolved or seek to affirm research traditions to which they themselves belong. Where such studies seek out transhistorically valid "findings," as a historian I explore knowledge claims and the contexts in which they emerged and sought to attain plausibility or authority. Where scientists turn to the past to study "advances" and "innovations," as a historian I am as interested in doubt, confusion, bias, misrepresentation, and failure as I am in successful steps toward the science of the present.[13] To give one example, I attend closely in this book to arguments over appetite cast in terms of "instinct," a term not much used by modern students of eating behavior. I do so not to show how the idea of instinctual eating was found wanting but to explore ways that thinking about instinct was tied to related assumptions and questions—about differences, for example, between animals and humans or problems of moral and social responsibility in eating behavior—that persist regardless of the shortcomings of the concept itself as gauged in modern science. Another way to think about this difference in approach is to ask what the implications are, for the science of appetite and eating in our own day, of setting aside concepts of instinct, which, historically, have been closely tied to ideas of "natural" eating. If an instinct for healthy eating is gone, what do we mean when we talk about natural foods and natural eating? Thus I take up arguments that are seemingly over not merely from historical interest but also from the conviction that language and concepts once historically salient are rarely definitively interred but remain part of a landscape of the past that always beckons. For good or ill, things once said are not easily unsaid, and it is only historical exploration that can reveal ways that speakers, in reverting to the past, often bring along implications and meanings that in our distinctly unhistorical age go largely unacknowledged.

Although the scientific and medical history of appetite remains largely unwritten, historians have, in recent years, illuminated many of the intellectual, social, and cultural developments that have invested the problem of appetite with new urgency. Indeed, what explains the outpouring of scholarly work on food and eating is, itself, a historical problem of serious interest.[14] My own path to appetite was not through the history of food but through the history of medicine. I came to the topic first in studying the history of vitalism, a framework for medicine and life science that historians long characterized as a kind of obscurantism, an intellectual stance that hindered the pursuit of positive knowledge.[15] From the early nineteenth century on, critics of vitalism decried its fidelity to what they cast as "metaphysical" principles: that is, that inherent differences divide living beings from the nonliving and that knowledge of life and the living must be gained in ways that differ essentially from the sciences of inert nature. Such critics sought to banish vitalist principles and procedures from life science and medicine, insisting on the universality of physicochemical constituents and processes in the living and nonliving. Yet vitalism remained an essential source of thinking about health and disease that deliberately emphasized questions of variability and individuality as opposed to regularities and normativities. In so doing, it sought to resolve hard distinctions between body and mind, somatic and psychic, material and immaterial, and thus laid special claim to medical domains in which the troubled interplay of mind, emotion, and body seemed to be manifest. In recent years interest in vitalism has revived, and historians have traced ways in which vitalist-inspired medicine has asked radically different questions from those posed by mainstream medicine, with its recourse to images of the "body-machine" and interest in parts over wholes.[16] Whereas the latter sought (and undeniably gained) certain advantages by eliding differences between the living and nonliving and seeking not psychosomatic interconnections but strictly physical causes of states of health and disease, vitalists, as Anne Harrington has written, sought "an idea of 'wholeness' that could . . . capture the deeper truth of every organism's functioning and striving."[17]

My awareness of appetite as a psychosomatic conundrum emerged as I investigated the place of medical vitalism in the history of psychiatry. Probing this history, I was struck by the fact that accounts of the role of vitalism in the emergence of psychiatry generally ignored the vitalist principle that psychic illness was often "seated" in the viscera and connected to troubles of appetite, eating, and digestion.[18] Raised, like others in my generation of intellectual and cultural historians, to take little interest in psychiatry that did not derive from Sigmund Freud, I first found amusing the idea that mental illness came from the stomach. Nonetheless, I wanted to understand the origin of such claims, and, because an obvious connecting point between psychiatry and the medicine of appetite and eating was the history of "eating disorders," I turned to that problem. In so doing, I came to see that appetite, while ordinarily a subject of lighthearted enthusiasm ("Bon appétit!"), was, for many, no laughing matter.[19] In some ways, finally, the puzzle that eventuated in this book was the realization that in modern psychiatry (as opposed to its nineteenth-century predecessors) many of those concerned with eating disorders came to agree that appetite—the desire or disposition to eat—was not at issue but, instead, that cultural pressures, psychosexual dynamics, and "the family romance" occupied center stage. How, I wondered, could troubled eating not be connected to the appetite for food? Among other things, this book is my effort to find the answer to that question.

Through the long history of appetite traced in this study certain themes and problems continually recur. The first of these is the question of the basic nature of appetite. Either implicitly or explicitly, all commentators on appetite conveyed the view that it was some kind of internal state or disposition, but in talking about it they struggled to establish a definite terminology. Aristotle defined appetite as a "faculty" or "power of the soul," as the "genus" of which desire, passion, and wish are the species. Others, less systematic, referred to appetite as a "sensation," a "desire," a "feeling," or a "longing," without specifying its source in body or soul. Still others attempted no substantive definition at all but simply referred to appetite as "that which" has a certain effect or is itself affected by some substance or influence, like the smoked herring and pickles noted in an eighteenth-century German dictionary.[20]

Efforts to define appetite have often entailed an attempt to differentiate it from hunger. This approach was already evident in Enlightenment commentaries, when observers laid out some of the differences between the two phenomena that would continue to be noted into our own time. Appetite was pleasurable; hunger, painful. Appetite was directed at specific foods; hunger, at food in general. Appetite involved desire; hunger, need. Yet others said that appetite and hunger were essentially the same, with appetite constituting the "first phase" of hunger, the pleasurable phase that was followed by discomfort and, finally, by pain. Like almost all the questions explored in this study, there is still no consensus on this matter. Some models of the appetite-hunger pair implicitly suggest a spectrum stretching from fleeting desire to implacable need, whereas others continue to posit strong differences between appetite and hunger. By now, the problem is usually situated operationally—appetite is like or unlike hunger depending on what we hope to do with it.[21]

Closely related to efforts to determine the relation between appetite and hunger is the problem of how and where appetite originates: is it, like hunger, a thing of the body, or does it entail activities of mind or soul or spirit? It might seem self-evident that the appetite for food is rooted in the universal material drive of living beings to obtain nourishment. This would connect appetite to the indubitably physical phenomena of ingestion and digestion. And yet through its history a good deal of commentary on appetite divorced it from the need for food, construing it instead as one of the basic impulses toward pleasure. In the premodern era, when appetite was chiefly the concern not of science and medicine but of theology and philosophy, this was indeed the dominant view of the "sensual appetites." Why, when science and medicine took over the problem of appetite from around 1750, observers largely set aside views of appetite as part of a general search for pleasure is itself a difficult question. The answer may lie in the fact that interrogation of pleasure was suspect and seemed necessarily to involve religious and moral judgments that unbiased inquiry into natural processes must avoid. Or it may simply be that pleasure (like appetite) was ineffable and that both science and medicine eschewed the ineffable in favor of the concrete. In any event, even though it was recognized that appetite had to do with the pleasure experienced in ingesting food, few efforts were made to understand the source or nature of this pleasure. The one exception was the effort to tie appetite to specific sensory qualities of foods—textures, flavors, and smells. But this move meant relegating appetite to the domains of "sensation" and "perception," and for a host of reasons, in the "history of the senses" those who broached these great problems focused their investigations on themes, especially those connected to vision, loftier than the humble sensory enjoyments provided by food.[22]

Still, in recognizing the specific pleasures of appetite, if not its relation with pleasure in general, commentators generally noted that while appetite had to do with materialities of food and nutrition, it was also clearly involved with mind or spirit. Although the language of appetite frequently targeted its shared corporal and psychic dimensions—"mind-body," "physical-moral," "psychosomatic"—such terms often simply masked tendencies to accent either its bodily or its psychic character in isolation. This was evident, for example, in the work of observers who emphasized variations of appetite in different human types. From antiquity onward, appetite was seen to differ not only in individuals—the chief concern of physicians—but also in accord with age, sex, level of "civilization," external environment, and other influences. Some of these influences were clearly physical in character: everyone agreed, for instance, that appetite demanded more and different kinds of foods in cold places than hot. But some were matters of temperament, spirit, or, in modern parlance, psychology. Of the typological differences accented through the history of appetite, none was of greater interest than the extreme variations of appetite said to be exhibited by girls and women, variations first established in ancient medicine in contrast to the norm of the temperate, self-controlled male. In Hippocratic medicine male-female disparities of appetite reflected humoral and anatomical differences, but when the modern history of appetite got under way in the eighteenth century, influential commentators on women's appetite held that it derived not from any single female feature but from the whole being of woman—body and mind, physical and moral.[23]

When such observers pointed to perceived differences in the appetite of women and men, moreover, they asked not only where and how these originated but also, often implicitly but sometimes explicitly, whether appetite could differ from the ideal or the ordinary without being pathological. Medicine from antiquity to the eighteenth century said yes. Appetite was individual; it developed and manifested in relation to myriad circumstances. It formed part of the infinitely diverse "nature" of individuals. To be sure, in looking at women and other human types whose appetite could run wild (savages, children), observers perceived patterns that tended toward or implied the presence of disease. But the idea that there was one, uniformly healthy appetite for all people beyond which disease was necessarily immanent—this was a claim that emerged only with modern life science and that took hold in scientific medicine only after long resistance by bedside physicians who extolled individual appetite as the surest guide to healthy eating.

Given the extraordinary break with traditional thinking that the notion of molding a uniform appetite represented, much discussion ensued of how this and related claims could be sustained. Hence the persistent preoccupation from the late eighteenth century onward with problems of method: how could appetite be known? As observers of all sorts noted, the most obvious way to learn something about appetite was to think about how it worked in oneself. This method of self-observation, sometimes supplemented by self-experimentation, was often used during the Enlightenment and into the early nineteenth century, yet by around 1850 its validity came under challenge.[24] Critics of introspection often emphasized the unreliability of self-recollection, the tricks that memory plays when we try to recall our experience accurately, but in respect to appetite and eating, there were other, more serious limitations to what could be known from pondering one's own experience. Virtually every nineteenth-century commentator on appetite was a well-educated male, the very prototype of the self-disciplined eater extolled since antiquity, and there seemed to exist an impassable gap between what such researchers could know by contemplating their own eating desires and what they sought to know about extremes or perversions of appetite. From the eighteenth into the early nineteenth century there was a vogue for confessional narratives by physicians who acknowledged their own anomalies of appetite, but after 1850 such accounts largely disappeared as a line was drawn between the educated and self-contained observer who had no experience of, or at any rate was disinclined to reveal, perversity of appetite.[25]

Even to physicians who sought insight from self-observation, the one method that was of unquestioned utility in studying appetite was bedside observation, which, since Hippocrates, had constituted the royal road to knowledge of the individual experience of health and sickness. Yet physicians argued extensively over how clinical observation should proceed and how its results should be interpreted. During the Enlightenment, for example, there was much dispute about the value and meaning of readings of the pulse. One might expect, amid these arguments, that appetite, given its tricky ontological status, would have evoked much dispute. But in fact it seems that appetite was a phenomenon whose value as a clinical sign was so ingrained in medical tradition that physicians did not question the language they used to describe it or to indicate how they knew the state of a patient's appetite. Instead, they employed an extensive but conventional vocabulary—"present" or "lacking," "strong" or "weak," "voracious" or "feeble"—and the matter of how they made such judgments did not come to the surface. One may presume that physicians judged the appetite largely on the basis of what patients told them, the subjective patient "narrative" whose significance drastically declined with the coming of the objective diagnostic measures of modern medicine.[26] Already in the eighteenth century, such narratives and other features of clinical observation were coming under challenge, and the search for more exacting investigative methods got under way, including those entailing purposely designed experimentation. By the early nineteenth century physiologists pursued two kinds of experiments on appetite. One involved cutting the vagus nerves that, since antiquity, had been thought to transmit the desire or urge to eat. The other entailed peering through openings into the digestive organs (created artificially or by wounds) to observe movements of the stomach that could then be compared to stated reports of appetite in humans or to apparent signs of the desire to eat in animals. Then, after 1850, experimental approaches tailored to both laboratory and "field" settings proliferated beyond physiology as physiological and animal psychologists turned their attention to problems of ingestion and digestion. By the early twentieth century virtuosity in experimental design had become essential to the standing of researchers throughout the sciences that claimed authority over appetite.

As different forms of experimentation took hold, technical issues specific to individual disciplines often dominated arguments among researchers. Historians of psychology relate that methodology became an idol in the early years when that discipline sought independence and legitimacy, but methodological contests also absorbed great time and energy in physiology and medicine, as when researchers in the 1920s and 1930s argued over the best route of entry to sites in the hypothalamus believed responsible for "regulating appetite."[27] Although appetite was a subject that escaped disciplinary boundaries, methodological arguments in separate disciplines tended to crowd out attention to how competing disciplinary approaches might be reconciled with one another. The one long-term contest that unfolded at a metadisciplinary level and concerned not procedural particulars but broad differences of perspective was that between experimentalists, who strove to establish closely controlled conditions of study, and clinicians, who held that appetite could be known only by observing its vagaries in individual patients. For some time historians of this experimental-clinical rivalry emphasized hostility between the two sides, but recently some have argued for an "integrationist" position, a meeting of minds either in specific investigative contexts or as a prevailing long-term trend. And certainly as scientific medicine took hold from the late nineteenth century onward, the old gap between experimental physiology and clinical medicine closed partially when at least some physicians came to be trained in methods current in the sciences and "clinical research" became a new reality. Yet in important instances the old division simply took new shape, as when medical holists of the 1920s and 1930s decried the limited vision of a medicine focused not on patients but on dysfunctional parts.[28] Appetite was an important topic for some of these physicians and, thanks partly to their efforts, in this sphere rivalry or mutual incomprehension persisted between laboratory and clinic. More significant is the fact that what had long been a two-sided conflict (between medicine and physiology) became increasingly complex as experimentation took hold in psychology, nutrition, and other fields. Thus, the methodological contests that unfolded up to 1950 set the stage for the multitudinous approaches to appetite of our own time—psychiatric, neurological, dietetic, endocrinological, gastroenterological, pharmaceutical—and thus for the persistent failure to achieve anything like a unified approach to appetite and its troubles.

For a time the close disciplinary focus of researchers on appetite seemed to give way to broader methodological and epistemological concerns when behaviorism emerged as the dominant framework of American psychology after 1900. Even in its early days behaviorism encompassed many varieties, but its overall import was to set a program for psychology in which investigation was limited to externally observable movements and activities while "mind" or "consciousness," indeed all "inner states" of observed organisms—including appetite—were set aside as objects of interest.[29] Subsequently, behaviorists devoted great effort to devising methods for observing stimulus-response patterns that could be quantified in a fashion comparable to data obtained in the "hard" sciences. This methodological triumph of the early twentieth century became a conceptual triumph, not just in psychology where it originated but also in medicine, physiology, and the broad span of the behavioral sciences. In the process appetite was undercut as a viable object of science and in its place new objects—"ingestive activity," "food intake," "mechanisms regulating ingestion"—took center stage.[30] As a result appetite, always an unstable object of science, disappeared for a time or enjoyed mention only when confined within demurring quotation marks.[31] In later behaviorist (or neo-behaviorist) formulations, even those who explicitly denied a role to inner states gauged the "strength" of stimuli in relation to a range of "motives" that included inherent physiological "drives" and, once again, "appetites."[32] Yet the subsequent history of the disciplines affected by behaviorist approaches reveals once again the force of specific disciplinary habits. In psychology itself it proved difficult to attempt an account of eating that dispensed with internal states of organisms, and "drives" and "appetites" made a (sometimes surreptitious) return.[33] Yet in physiology and a range of evolving medical specialties, such fine points of psychology went largely unremarked, and behaviorist dicta continued to supply a welcome rationale for limiting inquiry to readily observable and quantifiable activities, such as "food intake." Little wonder that when researchers from diverse disciplines gathered in the 1950s to assess the dominant investigative framework of the moment—the regulation of ingestion by assorted physiological mechanisms—some speakers ruefully noted that no coherent concept of appetite (or hunger) was agreed on even in closely related fields.[34] The psychologically astute among them knew that a strict behaviorist view that banished drives, appetites, and other inner states no longer commanded broad allegiance within psychology, whereas others, uninterested in shifting psychological paradigms, continued laying claim to behaviorist language in describing their own narrowly somaticist procedures. Thus, despite its seeming moment of high transdisciplinary impact, behaviorism did not supply the sought-after road to integration, and as I explore in the epilogue, efforts to weave together the complex strands of appetite perforce went on.

In the now-unrelenting battles over what and how to eat and what or who to blame for our predicaments of appetite and eating, two paired processes of homogenization are becoming apparent. One of them, the standardized eating fostered by the advent of industrial food and the elaboration of ever more complex profit-driven stratagems of the food industry, has by now been explicated in some detail by historians. The other, the drive toward homogenization of appetite and eating by and within science and medicine, requires equal attention. This second process, a drive toward uniformity justified by apparently unchallengeable values of health and longevity, has supplied the central dynamic in investigations of appetite for two and a half centuries. How and to what extent this facet of "nutritionism" has contributed to the discontents of appetite of our time can be illuminated only by historical investigation.[35] This study of how appetite developed as an object of scientific and medical study is my effort at a beginning. What is clear is that we are caught between two competing forces, one driving us to consume foods designed, with the help of sophisticated scientific procedures, to titillate the appetite, and, the other, to ignore the yearnings of our own appetite and to accept the discipline imposed by normative standards of "healthy eating." The extent to which we are bombarded by these conflicting directives is at times almost laughable. No sooner do we read, again, about the four or eight or ten rules of healthy eating, than we encounter images of "appetizing" concoctions to whose designers the health of eaters is of no concern. It would take a great philosopher of the human appetite to discern how this conflict has helped generate the discontents of appetite of our time, but that it exists and that it is intensifying are everyday realities. As a historian, I have nothing to recommend except that we become closely acquainted with the historical processes that landed us in this predicament—not just the economic, social, and cultural forces that led us to the era of big food, but also the unfolding within science and medicine of imperatives that, in ways that could never have been anticipated, too often help to undermine health. My own reading of these matters suggests that appetite is, like love or pain or anything else that is fundamentally human, integral to our whole being.

It is not only physical; it is not only mental; it is not some aggregate of disparate components of our nature. It is simply one of the modalities by which we establish who and what we are in the world around us. All efforts to substitute biomedical dictates for individual strivings of appetite come, it seems, at a cost. However these strivings are formed (a question still not answered), and whatever the consequences of acting on them may be, abandoning the care of our appetite to purportedly authoritative others diminishes the kind of health that, as Georges Canguilhem once observed, consists in freely deciding, for ourselves, what it means to be healthy.[36] Whether it is possible at this historical moment to restore any freedom to appetite and, in so doing, to take a different aim at its discontents, I do not know. But I do believe that history supplies a place to start.

PART ONE

Anxieties of Appetite: Created Needs in the Enlightenment, 1750–1800

Appetite [is] . . . a movement of the soul toward an object it perceives as a good although all too often it is a great evil. "FAIM, APPÉTIT," IN *ENCYCLOPÉDIE, OU DICTIONNAIRE RAISONNÉ DES SCIENCES, DES ARTS ET DES MÉTIERS* (1756)

Introduction to Part One

A good claim can be made that Western patterns of eating and drinking changed more dramatically in the eighteenth century than at any other historical moment, especially since the era brought the effective end of famine. Europeans had suffered starvation throughout the medieval and early modern eras, but for the most part famine came to an end with the arrival of the agricultural and commercial "revolutions" of the eighteenth century. Moments of severe privation continued to occur, along with the food riots that had frequently accompanied dearth, but for the most part, famine receded after the early years of the century.[1] Yet if the receding threat of famine was the most dramatic consequence of the changes in food production and distribution under way in the eighteenth century, it was accompanied by other enduring shifts, especially in the availability and consumption of new types of food and drink. Although historians debate the timing and extent of the dietetic transformations effected in these years, they generally agree that the eighteenth century saw the coming of what we might call "ingestive modernity," and that its arrival had a profound impact on the objectives of the modern state, class divisions, and the dynamics of social upheaval—in sum, the nutritional basis of Western life.[2]

Of the many alterations in Western habits of ingestion that occurred in the eighteenth century, one of the most significant for the future of eating and drinking was the arrival and increased availability in Europe of the so-called luxury substances, especially tea, coffee, and sugar. Some elements of the European population had already come to enjoy use of these substances, but it was only in the eighteenth century, once basic subsistence had been assured for large numbers of people, that such luxuries came to have ever-wider markets.[3] For a long time historians looked at this development—the increasing trade in and consumption of once exotic foods and drinks—chiefly as part of the growing reach of European commercial capitalism. More recently, scholars have investigated cultural processes accompanying such shifts in patterns of human sustenance, exploring how public entities—from the absolutist states of the early modern era to small-scale communities—come to agree on (or continually dispute) what constitutes human needs. Premodern societies are said to have been dominated by a sometimes-desperate struggle to acquire adequate clothing, shelter, and food, whereas surplus, the margin beyond need, has been seen as a prime characteristic of the modern. How this margin is perceived, who has the authority to delineate it, and how competing conceptions of needs win out over others are problems that, as historians have recently argued, set into play complex processes of political, social, and cultural negotiation and conflict.[4]

One interpretive construct that historians of the eighteenth century have found especially helpful in this regard is the Enlightenment notion of created need. In important ways historians who explore this theme simply take the lead of eighteenth-century observers themselves. As E. C. Spary has argued, the nature of need, the difference between need and desire, as well as the moral and political implications of "factitious desires" were all matters that evoked reflection, often anxiety, among the philosophes of eighteenth-century Europe. Jean-Jacques Rousseau (1712–1778) was the master builder of this discourse of need and desire, and his notorious conclusion that civilization brought with it false desires and new physical and moral dangers has served as a point of origin for arguments about what constitutes genuine need ever since.[5] The concept of created needs was essential to the cultural setting in which the nature and impact of the appetite for food and drink evoked intensified debate in the eighteenth century, both among thinkers like Rousseau and, the special concern of this study, among medical and scientific investigators who often regarded luxury substances as dangerous novelties and debated their effects on body and mind. To be sure, physicians and natural philosophers from antiquity onward had turned their sights on problems of appetite, and my discussion of eighteenth-century medicine therefore begins by looking at the legacy of the ancients in respect to ingestive desires and habits (chapter 1). The sixteenth and seventeenth centuries had also seen heated disputes about the how and why of eating, among both religious combatants who attacked or defended

Catholic teaching on mortification of the body and learned figures who counseled abstemiousness to achieve longevity.[6] In the early eighteenth century, moreover, there was much debate over what constituted healthy eating, the components and digestibility of assorted foodstuffs, and the like. The English physician George Cheyne (1671–1743), who at his most encumbered weighed in at 440 pounds, spent a lifetime worrying about his own egregious appetite and seeking ways to reduce his bulk. At roughly the same time, the physician who dominated European medicine up to his death in 1738, Herman Boerhaave of Leiden, explored the chemical nature of foodstuffs and their nutritional and therapeutic uses.[7]

Such examples could be multiplied, yet as I explore in chapter 2, it seems clear that concern over eating took on a new urgency in the second half of the eighteenth century. In these decades the how and why of eating drew the attention of new kinds of publications and institutions. "Medical topography"—physicians' descriptions of such features of particular locales as geographical setting, climate, mortality, and morbidity—included as a crucial element the foodstuffs available to and eating habits of local populations.[8] Hospitals and collateral institutions—the dispensaries that came to dot the landscape of London and other urban centers, the asylums for the mentally ill that emerged everywhere after 1750—routinely reported on ills associated with disordered appetite and eating.[9] Other developments are suggested by Michel Foucault's complex argument about the emergence of "biopower," which, despite the challenges of critics, continues to undergird much historical thinking about the preoccupation of the emergent modern state with ensuring a healthy population, an objective that entailed new attention to nutritional "needs."[10]

Of the public voices focused on matters related to food and eating in the mid-eighteenth century, physicians had the strongest claim to authority, given the centuries-long preoccupation of Western medicine with the effects on the body of things ingested. Early modern physicians routinely invoked ancient teachings on dietetics and the therapeutic use of food and drink. They were also leading contributors to the literature devoted to prolongation of life, although they faced competition from experts in the culinary arts who were dedicated to maximizing pleasure in eating. Ken Albala argues that although many Renaissance physicians saw medicine as allied to, or at least compatible with, the culinary arts, by the seventeenth century medical thinking was often antagonistic toward gastronomic indulgence.[11] Medical warnings about eating grew more dire in the eighteenth century: notwithstanding its heralded commitment to pleasure seeking, the Enlightenment saw increased fretting about what to eat.[12] Through this long evolution, one feature of medical counsel remained constant: a healthy diet consisted of what was appropriate to the individual on the basis of his or her age, constitution, and temperament; the influence of local climatic conditions; and, not least, particular eating propensities and habits.[13] After 1750, however, this tradition began to falter.

In part 1 of this study, I argue that the later eighteenth century witnessed not only increased anxiety about what to eat but also a new drive away from the individual focus of earlier medical counsel toward the view that food choices must conform to general standards of healthy eating. Breaking with physicians from Hippocrates onward who postulated no absolutes with respect to diet, scientific investigators of the later eighteenth century (examined in chapter 3) began arguing that, based on the material nature of foodstuffs and the uniform operations of digestion, certain foods were, for everyone, healthier and more "digestible" than others. This new perspective both reflected and gave added encouragement to inquiries into the nature of appetite itself as a crucial feature of the digestive process. By the end of the period such inquirers had begun to privilege experimental methods—used on themselves, inmates of institutions, and dogs and other animals—in an effort to displace individual appetite as a guide to eating and to establish general truths about the selection of foodstuffs and their effects on those who consumed them. Thus, it is to the later eighteenth century that the modern challenge to individual appetite as a reliable guide to eating traces its origins. To explore these beginnings, it is necessary first to examine what ancient physicians and philosophers held to be the nature of appetite and counseled about its management as the foundation of health. To many eighteenth-century investigators, challenging this ancient legacy was the first task to be accomplished in establishing a new science and medicine of appetite.

ONE

Why We Eat: The Ancient Legacy

For nearly two thousand years it was a truism of Western medicine that appetite was highly variable and that individuals ate in accord with an inner sense of which foods suited them best. A generally accepted corollary held that people desired to eat what was good for them. Certainly it was recognized that some people ate unwisely, consuming too much or too little or selecting foods that were "hard to digest." Errors of regimen were common and sometimes had serious consequences. But deviations of this sort did not belie the truth that beneficent nature or a providential deity had implanted in human beings (and in animals and perhaps even plants) an appetite for appropriate and healthful nutriments. The deep cultural roots of this faith in "natural appetite" are doubtless many and diverse. But in Western tradition the most powerful affirmation of appetite as a reliable guide to eating came from medicine, and, within medicine, from the teachings of the "divine Hippocrates."[1]

Diet and Health in the Hippocratic Corpus

Perhaps no medical system has ever been so dependent on thinking about food as that offered in the Hippocratic Corpus. Hippocratic medicine had little sense of the internal workings of the body and scarcely any sense of the functions of specific organs; rather, it was based on observation of inflows and outflows of the humors, the fluids that developed out of nutriments ingested and that governed both the maintenance of health and the onset of illness. Thus, an essential requirement of the Hippocratic physician was an understanding of the nature of food and its effects on the body. Fundamental to Hippocratic thinking was the notion that health depended on a balance of the humors and that disease resulted from imbalance, either surfeit or deficit. Food did not act alone in determining the state of the humors. Growth and exercise consumed the "power" imparted by food, and other activities, those eventually codified in Western medicine as the "non-naturals," were also essential to health. But, aside from breathing, no bodily activity was more fundamental than taking in and processing food. Similarly, no sign of bodily health or distress was more important than the desire to ingest food; its absence could presage mortal illness.[2]

Yet if in the Hippocratic Corpus the great importance of appetite and eating is unquestioned, little is said about where or how the desire to eat originates. The author of *Regimen I* recognizes the paradox that "the belly is without consciousness yet by it we are conscious of hunger and thirst," but says nothing more (*Regimen I* XII, Loeb IV:251).[3] Nor is there mention in the Corpus of how the physician knows the state of the appetite in a patient under his observation. Physicians seem to have judged the strength or weakness of a patient's appetite on the basis of his asking for or accepting nourishment, or on explicit declarations of how he felt about eating. In this respect appetite differed essentially from the physical signs on which Hippocratic physicians ordinarily relied—minute differences of texture and shape, look and feel, color and consistency of any part or product of the body. States of the blood, urine, phlegm, and sweat provided immediate evidence to the senses; blood, for example, might be thick or thin, dark or light, congealed or frothy. Visual evidence was primary, but touch, smell, and in some cases even taste were involved as well. Appetite was different; like thirst, nausea, or pain, it was a state to which the physician had no direct access.[4] This means that in the Hippocratic writings what would later be termed the "mystery" of appetite—inaccessible, changeable—was already apparent. Yet the lack of direct sensory evidence did not preclude medical judgments about appetite and its significance; rather, the physician compared the patient's report to other signs to judge, as Jacques Jouanna states, the "relative value of the observed signs" in the particular circumstances of an illness. Somewhat paradoxically, the hiddenness and changeability of appetite could be advantageous rather than detrimental, as disturbances of appetite reflected the infinite mobility of internal processes that could be seen only with the "'mind's eye.'"[5] The most dramatic representation of the place of appetite and hunger in this kind of "seeing" is found in *Prognostic*, the treatise that became the bedrock of medical semiotics, the study of the signs that allowed the physician to assess a patient's prospects for recovery, relapse, or death.[6] *Prognostic* presents perhaps the earliest known differentiation between appetite and hunger: appetite as a sign of hope ("Those chiefly recover who . . . quickly recover their appetite") and hunger as a state so wretched that the patient's aspect could be confused with the face of death (XVII, Loeb II:37).

Hippocratic physicians assumed the variability of appetite and recognized that no definite rules could be established for eating since individuals desired and ate foods in their own way. Hence the directive stated in *Aphorisms* that "food or drink which, though slightly inferior, is more palatable, is preferable to that which is superior but less palatable" (II:XXXVIII, Loeb IV:117–19). This view of the relation between ingestion and health seems to have derived both from philosophical principles, as when the author of *Nutriment* applied Heraclitus's "theory of perpetual change" to the problem of the assimilation of food, and from minute observation of individuals, the method that would make "Hippocratism" an enduring alternative to all ways of seeing the body based on universals or regularities.[7]

Reflecting such thinking, Hippocratic teaching on regimen urged the impossibility of establishing the exact mix of food and exercise necessary to ensure the health of an individual. The author of *Regimen I* observes that even if everything were known about a patient's constitution and age, the season of the year and changes of the winds, "a due proportion of food to exercise, with no inaccuracy either of excess or of defect" could not be determined (II, Loeb IV:229). Still, although exactitude in this regard was impossible, Hippocratic physicians did seek to describe general conditions that reflected disproportion between food and exercise, most significantly the condition called "surfeit." This term has varying meanings in different contexts, but some recommendations for alleviating surfeit suggest a resemblance to what we think of as "obesity": "Fat people who wish to become thin should always fast when they undertake exertion" and consume "rich [meats] . . . so the appetite will be satisfied with a minimum" (*Regimen in Health* IV, Loeb IV:49). Yet elsewhere, when the author of *Regimen I* considers the lamentable effects of "abundant flesh," the distance between Hippocratic surfeit and modern obesity is clear: "Abundance of flesh cannot fail to result in inflammation of the blood, and when this happens to a soul of this sort it turns to madness, as the water has been mastered and the fire attracted." Here surfeit was a kind of "inflammation . . . next door to madness" that could arise from different causes, including "intoxication, or . . . overabundance of flesh, or . . . eating too much meat" (XXXV, Loeb IV:291).

Was excessive eating ordinarily responsible for surfeit? The problem of how a patient came to suffer from this condition is complex. *Regimen in Health* sets forth a range of circumstances that could account for it and counsels the physician to attend to the patient's "physique"—fleshy or lean, dry or moist, soft or hard—and to "age, season, habit, land, and . . . the prevailing heat or cold" (II, Loeb IV:47–49). Thus, surfeit could result not only from individual choices about diet and exercise but also from a given type of body or from climatic and geographical influences. Such statements reveal a tendency within Hippocratic thinking to determinism—a person is fat if his given physique is fleshy—and an ambiguous sense of whether surfeit reflected a blameworthy indulgence of appetite or resulted from circumstances outside a patient's control. Fundamental to the problem of the patient's responsibility for conditions such as surfeit is the Hippocratic view of habit, yet here again the Corpus contains conflicting passages, some suggesting that individuals have the power to form habits, but others indicating that habits ineluctably arise in response to the interacting influence of age, season, terrain, and other circumstances.[8]

To explore the Hippocratic view of the formation of habits, some features of the larger cultural setting must be taken into account. Historians have suggested that Hippocratic physicians offered counsel principally to the class of Greeks made up of freeborn, male citizens.[9] Although the cases recounted in *Epidemics* show that Hippocrates and his followers treated women, children, slaves, and the occasional foreigner, it seems clear that physicians thought some patients—adult male citizens—capable of making clear judgments and forming healthful habits and others not. This element of Hippocratic teaching was to prove crucial to later thinking about healthy and perverse appetites in the Western tradition, which assumed that certain human types—children, women, "barbarians"—lacked the mastery of self required for control of appetite.

Some clues in the Corpus suggest the kind of patient addressed by the Hippocratic physician offering advice about appetite and eating. *Regimen in Health*, W. H. S. Jones tells us, was addressed to the "layman," the "ordinary, normal person, whose business," unlike that of the professional athlete, did not require "special diet and exercise" (Loeb IV:45n1). For this layman, the ideal was adherence to standards of moderation that assured harmony of body and mind and overall health.[10] The books on regimen generally promoted physical health, yet the value of bodily health blurred into larger cultural values of the honorable citizen. In his study of the sexual guidelines Greek philosophers and physicians offered to male citizens, Michel Foucault has shown that a principal virtue of the Greek citizen was self-mastery, an ideal that linked the capacity to govern one's own body, behavior, and speech to the capacity, all-important in the city-state, to govern others. The ideal of the male citizen thus required that he deliberately form habits central to self-governance.[11]

When the author of *Regimen III* described the individual who could best profit from medical advice, he made the state of the appetite an essential feature, addressing "men who appear to be in health, eat with an appetite, can take their exercise, and are in good condition and of a healthy complexion" (LXIX, Loeb IV:383). For such individuals recommendations for appropriate eating were often elaborate, taking into account a range of factors including the influence of the seasons (*Regimen in Health* I, Loeb IV:45). Though detailed, such recommendations generally included nothing that individuals could not implement on their own. Thus, while other elements of the advice on regimen called on the physician's expertise, the responsibility of the mature, self-governing male to moderate his own eating was clear.[12]

Yet the dietetic books also mention other categories of people who functioned outside the ideal of the self-governed male.[13] In Hippocratic thinking the four "ages" that differed by degrees of heat and moisture were highly significant for health and disease (*Regimen I* XXXIII, Loeb IV:279–81). Children were by nature moist and warm, and features of their diet followed accordingly. *Regimen in Health* presents a set of special rules for infants (VI, Loeb IV:53), and in *Aphorisms* a description appears of afflictions seen in children at different ages, including "very fat children" who were inclined to irritation of the gums, fevers, convulsions, and diarrhea (III:XXV, Loeb IV:131). Yet overall, relatively little heed is paid to children, a fact that is interesting in view of the close attention philosophers gave to training the bodies of young boys. Foucault argues that for such thinkers training in sexual comportment was primary, but he makes clear that rules for boys also extended to control of appetite. Xenophon set forth "the precepts and conduct of his master [Socrates] 'concerning eating and drinking and the pleasures of love,' " and Plato described the training of guardians to achieve "the threefold mastery of the pleasures of drink, sex, and food."[14] That such precepts were set forth by philosophers rather than physicians may mean that to the Greeks training of the body belonged to morals rather than to medicine. Yet while the Hippocratic authors seldom admonished male citizens-to-be in respect to eating, in other contexts they spoke directly to the value of exercising control over the desire for food when they contrasted the eating habits of Greeks and foreigners and, most tellingly for the future, of women and men.

Hippocratic thinking about the appetite and diet of barbarians appears chiefly in *Ancient Medicine*, which, as Jouanna has shown, portrayed the discovery of medicine, and specifically dietetics, as the foundation of civilization itself.[15] Perhaps the most famous ethnographic judgment in the Corpus was that made about the Scythian nomads whose monotonous desert climate and habits of food and drink produced a physique that was "gross, fleshy, . . . moist and flabby"; males were fat, lazy, and impotent, whereas the "fatness and moistness" of females were such that "the womb cannot absorb the seed" (*Airs, Waters, Places* XIX–XXI, Loeb I:123–25).[16] For the Hippocratic physicians, then, the failure to control diet was a clear marker of the barbarian, to whom "medicine was unknown."

The question of appetite in girls and women is more complex. The books best known for their consideration of women are the gynecological treatises *Diseases of Women* (I, II, and III), but on questions of appetite and eating, the text that exercised influence long after the classical era was the brief book *Diseases of Young Girls*, also known as *Diseases of Virgins*.[17] This text compares the peculiar ills of young women to the "sacred disease" (epilepsy) with its attendant "apoplexies and terrors," and asserts, in one of the most forceful Hippocratic statements about women, that these phenomena were worse for women because "female nature is weaker, and more troublesome."[18] In *Diseases of Young Girls* blockage of the menses is central to illness in postpubertal girls and women. The text describes how, with the inflow of nourishment, blood accumulated in the region of the womb and then, overflowing its natural locale, traveled upward, suffocating or obstructing the heart and diaphragm, "the place . . . critical for both mental aberration and madness" (251). The remedy offered was to echo through the centuries: such girls "should marry as quickly as possible" to ensure that uterine function was fulfilled. Yet the Hippocratic author also observed that even a woman who married and had sex might still suffer illness of this sort if she were barren. As Helen King demonstrates, this text reinforced the Greek custom of early marriage for girls and the more general tenet in Greek society that women's health required performance of the critical roles of wife and mother.[19]

Although physicians later adduced *Diseases of Young Girls* as the first Western text describing chlorosis, King shows that it actually says little about appetite and eating aside from a brief comment about blood accumulating with the intake of nourishment. Moreover, nothing in the text suggests that girls afflicted with the "disease of virgins" suffered because of perverse eating. Yet for Western physicians this text became the putative source for a long-lived tradition that associated girls and young women with two forms of disordered appetite and eating: a general aversion to food, in the words of Johannes Lange in the sixteenth century, "a turning away from food, especially meat," and the desire to eat strange, often nonfood, substances, described by the "English Hippocrates" Thomas Sydenham as "an unnatural longing for such things as are noxious, and unfit for food."[20] Other early modern texts specified that young girls substituted "perverse foods" for meat and other healthful substances, consuming earth, chalk, slate, and unripe grains to foster the pale greenish skin that signaled sexual desire.[21]

Although none of this rich detail is in *Diseases of Young Girls*, the text did supply the essential element of subsequent medical theorizing about appetite in girls and women—the connection between eating and reproductive life. It thus established a principle that for centuries went largely unquestioned: appetite and eating were essentially different in women and men. The connection between eating and reproduction is also made explicit in a brief section of *Regimen in Health*: "Women should use a regimen of a rather dry character, for food that is dry is more adapted to the softness of their flesh, and less diluted drinks are better for the womb and for pregnancy."[22] Other passages in *Regimen I* associated women with principles of moist and cold that in turn were linked to lack of vigor, even to imbecility and madness (XXXIV, Loeb IV:281; XXXV, Loeb IV:287). Given these fundamentals of women's nature—moist and cold, dominated by reproduction, tending toward madness—it is scarcely surprising that the Hippocratic books show no expectation that women could control their appetites or, in any of the ways expected of mature men, fashion a self. This view that women had little, if any, capacity to shape or control their appetite for food was to be a truism of Western medicine; disturbed appetite would figure as a central element of psycho-gastric ills, especially the "grand neurosis" of hysteria, to which women were thought to be exclusively or exceptionally prone.[23]

The teachings of Hippocratic medicine with respect not only to women and other inferior types but also to the ideal male are important for several reasons. The first is the extent to which Hippocratic physicians emphasized particularity of appetite even while noting patterns that could guide diagnosis and treatment. There is also an evident assumption of, first, a tendency to overeat, one that the self-controlled male citizen must master, and, second, of the peculiar appetites of particular human types, especially women, who were, in many respects, the prototype in Greek culture for everything wild, uncontrolled, even pathological.[24] This said, however, it is important to note also that some critics of the Hippocratic physicians believed them too little interested in the individuality of patients, especially in the necessity of fitting dietetic prescriptions to infinitely variable circumstances. Praxagoras (b. 340 BCE?), head of a school rivaling the Hippocratic physicians, was even more adamant that "general dietetics must be supplemented by individual dietetics" and that "what may in general be useful can in a special case be harmful."[25] Such criticism suggests a tension in ancient medicine between recognition of appetite as endlessly various and the imperative to devise rules of regimen and therapeutics that would be useful in daily practice. This tension is also evident in the work of the other great ancient physician whose work was to dominate Western medicine, the second-century Roman physician Galen. Before turning to Galenic medicine, however, it is necessary to look at the most important philosophical contribution to ancient teaching on appetite, which appeared in several works of Aristotle. While deeply indebted to Hippocrates, Galen was also a close student of Aristotle. His medical system was highly dependent on Aristotelian principles, and it played a large role in transmitting Aristotle's conception of appetite to later generations.

Aristotle and the "Psychic Power" of Appetite

The clinical approach of the Hippocratic physicians yielded not a definition of appetite but, instead, a sense of how it worked in health and disease in individuals. By contrast, Aristotle (384–22 BCE) explored the basic nature of appetite as one of the "faculties" or "powers" of the soul, providing a formulation that was to exercise influence well into the modern era.[26] Aristotle's treatise *On the Soul* (*De anima*) was required reading for the bachelor's degree of the medieval and Renaissance universities, and academic theses and disputations dedicated to its explication numbered in the hundreds.[27] As Fernando Vidal shows, learned individuals influenced by Aristotle regarded the soul as belonging both to science and philosophy ("physics" and "metaphysics"). Thus, early modern Aristotelians considered the soul and appetite from the perspective of the natural functions of living bodies while necessarily bearing in mind the spiritual viewpoint of the soul as the source of human rationality and moral choice.[28]

Aristotle's understanding of appetite forms part of his larger idea of soul, which in turn partook of a fundamental division between matter as "potentiality" and "form" as "actuality." In stark contrast to modern materialists, Aristotle established as the basic source of the nature of different living beings their immaterial "form" (412^a1, 1:656). Thus, it is the form of the human that makes us human, the form of the dog that makes a dog. This distinction between form and matter was also a way of distinguishing between soul and body while still insisting on their indivisibility. Aristotle describes this oneness metaphorically when he states that soul and body are like "the wax and its shape," and, elsewhere, that "the soul *plus* the body constitutes the animal" just as "the pupil *plus* the power of sight constitutes the eye" (412^b7, 1:657).

For Aristotle all living beings have a soul; soul is the "principle of animal life" (402^a7, 1:651) and the source of the functions or powers of living beings. Aristotle distinguishes between two forms of life: plants, which enjoy only the power of self-nutrition, and animals, which enjoy appetitive, sensory, and locomotive powers as well as the power of thought. It is in this context—classifying the powers of soul—that Aristotle defines the nature of appetite. In his scheme the primordial fact of animal life is the capacity to sense, and thus to feel pleasure and pain. In turn, sensations of pleasure and pain create "appetition" and guide the animal in its search for nutriment. "If any order of living things has the sensory," he wrote, "they must also have the appetitive; for appetite is the genus of which desire, passion, and wish are the species" (414^b1–3, 1:659). This formulation set the terms of discussion for a long time for those who sought to understand the nature of appetite through the operation of the external senses and the way that pleasurable sensations established recurrent desires. In Aristotle's account, we experience appetite—or indeed, any wish or desire—because we feel pleasure and pain, and we feel pleasure and pain because we are equipped with senses. For Aristotle, touch is primary in the desire for food because touch gives access to the knowable qualities of food—hot and cold, dry and moist. Although physiologists of the eighteenth and nineteenth centuries who debated the sensory foundations of appetite would reject Aristotle's emphasis on touch in favor of the powers of taste, his formulation remained into the modern era the point of departure in efforts to chart the relation between appetite and the senses.[29]

Stated in these terms, this Aristotelian way of thinking about appetite seems straightforward enough. But as Leon R. Kass has suggested, Aristotle's view places appetite at the core of the activities of living beings, in that "desire for" is the original motivator of all movements and actions, not just those involved in eating and nourishment. Kass argues that "simple awareness of some outside edible is insufficient to account for the action of self-nourishment"; instead, "something must energize the animal into action, . . . something like 'felt need' or 'appetite.'" In this sense, "appetite or desire . . . is the deepest principle of life."[30] Kass argues that an entire moral philosophy of appetite and eating is suggested by Aristotle's view that every living being is imbued with a particular form that guides the experience of desiring foods and eating and thus establishes the appropriateness of specific types of foods and ways of eating for the class of beings to which the individual belongs. In this regard Aristotle's thinking differs radically from modern investigations of appetite that take animal eating (that is, feeding) as the indispensable model for the human. Aristotle saw enormous differences between the appetite and ways of eating of animals and humans. These were differences not only of anatomical apparatuses but also of integral animal and human nature, the latter encompassing moral choices that had no place in the animal realm. Hence the approach to appetite that Aristotle developed in the *Nicomachean Ethics*, in which he proclaimed moderation the essential virtue while condemning excessive pleasure in food and drink as "brutish."[31]

Aristotle's explicit statements about appetite in *De anima* link it, then, to the operation of the external senses and sensations of pleasure and pain, but in other writings, especially *Parts of Animals* and *Generation of Animals*, he connects appetite not to the external senses but to the inner disposition of organisms to seek out and respond to nutritive materials in ways determined by their natural "inclinations."[32] Although Aristotle himself emphasized the role of such tendencies in the action of the viscera on nutritive substances, later authors would take his observations to mean that living beings had a natural sense of what kind of eating was appropriate to achieve healthful nutrition. The idea that ingestion, as the beginning point of the largely automatic process of transforming and absorbing nutrients, might itself be assigned not to voluntary choice or externally oriented appetite but to internal inclinations would hold special appeal for physicians committed to the principle of the healing power of nature. In such thinking the "tastes" or "desires" of the viscera provided guidance, unconscious or dimly perceived, to the kind of eating that could sustain or restore health. Here Aristotle's division of living beings into plants that enjoyed only the power of nutrition and animals that exercised one or more of the higher faculties (appetitive, sensory, locomotive, rational) would be transposed to the level of the individual organism in which healthful nutrition was assured by a "vegetative" automaticity blind to the play of sensation and voluntary action.[33] In this way Aristotle came to serve as the source for two opposed conceptions of the origin, functioning, and dependability of appetite. One emphasized exteriority, choices made amid the consciously experienced world of pleasure and pain. The other construed interior impulses from the hidden depths of the body as a special and superior kind of knowledge of organic desires and needs. Here, in outline, was the dichotomy, still salient in debates over how to know what to eat, between appetite as culture, formed by training, subject to conscious choice, and appetite as nature, given by the creator or "instinct" or, as would be said around 1900, the "wisdom of the body."[34]

It remains to ask how Aristotle's understanding of appetite—as a faculty of the "soul" of all living creatures possessed of capacities for sensation and movement, as the essential spur to directed activities required for nutrition and repair of the living body—relates to the medical view of appetite as a guide to ingestion that functioned in accord with the "nature" of individuals. As was true of his "science" overall, Aristotle's treatment of appetite exemplified an attempt to move meaningfully from the general to the particular, in this case to define appetite itself in the order of nature and then to explicate its exercise in plants, animals, and human beings.[35] Given the encyclopedic range of his intellectual concerns, Aristotle treated appetite from different perspectives, shifting from the metaphysical to the biological to the ethical, and thus wrestled with problems that lay largely outside the ken of medicine, at least in its Hippocratic form. Both in his considerations of the peculiar "form" of the human and in his specifically ethical injunctions to "moderation," Aristotle sought general truths about appetite, eating, and all natural functions that transcended the medical focus on healthy and sick individuals. Nonetheless, Aristotle's exploration of the peculiar nature and demands of human appetite differed profoundly from the generalizing and universalizing impulse that was to take hold with the coming of modern science. "Moderation" was an imperative for all human beings who were capable of ethical choice and action and who, exercising that capacity, stood apart from the beasts. But this imperative said nothing about the specifics of appetite or ingestion beyond the general counsel against extremes.[36] In this it was one with Hippocratic warnings against the dangers of surfeit and deficit, conditions that could not be precisely defined but that reflected the "nature" of the individual as ordained by nature at large. Thus, while moderation was a virtue all should strive for, it supplied no dietetic code. The same was true of the thinking of the Roman physician Galen, Aristotle's greatest expositor among physicians and the most influential systematizer of ancient medicine. In Galen's work appetite was again defined as a uniform feature of the physiology of humans and animals while nonetheless serving as a highly variable guide to ingestion in individuals mindful of health.

"Innate Appetites": Galenic Physiology and Dietetics

Hippocratic physicians accorded appetite an important place in diagnosis and prognosis, discussing it pragmatically in relation to the exigencies of preserving or restoring health. The other supreme authority of ancient medicine, Galen, broke new ground in seeking to elucidate the structures and processes involved in the drive to eat. Reflecting on what it is that "stimulates an animal to seek food," Galen held that "[nature] has granted to the stomach alone and particularly to the parts of it near its mouth [the cardiac orifice] the ability to feel a lack which arouses the animal" and urges it to seek nourishment, "each according to its own nature . . . when the stomach feels the need." Yet Galen was wary of according to the stomach any "innate power of perception," and he urged that "this faculty must flow in upon the stomach from . . . the source of sensation." He accorded this role to "a pair of nerves of considerable size [*nn. vagi*]" that, "descend[ing] to [the stomach] from above," form a complex network throughout the stomach "extend[ing] . . . as far as its lower end."[37]

These passages from Galen's physiological treatise *On the Usefulness of the Parts of the Body* (*De usu partium*) established essential claims about the nature of appetite. Perhaps most important was Galen's assertion that the desire to eat is localized in the opening to the stomach, where the individual "feels" the lack of food and the desire to ingest. Galen was acutely aware of the paradox contained in this assertion that the stomach could feel. His physiological system put the seat of sensation in the brain (not, with Aristotle, in the heart), and he was therefore obliged to explain how a visceral organ could experience sensation. The way in which he did so—postulating a link between the brain and the stomach via nervous connections—became a basic element of the physiology of appetite, the role of the vagus nerves.[38] Galen described the vagus nerves as stemming from the cranium and serving the esophagus, stomach, and other viscera. This function fit within his tripartite physiological system based on the three central organs of brain, heart, and liver; the three powers or faculties that enabled their functions—the brain's work of ratiocination, subserved by feeling and movement; maintenance of vitality or "innate heat" by the heart; and the nutritional activity of the liver—and the three varieties of pneuma, an "air-like substance" that linked these organs to the rest of the body. The vagus (also called the pneumogastric) nerve carried "psychic pneuma" from its origin in the cerebral ventricles and what Galen and other ancient physicians postulated as a "net-like plexus at the base of the brain" to the parts they served.[39]

Within this physiological scheme, it is not clear which of the two bodily parts—the cardiac orifice itself or the psychic pneuma that moved through the nerves and caused the orifice to feel a lack of and desire for food—was the more important. What is clear is that Galen's work characterized the prompting to eat as at least partially a nervous phenomenon. Of equal importance was his designation of this nervous activity in the stomach simply as a feeling of lack. Although in *De usu partium* Galen did not delve further into the nature of this lack, he did so elsewhere, especially in his writings on hygiene, where he urged that, "from the very beginning Nature has given, not only to animals but also to plants, *innate appetites* for what is always escaping." That the appetites were inborn was evident: "For we do not learn from anyone how to eat or to drink or to breathe, but from the beginning we have in ourselves the power of doing all these things, without instruction."[40] With this line of thinking Galen explained why, as many ancient and later physicians asserted, digestion was the most important bodily function.[41] The living body existed in a state of "perpetual flux" in which it continually lost components necessary for life and health. These losses could be repaired only by the intake of materials from outside the body and their transformation and "assimilation" through the power of digestion. Thus, digestion mattered most not because it was the body's highest or, as Galen termed it, following Aristotle, most "noble" function—this honor was reserved for rational thought—but because life itself could not continue without the intake and processing of restorative materials.

Galen developed his view of the primacy of digestion within the teleological framework that he took from Aristotle and that informed all of his thinking about living beings. The principle of teleology posited that everything in creation manifested the wisdom of nature or, as Galen frequently termed it, the Creator. Large sections of his physiological writings were devoted to celebrating the wisdom, justice, and beneficence of Nature.[42] It is one of the paradoxes of ancient thought that Galen, who was both a philosopher of the body and a practicing physician with extensive experience of human frailty and disease, nonetheless held that the human body, like everything else in creation, was created perfect by a "Nature" that did not err. In respect to nutrients this meant that nature assured that living creatures—human beings, animals, plants—innately desired those foods they required for life and health.[43] No creature, Galen believed, had to be taught what to eat. Hence the lack felt at the cardiac orifice and conveyed via the vagus nerve served as the anatomical substrate for inherent eating desires. As Robert J. Richards has shown, a remarkable facet of Galen's work on appetite was his use of a method to investigate animal eating that became standard only centuries later: isolating a newborn animal to see whether it could and would choose foods appropriate for its nourishment. Reporting an experiment on a newborn goat taken by caesarean section, isolated from its mother, and presented a choice of foods and drinks, Galen described how the observers, seeing the "kid . . . sniff each of the bowls in the room, and then from among all of these," smell and lap up the milk, "gave a yell; seeing realized what Hippocrates had said: 'The natures of animals are untutored.'"[44]

The fact that Galen asserted an instinctual knowledge of healthy eating does not, of course, mean that he precluded the possibility of eating "errors." As a physician, he had a wide experience of patients who suffered from disordered appetites and unhealthy eating, and he offered detailed recommendations to correct mistakes of regimen. Much more than Hippocrates, Galen commented on vagaries of appetite and on extremes of bodily weight and shape with clear prophylactic and therapeutic intent. In his treatises on food, Galen offered assorted correctives for "jaded" or "weakened" appetite, including a remedy of his own devising concocted of apple juice, honey, vinegar, ginger, and white pepper simmered to "the consistency of honey."[45] But Galen's primary work in this respect was his treatise on hygiene, where he took up the question of whether human beings can enjoy a perfect body or a perfect constitution. Despite his commitment to a sublime, unerring Nature, Galen concluded that everyone deviates slightly from perfection and harmony. Rejecting the idea "that health and good constitution are one and invariable," he insisted instead on changeability and variability. Individuals varied by constitution and age and also "in respect to slenderness and corpulence."[46] Galen stressed that no bodily constitution or form is unchanging but, instead, alters over time in accord with climate, age, "customary vigor" in exercise, and the action of all the "non-naturals."[47] Thus, the successful physician knew that no single "plan of life" could be devised for all patients (*Hygiene*, 223). In relation to food and drink, Galen stressed the impossibility of absolutes. The amount of food to be ingested varied with the amount of exercise performed. Wine was not good or bad but suitable or unsuitable according to circumstance (165). Some foods might suit certain ages better than others, but the essential question was always whether, for an individual, a certain food was easy or hard to digest (207–8).

In view of Galen's emphasis on variability, it may be asked where he drew the line, if in fact he did, with respect to eating and the weight and shape of the body. Here he did not give many clues. In offering programs of weight reduction or gain, he described particulars (for the obese, rapid running, massage, and "abundant food of little nourishment" [*Hygiene,* 257]) but not his general rationale or objective. Nor did he explain the dangers of the opposite extreme of "emaciation" (*Hygiene,* 25, 276). Thus, although Galen evidently believed that extremes of thinness and corpulence could and should be treated, he set forth no clear guidelines for defining these states or judging when therapeutic action was necessary. Whether he acted because he himself observed that a patient suffered discomfort, pain, or disability as a result of these conditions, or because the patient himself (and I use the masculine pronoun advisedly) was unhappy with his state is unclear. Still, on the question of why people ate in ways detrimental to health when they should instinctually know what and how much to eat, Galen was clear: they failed to heed both nature and the medical art.

Unlike Hippocrates, Galen stated outright that a patient's failure to maintain health resulted from a failure to live in accord with nature. Optimistically, Galen asserted that most people did in fact heed nature's guidance, and if they strayed from nature's path and were "injured by unsuitable things, they have abjured them." But others, "overcome by pleasure or through excessive folly . . . persist in their vicious customs" (*Hygiene,* 225). Here Galen stated what was to be a perennial claim of physicians: an inappropriate desire for pleasure—"folly"—distorted both the perception of what was good and choices based on that perception. On occasion Galen was severe, remarking on the "shame" that attaches to the man who does not act in accord with nature's dictates, who "need[s] someone else to put . . . food into his mouth and attend to his toilet necessities." Galen's ultimate judgment on such individuals was harsh: "For it were better for anyone not an utter coward to choose ten thousand times to die than to endure such a life" (*Hygiene,* 189).

Despite Galen's pronouncement that "most people" obey the dictates of nature, it is also evident that, like the Hippocratic physicians, he believed that only certain kinds of people enjoyed sufficient powers of will and self-control to benefit from the "hygienic art." The "weak-willed and undisciplined" succumbed to the "sweetness of present pleasure," while those consumed by ambition or vanity neglected to eat appropriately and shunned remedies for unhealthy leanness (*Hygiene*, 256). Thus, Galen addressed himself only "to the educated," lamenting that others would always eat "as dumb animals do," failing "to judge by experience what foods and drinks are injurious to them" (*Hygiene*, 275). Here was a striking divergence from the view of animal eating that Galen articulated in admiring nature's guidance even to the "untutored" kid. With hygiene rather than teleology as his theme, Galen portrayed animal eating as blind and heedless, the obverse of human ingestion, which was guided by experience and the medical art. Pursuing this theme, Galen specifically excluded certain types from the benefits of hygiene. Slaves, even if given time to cultivate their health, would still eat "injudiciously," whereas Germans and "other savage or barbarian people" would profit no more from hygienic advice than "bears, boars, [or] lions" (*Hygiene*, 255, 32).

"Barbarians" failed to follow the dictates of hygiene because they ate like beasts. But what of women? Although there is nothing in Galen's physiology to suggest that the mechanism of the drive to eat—the sense of lack felt in the cardiac orifice—differed in women, the advice on regimen that Galen offered in *Hygiene* was exclusively aimed at men. He discussed eating in relation to kinds of exercise and work performed by men, and he made specific recommendations for "lads" in their first, second, or third stages of "seven years." Of these boys and young men he stated emphatically that not only must their bodies be formed but their minds and habits as well (*Hygiene*, 37–38, 90, 92). In contrast, Galen referred to women only to mention which foods were helpful in stimulating menstruation or to decry women's cooking or their childish taste for "jujubes."[48] Thus, for Galen, managing the appetite was a strictly masculine affair.

This discussion of the place of appetite in ancient medicine emphasizes essential claims made in those texts that remained the most readily accessible and the most commonly referred to by later physicians. Vivian Nutton, in his recent survey of ancient medicine, observes that the most striking feature of medical history in the era from roughly 800 BC to AD 620 is the remarkable diversity of thinking, a diversity that "gives the lie to the notion of a single medical system in Antiquity." At the same time, he documents the fact that, progressively from the ninth century onward, the corpus of ancient medical works was steadily reduced—by physical destruction, linguistic change, new material processes of publishing and preservation, and many other factors.[49] Hence, when ancient medical texts began to be published during the Renaissance, a kind of fixed corpus was established, and it was to these canonical texts (principally of Hippocrates and Galen) that physicians and scholars of later years resorted when they searched for guidance. The texts surveyed here should suffice to suggest that ancient medicine, despite its search for reliable patterns in the effects of eating habits and specific foodstuffs, always returned to the truths that health and illness unfolded in individuals and that all medical measures, including the management of appetite, must be suited to individuals. It was precisely this element of the ancient legacy that had to be overcome for a modern science and medicine of appetite to emerge, a process that began in earnest in the eighteenth century.

TWO

"False or Defective" Appetite in the Medical Enlightenment

In the early eighteenth century the attitudes of physicians toward ancient teachings on appetite, ingestion, and digestion were divided. Many physicians continued to extol the Hippocratic tradition, with its emphasis on particularity in health and sickness and the supreme value of bedside observation. Indeed, from the late seventeenth century a "Hippocratic revival" that gave new force to such views got under way.[1] Self-declared Hippocratists who regarded appetite as highly changeable described its fleeting states in detailed case reports modeled on accounts in the Hippocratic Corpus of illness in individual patients.[2] In contrast, from the sixteenth century onward, medico-philosophical "systems" that questioned ancient authority or explicitly sought its overthrow had proliferated. These frameworks raised new questions about both the bodily processes involved in appetite, eating, and digestion and—as modern forms of mind-body dualism took hold—the relation between psyche and soma in governing these activities. Thus, arguments unfolded about the nature and work of bodily agents ranging from chemical "ferments" to nervous filaments to "hydraulic" and other mechanical "forces." At the same time, the links between corporal function and the moral and psychological dimensions of appetite—its functioning as dictated by providence or as derailed by the power of passion or desire—received attention in formulations that uneasily blended medical tradition, Christian dogma, and philosophical innovation.[3] In crucial ways these new medico-philosophical frameworks prepared the ground for the full-scale assault on ancient teaching on appetite that was to come in the eighteenth century.

Chemistry and Mechanics of Appetite

In the sixteenth and seventeenth centuries, questioning of the ancients gained momentum among physicians committed to the self-consciously innovative frameworks of "iatrochemistry" and "iatromechanism." The first of these emerged from the teachings of the medical iconoclast Paracelsus (1493–1541) and his disciple Joan Baptista van Helmont (1580–1644), while iatromechanism was associated first with the philosophy of René Descartes (1596–1650) and later with Newtonianism.[4] Iatrochemists shifted attention away from the influence of overall constitution in dietary choice and digestion, seeking instead to identify the action of "ferments" and other chemical agents unknown to the ancients. Helmont, for example, stressed the special appetizing quality of sharp substances and said of salt and saltpeter that " 'it hath a spirit in it that causeth hunger.' "[5] The Leipzig iatrochemist Michael Ettmüller (1644–1683) defined appetite as a "desire of Solid Food, caus'd by the gnawing of a Volatil Acid in the Mouth of the Stomach" and held that different animals "covet[ed] their peculiar Food" thanks to the work of ferments peculiar to their type. Ettmüller also traced a chemical origin for the pica of pregnant women and those suffering "Melancolic Hypocondriac fits": "corrupt matter" in the stomach caused "the fancy to pursue and entertain unusual objects."[6]

In the seventeenth century iatromechanists throughout Continental Europe explored appetite and hunger within the framework of Descartes's mechanical philosophy. Descartes himself described hunger as an "idea" of the soul, itself "seated" in the pineal gland, occasioned by the action of stomachal fluids. In the stomach liquids "turn against the stomach itself and, agitating the filaments of its nerves more strongly than is usual, . . . cause motion in the parts of the brain from which the filaments come. This will cause the soul, when united to this machine, to conceive the general idea of *hunger*." Appetite was roused in a similar fashion but had a more precise aim: liquids that were "so disposed as to employ their action against certain . . . foods rather than others . . . act in a particular fashion also against the nerves of the stomach," causing "the soul . . . to conceive an appetite to eat certain foods rather than others."[7] Although Descartes, like Galen, emphasized the role of the nerves, his physiology eliminated the "innate instincts" that in Galen's system guided ingestion. He posited instead the work of "internal senses" that arose from nervous movements but were experienced as sensations only thanks to their reflection in the soul.[8]

From the 1690s onward, iatromechanism in Britain developed "in the shadow of the *Principia*."[9] Physicians inspired by Newtonianism sought to show the lawlike workings of appetite and hunger. The Scottish physician Archibald Pitcairn (1652–1713), leader of a "hydraulicist" school indebted to Newton, attributed appetite and hunger to the quantity and composition of the gastric juices and their effects on stomach fibers: a "greater Quantity of Bile . . . destroys the Sense of Hunger" and "a load of Viscidity at the Stomach" causes "a Dejection of the Appetite."[10] Pitcairn also focused on body fluids in analyzing the consequences of "a Stomach too full," which caused "an *Anorexy*, or a Loss of Appetite" and a "Stomach too empty," whose lessened viscidity could result in increased appetite or, in extreme form, "dog-hunger" ("Boulimy") (241, 258–61). Newtonians like Pitcairn sought to banish the uncertainties of ancient medicine by viewing the body as a web of vessels and fluids that obeyed "mechanical laws."[11]

The extent to which these new medical systems encouraged a search for uniformities of appetite varied. Iatrochemistry, with its origins in alchemy and diverse forms of mysticism, continued up to the eighteenth century to favor a sense of the spontaneity of life processes. Iatromechanism, in contrast, promoted—as the example of Pitcairn indicates—a search for the lawlike in the organic as in the inorganic realm. Even so, physicians who employed mechanistic explanations of body processes often still recognized a high degree of variability in appetite, eating habits, and dietetic remedies out of respect for medical tradition or practical experience, their own or others. Such was the case with the Halle mechanist Friedrich Hoffmann (1660–1742), whose rationale for recommending that old men consume ass's milk to prolong their lives rested not on iatromechanical theory but on favorable reports of this prophylactic for aging by the seventeenth-century physician Guy Patin.[12]

For physicians committed to the man-machine construct, the matter of whether appetite reflected uniform processes was embedded within the related and vexed question of how a phenomenon originating in the lowly stomach could become a matter of consciousness, something on which the individual could act with moral choice and volition. Indeed, this difficulty made appetite a paradigmatic problem of the relation between body and mind or soul. Descartes avoided this question when dealing with appetite and hunger, referring simply to "the soul, when united to this machine" but not attempting to explain just how this union was accomplished. Yet despite such vagueness, mechanist dualism established the terms of discussion of appetite for ensuing decades, and, in key respects, into our own day. One effect of mind-body dualism was to encourage the emergence of a separate science of psychology that, though construed by its advocates as part of natural science, treated appetite and kindred phenomena as operations of soul largely unrelated to corporeal physiology.[13] Even a figure such as David Hartley (1705–1757), a physician by occupation if not by training and one frequently associated with mechanist materialism and "physiological psychology," viewed appetite chiefly from a moral perspective. In his influential *Observations on Man* Hartley attributed "natural hunger" to "vigorous vibrations" resulting from "increase[d] contractions of the muscular coat of the stomach," but when he explored "the appetites," he emphasized their transformation by immoral inclinations and choices. Hartley held that in children the appetites were "the best guides . . . in respect of quality and quantity" of food until "their tastes and appetites [were] perverted . . . by customs and practices derived from our corruptions."[14] Only the work of divinely ordained conscience could overcome "the irregularity and inordinateness of the bodily appetites" grown "excessive through unlawful gratification" (2:226).

Other godly mechanists also saw appetite and digestion as operating under the guidance of a benevolent providence. The Jansenist Philippe Hecquet (1661–1737), author of a major treatise on digestion, developed a view of appetite that corresponded to his general theory of digestive function. Denying the role of obscure chemical processes, Hecquet instead attributed digestion to a straightforward process of "trituration," an ensemble of grinding and mashing actions in the stomach. An austere vegetarian, he saw trituration as most successfully exercised on the "raw vegetables . . . [that were] formerly the ordinary nourishment of men."[15] For a figure such as Hecquet, trituration fit within a wider philosophy of the body that saw life and the organization of matter as directly dependent on the guiding action of the "Great Designer" and thus all bodily activity, including trituration, as the result of movements imparted by God.[16] In this way of thinking, appetite too was governed by providence: beneficial as given, it was easily corrupted by wayward desires. The Calvinist Herman Boerhaave (1668–1738), often regarded as the exemplary mechanist of the early eighteenth century, emphasized not wayward desire but divine beneficence when he counseled that patients with seemingly unhealthy appetites must be allowed to eat as they desired for "this automatic motion or want . . . happens through 'the will of the great Author of all Things.'"[17]

Around 1700, so-called animist physicians shifted away from mechanist constructs of body-mind relations toward theories that saw corporal activity as directed by the soul. A leading figure in this shift was the Halle professor G. E. Stahl (1660–1734), who viewed the soul as fully united with the body and the seat of "all perceptual processes," intellectual and emotional. Stahl viewed hunger and thirst as "determinations" of the soul, which guided the organism in the choice of substances to satisfy bodily needs; erroneous choices resulted from wayward appetite, which could undermine health by habitually heeding "false desires."[18] In exploring the nature of appetite, animists both drew on and contributed to "psychologia." The Montpellier medical professor François Boissier de Sauvages (1706–1767), for example, followed Stahl in seeing disorders of appetite as primarily mental and spiritual in origin and in postulating that a soul disordered by passion could be the direct cause of physical maladies.[19] Medical animism thus served as one of the powerful currents that fostered interest in how desires and feelings could effect "movements" in both body and mind.

Animists' advocacy of the soul as responsible for both normal and disordered bodily function coincided with work on the anatomy and physiology of the nerves that intensified from the later seventeenth century onward. Working within the Galenic model of the nervous system that gave the brain and spinal cord (regarded as an extension of the brain) direction of the material activity of the body, students of nervous structure and function, from Francis Glisson (1597–1677) and Thomas Willis (1621–1675) onward, sought to define the relation between the substrate of brain and nerves and the mental and emotional activities previously assigned by theologians and physicians to the soul.[20] These developments meant that by around 1750, discussions of appetite came to be tied to larger inquiry into the nature of feeling itself, whether conceived as sensations dependent on the work of the external senses, "internal senses" arising from obscure visceral origins, or—and here a complex vocabulary developed—activities enabled by the "power," "faculty," or "property" of "sensibility." In the eighteenth century these problems increasingly became the special province of medical physiology, but they also figured in the medical project of "nosology," the effort to bring new certainty to medicine by ordering and classifying diseases as distinct entities comparable to other natural phenomena.[21] With these efforts, medical thinking about appetite shifted in important ways. Whereas in medical tradition the state of the appetite served chiefly as a sign of general health and disease, the nosological effort to concretize disease gave troubled appetite enhanced importance as a distinctive kind of illness with its own varieties.[22]

Maladies of Appetite in Eighteenth-Century Nosologies

The most influential Enlightenment nosologists were the Montpellier physician Sauvages and his Edinburgh successor William Cullen (1710–1790), both of whom sought to identify and give advice on treatment of every known disease (for Sauvages, more than 2,800 of them).[23] In these efforts to provide a stable classification of all diseases, appetite assumed a new status: no longer simply a sign, it appeared as a phenomenon subject to its own identifiable disorders. Sauvages conceived of appetite in terms supplied by eighteenth-century psychology, presenting it along with imagination and judgment as components of reason. Accordingly, he classified disturbances of appetite not with bodily disorders but with those that "trouble the reason." These maladies included variants of "voracious appetite" ("canine hunger," "bulimia") as well as "antipathy to food" and the "depraved appetite" evident in pica.[24] Thus, in considering the *bizarreries* afflicting appetite, Sauvages first identified the particular "species" of disease and then described its causes, signs, and symptoms, as well possible treatments. Pica, for example, he divided into varieties seen in children, young women suffering from *pâles couleurs* (chlorosis), pregnant women, those suffering from scurvy, those who indulged "envies bizarres," and the "simulated pica" of street entertainers who sought to impress spectators by eating stones.[25] Except for the form associated with scurvy, all of these were disorders of the will suffered by patients of limited rationality—children, women, street entertainers—whose behavior was "bizarre, capricious, [or] willful." Such people, "lacking reason, constantly desire that which is not a true good and feel disgust for what appears as bad but is in fact a genuine good" (2:665).

In his nosology Cullen drew heavily on the prior work of Sauvages. He too included "want of appetite" as a symptom characteristic of many diseases while reserving his principal discussion of troubled appetite to his fourth class of disease, "*locales*," for which he tried to identify diseases as "affection[s] of a part, not the whole body" (154). Cullen looked closely at the order of disease he labeled "dysorexiae" or "false or defective appetite," situating within it conditions Sauvages had termed

1 Marshall Hall, *Commentaries on Some of the More Important of the Diseases of Females* (London: Longman, Rees, Orme, Brown, and Green, 1827), plate 7. In the nineteenth century the term "chlorosis" largely replaced Sauvages's *pâles couleurs*; whatever its label, the condition was marked, as Hall observed, by an "appetite . . . generally impaired and capricious, and even depraved" (57). Wellcome Collection.

bizarreries or *morosités*, including forms of anorexia, bulimia, and pica ("a desire of eating what is not food" [163]). Yet although Cullen followed Sauvages in many respects, he criticized his predecessor for classifying these diseases among the "vesanias" when in fact "almost every species of Dysorexia is . . . an affection of a part, rather than of the whole body" (162n). Thus, he denied, for example, that different forms of anorexia constituted distinct diseases, stating, "Every Anorexia seems to me to be symptomatic, and to vary only according to the disease it accompanies" (164). Still, Cullen himself was uncertain about where, finally, ills of appetite should be placed. His main discussion of appetitive ills appeared with "local diseases," but he also listed them among the "neuroses" (the term, of his coinage, designating nervous disorders).[26] In the end, then, Cullen was unable to decide whether ills of appetite were mental or corporal, "whole body" or local, "idiopathic" (independent) or symptomatic.

Despite the strenuous efforts of Sauvages and Cullen to clarify the myriad manifestations of troubled appetite, their nosologies indicate a divide in medical thinking about the origin and nature of appetite that was to deepen from the eighteenth century onward: was appetite a mental phenomenon, as Sauvages asserted in defining it as a component of reason, or was it a bodily phenomenon, as Cullen suggested when he criticized Sauvages for including appetite disorders within the class of *folies*? In this fashion, nosologists approached, without doing much to resolve, questions that from this point onward have bedeviled inquiries into appetite, not only whether appetite was bodily or mental but also the therapeutic and moral question of whether appetite was under the control of reason, judgment, and will or, instead, was a corporal phenomenon functioning largely outside the reach of mental operations. As will be seen, such questions would influence not only choices of treatment for patients suffering appetitive ills but also the choice of methods and future direction of research on the origin and nature of appetite.

Maladies of Appetite in Practical Medicine

While raising such issues as the bodily or mental character of appetite in their nosologies, Sauvages and Cullen also suggested types of appetite disorder that most concerned physicians in practice. These included the two extremes of over- and undereating but also such diffuse and elusive conditions as dyspepsia and, a variant much discussed in Continental literature, the syndrome known as *maladies des gens de lettres*, a pathology afflicting the learned that almost always encompassed disturbances of appetite and digestion. Given the enticements of late eighteenth-century food culture—the greater accessibility of once-luxurious food and drink and the spread of sophisticated modes of preparing food—physicians worried especially about the dangers of excess and the harm caused by consuming overrefined comestibles. Up to midcentury, medical worries of this sort were often intermixed with religious condemnations of gluttony that had picked up force since the sixteenth century as Protestant and Catholic reformers articulated austere codes of morality emphasizing individual responsibility for controlling the appetites and all forms of bodily comportment.[27] But there is ample evidence that concerns about overeating changed character as the eighteenth century wore on. Few physicians explicitly challenged religion as a guide to appropriate restraint of appetite, yet specifically medical concerns, embedded within a larger discourse of "diseases of civilization," steadily displaced older views of gluttony that associated excessive eating with sin.[28]

Since the late seventeenth century, "corpulence" had gone through a gradual process of medicalization, and Enlightenment physicians believed the condition to be ever more widespread.[29] Yet why people grew fat was much debated. Medical descriptions of "canine hunger" and "bulimia" that appeared around 1750 anticipated the efforts of nosologists to sort out physical and psychic etiologies for different forms of voracious eating. In an entry in the *Encyclopédie*, Louis de Jaucourt (1704–1779), a student of Boerhaave, emphasized physical causes of such maladies when he characterized canine hunger as a form of excessive eating accompanied by constant vomiting and also defined bulimia as marked by oppression of the stomach, difficulty in breathing, a feeble pulse, cold, and fainting.[30] Although these ills had different symptoms, both resulted, Jaucourt argued, from physical problems of the stomach, ranging from acrid humors to excessive sensibility of the stomachal nerves. Another entry likely written by Jaucourt represented these conditions as aberrations of hunger, which he distinguished from appetite by noting that hunger was connected "to need alone," whereas appetite was linked to "taste and pleasure," and that hunger concerned only food, whereas appetite often involved desire for objects that proved "a great evil."[31] The implication was clear: disorders of hunger were physical, and deranged appetites were born of self-delusion and faulty judgments.

As did Jaucourt in France, British physicians distinguished obesity caused by the indulgence of appetite and excessive eating from that caused by physical disturbances or external influences. Physicians condemned overeating, especially of soft, smooth, oily, and fatty foods, and charged the obese with "sloth and idleness." But they also adduced as causes of corpulence bad air, defective "transpiration" or evacuation of waste materials, an abnormal state of the blood, or, a cause some regarded as hereditary, a laxity of solids, especially the "cellular or fatty membrane."[32] While reproving those who ate too much, the physician Matthew Flemyng nonetheless noted the puzzling fact that not all obese people were big eaters or all thin persons abstemious. All these physicians urged patients to change their habits—to exercise, consume a moderate diet, and reduce or eliminate the consumption of "spirits."[33] But they also prescribed "vomits," purgatives, and other drugs to resolve physical problems outside the patient's control.[34] Thus, medical thinking about corpulence, though linked to the larger moral discourse of luxury and excess, often had little to do with indulgence of appetite, emphasizing purely physical factors instead.[35]

Appetite occupied a similarly ambiguous place in medical thinking about other alimentary and stomach ills. Loss of appetite headed Cullen's list of the symptoms characterizing dyspepsia, a term increasingly favored for indigestion and diffuse stomach upsets.[36] Yet why and how appetite was disturbed in dyspepsia was unclear. Cullen tended overall toward the view that the loss of appetite in dyspepsia was nervous in origin, but he said nothing specific about the location or work of the nerves involved. His lack of specificity on this point reflected a general shift, under way from the 1760s, toward viewing dyspepsia and related ills as physico-moral maladies that depended not only on the state of the stomach but also on the general condition of the nerves or, as nervous influence was increasingly characterized, nervous "sensibility." In this respect there were interesting national differences. While "dyspepsia" was a label favored in Britain and North America, in Continental Europe the Swiss physician Samuel-Auguste Tissot's (1728–1797) *maladies des gens de lettres* was perhaps the most prominent syndrome in which troubled appetite resulted from a general disturbance of nervous sensibility. In scholars and other sedentary individuals who engaged in excessive mental work, Tissot saw the stomachal nerves "sympathizing" with an irritated brain and depleting the body's "vital forces."[37]

In ways that I explore later, concepts of heightened sensibility and deficiency of vital force gained ever more adherents in the late Enlightenment with the spread of medical vitalism.[38] These concepts undergirded the increasing tendency of physicians to see maladies like troubled appetite as highly individualized yet at the same time as marking specific physico-moral types. For a complex of reasons, the type that especially fixed the attention of late Enlightenment physicians was the

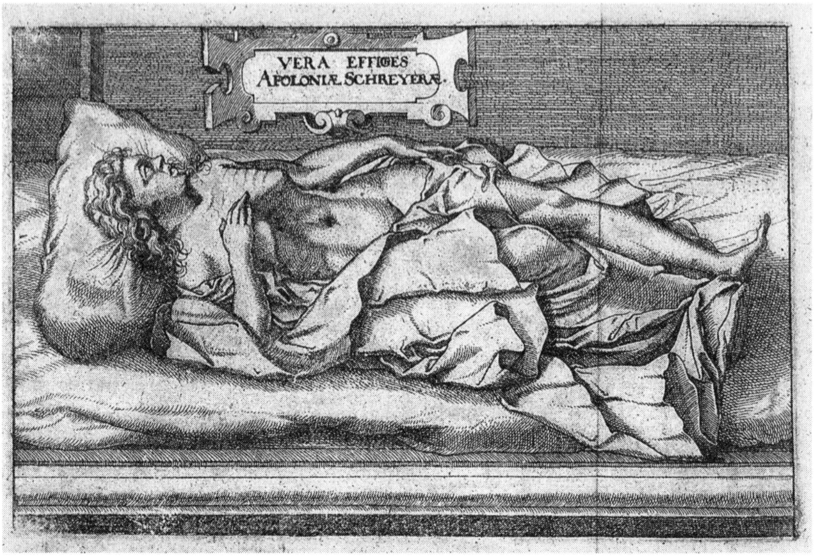

2 Apollonia Schreier, etching of a young woman after a fast of eleven months, near Bern, Switzerland, 1603. Wellcome Collection.

female. Medical concern with female pathologies of appetite—girls or women who would not eat or who craved strange substances—seems at first glance strikingly at odds with the anxieties about excess manifested in "diseases of civilization." But the attention given to these peculiarly feminine pathologies intersected in important ways with claims that well-managed appetite was a mark of civilization. In Catholic countries medical discussion of the pathological eating of girls and women often formed part of an attack on practices of mortification of the body through dietetic asceticism. Late Enlightenment physicians eagerly dredged up old cases or discovered new examples of "fasting maidens," holy women, and female saints who claimed the capacity to subsist for days, months, or years without eating. Except for the occasional adolescent boy, such cases concerned women and girls exclusively, and they demonstrated a host of lamentable female characteristics encouraged by such religious practices—credulity, vanity, and mendacity chief among them.[39]

Physicians who ridiculed the superstitions undergirding such beliefs sought to replace ascetic eating practices with the norm of moderation taken from the ancients but by the early modern era developed in concert with ideals of polite, urbane behavior that signified the advance of civilization.[40] At the same time, physicians in all countries, Catholic and

Protestant, participated in a broad antifeminist drive to reaffirm essential differences between the sexes and to undercut feminist arguments for women's intellectual and moral equality that had gained ground since the late seventeenth century.[41] In this context the putatively disordered appetites of women were a frequent target. From around 1760, physicians increasingly stressed the links between women's appetite for food and sexual and reproductive processes. In some cases this meant simply reemphasizing the dogma that women were "dominated" by the uterus and reproductive system.[42] More innovative was the view promoted by physicians in the vanguard of the late Enlightenment "cult of sensibility" that women were women not in one organ or function alone but, thanks to their peculiar nervous sensibility, in their whole being.[43] Whatever the reasoning, this renewed disposition to stress female difference meant that physicians diagnosed distinctive pathologies of appetite in girls and women that reflected what were perceived to be common female failings—weak reason, wild passions—that explained their resemblance to other ungovernable types, including children, savages, and people made mad by hunger. Medical representations of the "bizarre" or "monstrous" cravings of pregnant women and nubile girls for substances such as chalk, clay, dirt, and unripened grains recalled the crazed eating that in proverbs and folk tales marked the behavior of the hungry amid prolonged dearth—the consumption of bark, leaves, boiled bones, and, most horrifying, human flesh. Even in times of plenty women were given to "harmful excess" when deranged by "dangerous passions."[44]

Chief among new medical preoccupations was the link between disordered eating and "nervous" afflictions—the blend of psychic and physical distress commonly called the vapors—that physicians believed to be constantly on the rise, products of the "false appetites" stimulated by hectic and pleasure-driven urban life.[45] Early in the century both men and women suffered vaporous and nervous affections, and Cheyne and other physicians regarded hypochondria (usually male) and hysteria (usually female) as "largely interchangeable" illnesses. Yet by the late Enlightenment, the vapors and hysteria, the most frighteningly debilitating of nervous ills, had come to be regarded as the special province of females, despite the persistence of a gender spectrum that occasionally included effeminized males.[46] In the late-century constellation of symptoms, moreover, it was girls and women whose special nature made them susceptible to long-term, sometimes even fatal, troubles of appetite. Certainly sedentary and cerebral men suffered nervous-based eating and digestive problems, but increasingly these tended more to winds, "eructations," and constipation than to a persistent loss of appetite or aversion to eating.[47] In contrast, cases of chronically undermined appetite, with attendant loss of general vitality, were almost always female.[48] By the 1790s cases described by the founder of mental medicine, Philippe Pinel (1745–1826), suggested that if a man lost interest in eating and died of inanition, severe mental illness must be the cause.[49]

In the late eighteenth century physicians who made claims about appetite did so largely on the basis of time-honored methods. If they were philosophically inclined, they might rely on introspection, asserting with Antoine Le Camus (1722–1772) that self-examination of taste could illuminate how individuals experienced their peculiar pleasures and aversions for wine and other substances.[50] Other physicians, including the celebrated nosologists, gauged the limits of appetite and hunger by culling strange tales from history, travel literature, and medical tradition itself, with its panorama of bizarre, exotic, and monstrous bodily phenomena.[51] Yet the method that gave medicine its special distinction with respect to appetitive and all other ills was bedside observation, regarded by some physicians as the only reliable way to gain knowledge of human health and disease. Assessments of appetite were a constant of the practitioner-patient relationship, and although little discussion is to be found in illness narratives of just how to gauge the state of a patient's appetite, this seems to have been the case precisely because physicians had full confidence in their capacity to know a good, bad, failing, weak, or reviving appetite when they saw one.

Thus, despite the increased disposition to see distinctive diseases or fixed patterns in manifestations of troubled appetite, many under the influence of gender, eighteenth-century physicians continued overall to emphasize the individuality of appetite, as seen both in eating habits and in episodes of disturbed appetite that could shift in an instant depending on changing circumstances. This "Hippocratic" approach to illness retained great prestige in the eighteenth century, with physicians from every camp declaring allegiance to the "father of medicine."[52] Nonetheless, ways of conceiving the body and its maladies that downplayed variability and sought uniformities or universals in the nature and activities of living organisms steadily gained force from midcentury onward. This search for uniformity entailed a new effort not simply to discern how appetite manifested in different categories of patient and particular forms of illness but also, more fundamentally, to explain the origin and nature of the phenomenon of appetite itself. Physicians were prominent among those who took up this task, although they did so less in their capacity of ministering to sick patients than in the increasingly honored role of the "physiologist," who employed modes of investigation beyond the bedside to illuminate the nature of vital functions.[53] Physiology long retained its connection to medicine and remained a science focused on the human body and its functions. Yet by the mid-eighteenth century, leading physiologists, in company with natural historians and other investigators, began to urge that human characteristics could be grasped only in relation to those of other living creatures. While "man" remained the measure of all things, knowing the nature of animal and plant life was increasingly seen as indispensable in setting the boundaries of the human, in appetite and eating as in all essential functions. The bifurcation between the tradition-minded physician who followed the ancients in stressing the individuality of human appetite and innovators who sought new methods to establish processes universally at work in organic nature is the subject of the next chapter.

THREE

Human and Animal Appetite in Natural History and Physiology

Throughout the Middle Ages and the early modern era the eating ways of animals—their avid desires, their clever styles of searching for and getting food, their competitions and acts of aggression—held an important place in European folklore as well as in the meditations of philosophers, theologians, and poets.[1] Nonetheless, the systematic observation, certainly the deliberate manipulation, of animal eating began only in the eighteenth century as investigators trained outside of medicine and supported by omnibus scientific institutions like the Royal Society in England and the French Académie des Sciences dedicated ever-greater attention to the "sciences of life."[2] Such activities entailed a new determination to achieve an ordering of life phenomena previously regarded as spontaneous, even wild or monstrous in character. Thus, natural historians strove to illuminate appetite and related phenomena in relation to the broad canvas of organic beings and their nutritional activities. Experimentalists went even further, attempting to reveal the workings of ingestion and digestion, human and animal, by staging experiments that achieved uniform and replicable results. By century's end, such procedures, and the larger disposition to study human appetite in relation to that of animals, had come to be widely seen as indispensable to a new, "scientific" mode of interrogating appetite.

The Natural History of Appetite and Ingestion

A landmark work that exemplified new approaches to appetite and eating was that of the French naturalist René-Antoine Ferchault de Réaumur (1683–1757), who sought, in investigating animal eating (especially in birds), an unassailable method for settling long-standing debates about the digestive process. Réaumur devised an ingenious test to determine the relative importance during digestion of chemical action and trituration by having birds ingest food enclosed in glass and metal tubes that prevented trituration, and then examining the remains of the digestive process. Finding in some birds that digestion proceeded unimpaired even though mashing and grinding operations were prevented, Réaumur proved, to his own satisfaction and that of many who commented on his work, that digestion was at least partially a chemical rather than mechanical process.[3]

Réaumur's effort to define the digestive process with new certitude by experimenting on selected animals continued work he had previously undertaken to demonstrate the efficacy of new methods of hatching and raising domestic poultry.[4] Arguing on behalf of rapid hatching achieved through the use of heated ovens, he responded to objections from critics who believed that poultry nurtured "in a different way from the ordinary" would not be of the same quality as those raised in the usual fashion. He was particularly concerned to show that his chickens could be fattened and would have the same good flavor: "Why should they not have the same appetite as the others! And why would they digest more poorly?" (2:251–52). In presenting his conclusions, Réaumur did not draw out all the implications of such investigations, but he defended two claims of great interest in changing views of appetite. One was that experimentation did not interfere with "natural" processes of feeding and fattening, including the animals' choice of foods according to their age, size, and sometimes rapidly altering tastes (2:200). Second, he urged that these processes were, in ways that experimentation could show, uniform from one individual of a species to another, even when interfered with by manipulative practices of the kind he employed.

Réaumur's work was much admired, but no naturalist of the period wielded influence equal to that exerted by Georges-Louis Leclerc, comte de Buffon (1707–1788), whose massive *Histoire naturelle, générale et particulière* not only included extensive detail on the eating habits of diverse animal species but also made appetite central to his speculations about human and animal nature. Buffon approached the problem of appetite systematically in the "Discours sur la nature des animaux" that opened the fourth volume of the *Histoire naturelle,* an essay that considered at length the differences separating human beings from animals and the reasons for human superiority over animals in the order of nature.[5] Buffon sought to distinguish, first, functions that characterized the animal from those of the vegetable order and, then, animals from humans. The former distinction rested on the idea, taken from Aristotle, that vegetables experienced only "interior" functions of nutrition, growth, and reproduction, whereas animals were equipped with structures and faculties that allowed them to engage with the exterior world.[6] Chief among these was the capacity to take in impressions from the outside and to act in response to knowledge gained by this means. And just as the capacities of feeling and movement divided animal from vegetable, so specific modes of feeling and movement divided animal from human. Although in important ways Buffon was committed to the Cartesian view that animals lacked a soul and operated essentially as machines, he embraced Lockean principles in positing that all animal beings experienced sense impressions as either "agreeable" or "disagreeable" insofar as they helped the animal to fulfill its "natural needs."[7] Thus, if an animal consumed a foodstuff that met its need for nourishment, it experienced pleasurable sensations. These sensations in turn created an "appetite" or "desire" for nourishing foodstuffs that persisted as part of the animal's "interior sense," which subsequently guided its search for and ingestion of food. In animals, Buffon held, this interior sense was tied particularly to exercise of the senses of odor and taste, which were so exquisitely developed in animals that their choice of food was "purely mechanical" and virtually error-free.[8]

The nature of appetite was equally crucial to Buffon's delineation of human nature. Human beings also took in sense impressions that they experienced as agreeable or disagreeable, but the sensory impressions of greatest moment to human activity derived from the more complex and "elevated" senses of sight, hearing, and touch. This meant that in the exercise of appetite human beings were much inferior to animals and given to error; although in the ordinary course of things an animal would never eat anything harmful, human beings would, under many circumstances, as readily consume poisons like hemlock as they would nourishing substances like parsley (4:32). Yet the distinction Buffon drew between human and animal sensory powers was by no means his central consideration in differentiating human and animal appetite. Man's more complicated and variable appetite differed from that of the animal primarily because—as with all human actions—its functioning was determined not only by the sensory apparatus but also by the operations of the "soul." While animals were guided by a "purely material" interior sense, human beings were endowed with a "spiritual substance" whose guidance was not automatic but instead unfolded in accord with the dictates of uniquely human capacities of imagination and passion, judgment and reason.[9] In this difference lay not only human nobility but also the source of human error and suffering. Enticed by imagination, itself made possible by the strictly human capacity to compare remembered sensations, or "deranged by passion," human beings sought to increase their pleasures beyond what nature provided in abundance and, in so doing, brought on themselves miseries born of desire, excess, and debauchery. By this means Buffon arrived at the paradox that although human beings were superior to animals in "dignity," they were inferior to animals in health. Human beings knew far more illness and physical suffering than animals thanks to the agitation caused by the search for pleasure and the uncertainty, in humans, of the sensory powers connected to need and appetite (4:32–34, 43–48).

There is still another dimension to Buffon's thinking about appetite that complicates even further this account of the certainty of animal eating and its contrast to human appetite. In the essay on the "Variétés dans l'espèce humaine," Buffon offered lengthy descriptions of the diverse human types observed by European voyagers, and he attempted to explain the nature and origin of their striking differences.[10] These anthropological reflections led him to postulate that human beings differed among themselves primarily because of the influence of climate and diet. In this context Buffon explained appetite in accord not with eating desires but with the availability of given foodstuffs in particular regions and concomitant dietary habits. Here Buffon celebrated the ways in which human beings of superior capacities shaped their environment, propelling themselves beyond the "state of nature," and he lamented the failure of barbarous and "degenerate" peoples around the globe to do anything more than respond to basic needs. The monotonous diet seen among Lapps, Greenlanders, and other "gross . . . [and] stupid" peoples (3:373–75) was, therefore, a sign of lack of imagination, the very capacity that, when heightened among civilized Europeans, prompted dangerous excess. In sum, there was, to Buffon, no dietary noble savage whose eating instincts could serve as a model for the European appetite that so frequently went astray. Again following Aristotle, Buffon argued that the only healthy appetite was that based on reasoned, civilized moderation.

The Physiology of Appetite: Mechanists and Vitalists

While Buffon's natural history attempted to be comprehensive in covering the history and traits, structures, and activities of living beings, the new science of physiology emergent around 1750 took as its special domain problems of bodily function. In exploring the nature of appetite, physiologists shared with Buffon the complex vocabulary of sensation, the sensory, sentiment, and sensibility, but their goals differed from his.[11] Buffon was intent on mapping differences and similarities in human and animal nature and on charting the appetites of varied species. Physiologists, in contrast, sought to understand human appetite as part of the processes of ingestion, digestion, and nutrition that humans shared with many species; when they studied animals, they did so principally to illuminate human functions. To be sure, the value of human-animal analogies was itself a matter of dispute, with some urging that the greater complexity of human phenomena limited the utility of such comparisons.[12] The perceived complexity of the human drive to eat was apparent in the shifting lexicon that physiologists drew on to characterize it: in French, *besoin, désir, fonction naturelle, goût, pente de l'âme, sensation, sentiment* or *sentiment intérieur*, and, in English, "desire," "natural inclination," "need," "sensation," and "sense." This unstable terminology arose in part from physiologists' efforts to explain the difference between feelings distinguished in common parlance as "appetite" and "hunger." This they ordinarily did by defining both phenomena as sensations but then differentiating the two on the basis of pleasure and pain. Physiologists described hunger as a "need," characterizing it as "pressing," as born of "abstinence" and deprivation, as "voracious" and indiscriminate. Appetite, however, they spoke of using a language of pleasure. Allied to taste and desire, appetite was satisfied only by appealing foods. Delicate, soft, and easily disturbed, it was associated with good digestion and overall health.[13]

While physiologists agreed in positioning appetite and hunger within the nexus of pleasure and pain, they were strongly divided on the origin of these differing sensations and the implications of the pleasure-pain opposition. As with many physiological investigations of this period, views of appetite and hunger differed depending on the authors' relative allegiance to the theoretical postures of mechanism and vitalism. Mechanist medicine whether Cartesian or Newtonian posited that the human body was governed by the same forces and laws that governed the movements of all material bodies in the universe. Mechanists attempted to explain bodily functions with analogies to mechanical tools and actions, focusing more on the operation of individual parts than overall bodily processes. Vitalists, in contrast, insisted on the uniqueness of living organisms, which were governed by a "vital principle" or "vital forces" that acted in ways fundamentally different from the forces and laws ruling "unorganized" or "brute" bodies. Thus, vitalism was, to use a twentieth-century term, "holistic," stressing interdependencies of functional activity, and also teleological, seeing in living organisms innate self-regulatory processes that served to maintain health and life. Vitalist critiques of mechanist thinking got under way in the 1740s, and by the later decades of the century were vigorously pursued across Europe, most notably by the Montpellier school in southern France.[14] Although on some questions differences between mechanists and vitalists gave way as the eighteenth century progressed, divisions related to appetite and hunger and the larger study of ingestion and digestion remained strong. While mechanists sought specific structures and identifiable somatic causes to explain these phenomena, vitalists referred all phases of digestion—of which appetite and hunger were the first—to the work of an overarching "principle of life" or to a specific "digestive" or "assimilative" force.[15]

It is little wonder that mechanist and vitalist explanations of appetite and hunger radically diverged, as the most divisive issue separating these camps had to do with the fundamental nature of sensation and sensibility. Here the work of Albrecht von Haller (1708–1777), the Swiss physiologist whose standing in his domain equaled that of Buffon in natural history, was pivotal. Although on some questions Haller bridged the divide between mechanists and vitalists, he differed markedly from the latter in holding that sensibility, and the related but distinct property he identified as "irritability," were properties lodged in specific kinds of tissue—sensibility in nervous tissue, irritability in muscle tissue—and he worked intensively to gauge the effects of specific stimuli and irritants on the different tissues.[16] Vitalists, in contrast, saw these forces as spontaneously inhering within living organisms and as functioning independently of specific structures, nervous or otherwise. Although they did not discount the role of nervous stimuli, they denied that nervous structures were necessarily involved in vital action, emphasizing, instead, the general and freely functioning sensibility of the organism.[17] Accordingly, vitalists conceptualized "internal sensations" of appetite, hunger, and thirst as inherent urges that arose from deep within the organism and drove it to the performance of functions essential to the maintenance of life.[18] Thus, mechanists and vitalists differed strongly as to whether or not sensations of appetite and hunger required a describable apparatus or mechanism. Mechanists searched diligently for "organized" structures to explain these phenomena, and vitalists held that such efforts missed what was most important about internal sensations: their spontaneous functioning to preserve the general life and health of the whole. Of the many mechanist hypotheses proposed, the most influential and long lasting were those offered by Haller in his physiology textbook of the late 1740s, and then, some decades later, by the Italian naturalist Lazzaro Spallanzani (1729–1799).[19]

In his inquiry into what drove ingestion, Haller focused chiefly on hunger rather than appetite, defining it as a "sense of pain," an "acute," "intolerable" sensation, created by the "grating" of irritated nerves and tissues against one another:

> We are solicited to take food, as well from the sense of pain we call hunger, as from that of pleasure, which is received by the taste. . . . The first of these proceeds doubtless from the sensible folds or wrinkles of the stomach rubbing against each other by the peristaltic motion, of which there is an acute sensation, joined with a pressure from the diaphragm and abdominal muscles, by which the naked villi of the nerves on one side grate against those of the other, after a manner intolerable. Thus we are effectually admonished of the dangers ensuing from too long abstinence or fasting, and excited to procure food or nourishment by labour and industry. To this sense perhaps the gastric liquor or juice of the stomach, collected and sharpened after feeding, does in some measure conduce, unless it becomes putrescent.[20]

Thus, Haller held that hunger resulted from movements occasioned in the stomach by muscular irritation and by nervous sensations operative in the stomach's abundant network of nerves, although he was unable to determine with any precision the relative importance of irritability and sensibility in the work of the stomach. Similarly, in brief remarks he made on appetite, he proposed a straightforward mechanical explanation for its appearance, suggesting that appetite resulted from muscular pressure that caused saliva to pour into the mouth.[21]

Haller's characterization of hunger as the result of stomachal movements was still approvingly cited in the early twentieth century.[22] But his formulation left many unanswered questions, including, as he himself indicated, the role of the "gastric liquor" in the progressive phases of digestion. The Italian naturalist Lazzaro Spallanzani, who had already gained celebrity for his work on generation, took up this question in a series of essays published in Modena in 1780 and translated into French in 1783.[23] Spallanzani rejected explanations of digestion based on trituration

Herrn Albrecht von Hallers
Anfangsgründe
der
Phisiologie
des menschlichen Körpers.
Aus dem Lateinischen übersezt
von
Johann Samuel Haller.

Erster Band.
Die Faser; die Gefässe; der Umlauf des Blutes;
das Herz.

Berlin,
bei Christian Friedrich Voß, 1759.

3 Albrecht von Haller, *Anfangsgründe der Phisiologie des menschlichen Körpers*, vol. 1 (Berlin: Christian Friedrich Voss, 1759), title page. Engraving by P.-F. Tardieu after Charles Eisen, celebrating knowledge gained by dissection and vivisection. US National Library of Medicine.

as well as the eclectic view, which he attributed to the "celebrated" Boerhaave, which rested on a mix of trituration, the action of stomachal heat, and chemical fermentation (2, 255–56). His own investigations were inspired by the work of Réaumur, whose approach he extended by performing experiments on animals possessed of the three main types of stomachal structure—muscular, membranous, and "mixed." Following Réaumur, Spallanzani first had his animals swallow metallic or wooden tubes filled with assorted foodstuffs; as in Réaumur's experiments, the tubal casings prevented the mechanical compression required by triturationist theory. He then waited for varying spans of time while digestion proceeded before slaughtering the animals, retrieving the vessels, and studying the remains left in the stomach. These experiments, all of which showed that the substances the animals ingested were at least partially digested without the aid of mechanical action, convinced Spallanzani that digestion was either solely or principally (depending on the stomachal structure) the work of the gastric juice (251–53).

Spallanzani went beyond Réaumur's work of the 1750s by extending his inquiries to human digestion, which he studied by undertaking often painful self-experimentation. Following the same procedure he had used with animals, Spallanzani swallowed cloth bags and wooden tubes containing different foodstuffs to assess the effects on these vessels of passage through the digestive tract, the time required, and the state of the materials after excretion. In other experiments he collected his own gastric juice (after self-induced vomiting) and, by testing its effects on a variety of foodstuffs, demonstrated the success of "artificial digestion" achieved outside the body.[24] In these descriptions Spallanzani attempted to provide a convincing explanation of digestion, but he also, evidently against the grain, offered fascinating insight into the attitude of a mechanist-minded researcher toward his own feelings and reactions to such an unusual pattern of ingestion. He acknowledged the "repugnance" he sometimes felt at his own procedures and lamented the "weak stomach" that constituted an obstacle he had to overcome (236, 247). Yet, overall, he described his own bodily experience amid these experiments in a tone of stoic detachment differing little from that he adopted in discussing his animal subjects. Emphasizing the size and shape of the vessels ingested and the qualities of the foodstuffs they contained, he paid little heed to the apparent disposition of his animals toward this kind of eating. Whether they exhibited signs of appetite at the outset or lost it along the way he did not say. Indeed, the only points at which Spallanzani brought appetite and hunger into his account were moments when these proved useful in discrediting competing theories of the digestive process. Discussing the function of the "small stones" found in the stomachs of different species of birds, for example, he denied that these parts aided in trituration and suggested, instead, that their movements might serve to augment the birds' appetite (27–28). Similarly, he made use of hunger in arguing against the role of putrefaction in digestion. Noting that many animals had an aversion to putrefied meat, he observed that others, beasts like the vulture and the hyena, willingly nourished themselves with "corrupted matters" (314). Still others had a horror of such substances but could be habituated to their ingestion little by little so that, when pushed by hunger, the animals consumed them voluntarily (314–15). This accommodation to rotten flesh he associated with the fact that such substances were "restored" in the stomach by the "antiseptic" action of the gastric juice.[25] Thus, ingestion proceeded thanks to some combination of strong need (expressed in hunger) and an inner readiness to consume substances that, though initially undesirable, were rendered acceptable by the animal's innate body chemistry.

Spallanzani never asked directly whether it was gastric juice that stimulated the desire or urge to eat, and, as just noted, he thought it possible that in certain birds at least appetite was encouraged by the mechanical action of little stones in the stomach. Still, his readers, including his translator and commentator Jean Senebier (1742–1809), saw in Spallanzani's account the most "natural" explanation for the origin of hunger, "a sentiment whose intensity must increase with the duration of the action" of the gastric juice on the glands "responsible for its preparation."[26]

Others were not so sure. Although leading physiologists of the early nineteenth century credited Spallanzani for ending useless speculation about the mechanisms of digestion, others contended that his results could not be applied to human beings, in whom digestive capacity varied enormously.[27] Nonetheless, despite continuing argument about the role of the gastric juice, Spallanzani's work helped both to ensure that chemical theories of digestion gained a legitimacy they had not enjoyed since the heyday of the old iatrochemistry and to undercut objections to the intrusion of chemistry into physiology lodged by vitalists who posited an absolute distinction between vital and physicochemical processes.[28]

An essential point to make here is that, in dealing with the urge to eat, both Haller and Spallanzani attended principally to hunger rather than to appetite. As would remain the case into the twentieth century, physiologists who sought specific structures or incitory processes to explain the drive to eat focused on hunger as the sharper and more distinct of the two phenomena.[29] This stance encouraged some physiologists to assign appetite a position as the first "phase" or "degree" of hunger, a vague feeling that was experienced pleasurably but hard to pin down.[30] Mechanist physiologists found appetite difficult to explain in part because it was associated with the pleasure that came from specific foodstuffs, a matter they sought to illuminate by reference to the physical apparatus of gustatory sensations. Haller led the way in this respect: in characterizing the difference between the painful and pleasurable sensations that drove ingestion, he emphasized not the distinction between hunger and appetite but that between hunger and taste, a maneuver that allowed him to concentrate on the sensory structures of mouth and tongue in seeking a definite physical source of the pleasurable sensations in question (1:258–65). Still, the origin of taste was not a problem that mechanist physiology was well equipped to handle, as Haller himself admitted. In discussing varieties of taste sensations, he recalled his initial inclination to tie the perception of different tastes to the figures, shapes, and sizes of minute parts of foods ingested, and his eventual realization that such an approach was inadequate. After discussing the shape of various salts, he concluded that tastes are "too inconstant or changeable, to allow such a theory. . . . The mechanical reason, therefore, of the diversity of tastes, seems to reside in the intrinsic fabric or apposition of their elements, which do not fall under the scrutiny of our senses."[31]

In regard to matters of taste, mechanists were in much the same quandary as the one they faced in attempting to explain glandular secretions by means of the compression of glandular bodies by surrounding muscle and bone. After 1750, it was widely recognized that this approach did nothing to explain why particular glands secreted particular fluids. Indeed, it was in addressing this problem that vitalists first made inroads against mechanists, denying the explanatory power of such a model for glandular action and substituting for it a view based on "internal sensations," specifically the "taste" or "desire" of the gland that determined which components of blood it drew to itself and acted upon in furtherance of its specific function.[32] Building on this view of the body's internal wisdom, the Montpellier physiologist J.-C.-M.-G. Grimaud (1750–1789) developed a view of appetite and hunger, eating and digestion, that emphasized not anatomical structures but the work of the "peculiar sensibility," the specific "tastes," "odors," and "savors" of the organs and systems engaged, all of them guided in their choices by the vital or "intelligent principle" that acted to conserve the life of the organism. For Grimaud, hunger was "a conscious sentiment of the need for food: I say *conscious* sentiment in order to distinguish it from the *natural* sentiment that regulates the nutritive acts of plants and of each part of the body without the one or the other having any way of perceiving or recognizing these acts. . . . The sentiment of hunger is immediately determined by the principle that animates the body, which has knowledge of its needs and which applies this sentiment exclusively to substances capable of satisfying these needs."[33] Although, according to Grimaud, specific structures of tongue, mouth, esophagus, and stomach had important roles to play in the desiring, taking in, and digesting of food, these processes engaged the bodily economy as a whole and involved not only the obvious anatomical structures that drew the attention of mechanists but also the skin, the connective tissue, and, most important, the architectonics of sympathies and synergies that united the organism.

This gulf between mechanist and vitalist styles of explanation suggests that questions surrounding appetite and hunger were tied to debates at the heart of late Enlightenment physiology. This linkage is evident in the unstable appetite-hunger lexicon, in which appetite figured as urge or desire and hunger as need. The use of these freighted terms indicates the extent to which physiological debates were still enmeshed with moral ones: as a need, hunger could not be judged or denied, but as an urge or a desire, appetite raised questions that were inescapably joined to issues of choice and free will. As such, the appetite-hunger discourse was tied to that of the ultimate "ends" of the human organism, its fashioning and purpose in the larger scheme of the natural and moral universe. An especially revealing theme in this connection involves the connection between appetite and "instinct," a problem that spoke not to the mechanisms or forces that drove eating but to its larger moral significance. In this context appetite was no longer subordinated to hunger as its mysterious partner but instead emerged, with all its vagaries and dangers, as central to any understanding of the urge to eat.

Eating by Instinct

As observed earlier, Buffon argued that the distinction between instinctual ("purely mechanical") action and reasoned choice defined the difference between animal and human life and also established the superiority of humans over animals. Though healthier, the instinctual eating of animals reflected their lack of the higher mental faculties that established human preeminence and that, paradoxically, led to the ingestive evils so widespread in the eighteenth century. Spallanzani's statements on this matter were more oblique but ultimately no less telling. When he suggested that the stones found in the digestive tract of barnyard fowl existed to augment the animals' appetite, he reflected on what relation there might be between this peculiar anatomical feature and the instinct that apparently guided the birds' eating habits. Noting that when first born, the chicks, whether famished or well fed, ran "greedily" to eat "everything . . . little stones, fragments of brick, hardened plaster," he observed that the presence of the stones in the stomach might be thought a result of the animals' "stupidity" in eating indiscriminately. But this, he continued, seemed not to be the case: the stones were in fact a natural feature of the birds' anatomy, not a result of "stupid" eating. As they matured, the chicks stopped eating inedible substances, developing instead "the instinct that was sleeping in their infancy" and "chang[ing] in this regard, as in several others, their mores and their character" (29–30). Put in twentieth-century terms, Spallanzani's response to the question of whether and how instinct figured in eating choices blended elements of nature and nurture: his chicks enjoyed instinctual knowledge of what to eat, but that instinct was latent and had to be developed, as the animals' "tastes" were refined by experience. How this happened he did not say, but his reference to the animals' mores and character suggested maturation through processes of learning or training that were comparable at least in some ways to human experience.

Because Spallanzani's investigations of digestion ultimately worked to explicate human physiology—as his extensive and painful self-experimentation indicated—this passage on the instinctual basis of eating implicitly raised the question of whether human beings, as opposed to barnyard chicks, know naturally or instinctually how and what to eat. And that question raised in turn the larger one of whether the human organism functioned naturally to its own benefit, or, if not, what the source was of its "errors" or waywardness. Transferred to the human domain, the problem of nature's wisdom or fallibility in guiding eating was fundamental both in medicine and in all the sciences that presumed to speak of bodily processes impinging on issues of human choice and action. Thus, when inquirers identified hunger as an innate need, often a painful one, while situating appetite somewhere in the realm of desire, inclination, or pleasure, they accorded hunger a basic legitimacy—that which indicated the organism's demand for foodstuffs that would keep it alive—and identified appetite as a potential source of error and illness.

Yet on this question there was, again, a mechanist-vitalist divide. Mechanists tended to focus on apparatuses that enabled functions and, when these failed, to blame immediate mechanical causes while seeing as the fundamental cause of disorder voluntary errors of regimen, often, if not always, made in response to the siren call of pleasure. Hence the contrast Haller drew between the jaded appetite of the "rich idler" and the pleasure taken in simple foods by the plowman and the hunter.[34] In his handling, the "internal sensations" governing eating covered perception, imagination, memory, attention, judgment, and will, and not the deep-seated urges to which vitalists like Grimaud attended. In Haller's scheme, if an eater erred, he did so because of poor judgment and bad choices.[35] For vitalists, however, the beneficial role of instinctual choice was essential given their view of the organism as a unity. While vitalists recognized that some phenomena were governed by conscious choice and others directed by the internal wisdom of the body, they tended to underplay that distinction and to emphasize that health must be guided by the dictates of the vital principle. This was especially true in the realm of eating and digestion, as it was these functions that allowed the organism to repair constant losses of force and to ward off the destruction that threatened living beings at all times. For vitalists, in short, eating was too important to be left to the conscious choice of the individual; rather, health and vitality rested on the eater proceeding in response to the inner sensations that differentiated good foods from bad and that shaped feelings of appetite, hunger, and satiety. Hence the insistence of vitalists that people who ate naturally ate well. Grimaud stated a medical truism when he observed that digestion works better when appetite is present, but he asserted a vitalist principle when he urged that, even if foods were "contrary in themselves," they were of value to patients if consumed as objects of desire. Repeatedly, moreover, Grimaud emphasized that "the people," close to nature and in many respects more like animals than the civilized folk of the city, knew automatically what and how to eat.[36] In this sense vitalists saw appetite as a general disposition to act that replicated the "appetites" or "tastes" of the inner organs, which "knew" which substances to assimilate or reject in keeping with vital "needs."

Where does this leave Spallanzani's little chicks? As a mechanist-minded investigator, Spallanzani considered appetite at least partly dependent on learning through sensory experience rather than an organismic given of the kind posited by Grimaud and other vitalists. These different views on appetite and instinct had wide-ranging implications for human eaters and the physicians who treated them. It might be thought that in practice none of this made much difference: whether an individual was guided on what to eat by instinctual appetites or had to learn to make rational choices of food to preserve good health, all physicians recognized that errors would be made and proposed means to correct them. But in fact the practical consequences of theorizing of this sort were profound. The vitalist dictum that digestion was optimal when individuals ate what they desired meant that there could be no absolutes in respect to diet: to be healthy, individuals should consume what their peculiar appetites urged them to eat. In contrast, Spallanzani was among the earliest to suggest that, given the material nature of foodstuffs and standard operations of digestion, certain foods were, for everyone, healthier and more "digestible" than others. Accordingly, he ranked foods from easily digestible, such as veal, to the "acidic and salty foods that are ill suited to digestion."[37] Spallanzani did not go very far with this line of investigation, but his findings on the actions of gastric juice did a great deal to cast doubt on the individuality of the digestive process, from the first prompt of appetite to evacuation.

How and why this normalizing strand of eighteenth-century thinking about appetite began to gain force is a complicated matter, but an important element of the story involves the relative promise and prestige of the methods employed in the assorted sciences of appetite. By 1800 the decisive way in which animals came to figure in the investigation of appetite was not by the peaceful observation of naturalists like Buffon but through aggressive experiments devised by physiologists, who, asserting a new methodological clarity, gained special influence in investigating the "obscure" phenomena related to digestion.[38] Before 1800, isolated investigators had studied appetite and hunger by subjecting animals to starvation or feeding them particular foods in isolation to gauge the effects of distinct diets. The Italian naturalist Francesco Redi (1626–1697) had done so in the seventeenth century, and Haller reported results from such experiments in his physiological treatise of 1757–1766.[39] But by the end of the eighteenth century, many physiologists saw such experimental use of animals not as optional but as essential in seeking to differentiate true from factitious needs. Certainly the extent to which such findings about animal eating could be applied to humans remained a matter of contention, with vitalists such as Grimaud generally hostile to the idea of extrapolating from knowledge of animals to human beings. The vitalist conception of organismic unity—the interrelation of physical and moral factors including habit, occupation, temperament, and the all-important "passions"—meant that human life and health had a specific character not shared with animals. Such criticisms notwithstanding, the study of animal diet and eating habits, as well as the structure and work of teeth, mouth, gullet, and all parts of the alimentary tract steadily gained importance through these years, and their value was widely recognized even if, as Spallanzani himself saw, such comparisons carried only the weight of "analogy" and could not take the place of direct investigations of human phenomena.[40] Even the vitalist Charles-Louis Dumas (1765–1813), author of a four-volume physiology textbook published in 1800–1803, moved to supplement vitalists' traditional reliance on clinical observation with experiments on dogs that were intended to gauge the extent to which hunger could be mitigated by administering opium and other substances known to damp down appetite.[41] Dumas was careful to conclude his discussion by urging that his experimental results were in full conformity with the findings of clinical medicine, but the fact that he, as a vitalist standard-bearer, took up such methods indicated that animal experimentation was increasingly thought to be of peculiar value to investigations of appetite and hunger. The point here is not that experimentation itself was new, whether done on animals or on humans, but that the conviction had taken hold that phenomena once believed to be highly subject to individual variability were in fact lawlike in character, or, put differently, determined by universal physical and chemical processes. Where this left appetite was a problem for the future.

PART TWO

The Elusiveness of Appetite: Laboratory and Clinic, 1800–1850

In all organic bodies there is a tendency towards the exterior, and an effort to preserve their own integrity. Even plants are the subject of such a striving, . . . [which has] even in man a much wider basis than the brain, and is an action manifesting itself throughout the whole organism. . . . [T]his organic *appetitus* is . . . felt as an obscure power of the individual declaring itself in the manner of an instinct. JOHANNES MÜLLER, *MÜLLER'S ELEMENTS OF PHYSIOLOGY* (1843)

Introduction to Part Two

During the Enlightenment debates over appetite and eating reflected broad cultural anxieties, but the stakes in these arguments rose still higher during the upheaval unleashed by the French Revolution of 1789. Indeed, one way to view the Revolution is to regard it as a struggle over unsatisfied appetite. It has long been argued that a crisis of consumption set off this great political cataclysm, in which the hungry sansculotte is a stock figure. But in recent years historians have proposed that the political strife of the era was less about hunger than about the failure of state and economy to satisfy demands for such "necessary" commodities as white bread and café au lait.[1] Either way, the importance to revolutionary politics of desires and demands for food is clear. The Revolution also gave new urgency to issues of valuation and control of intimate human desires that were already prominent in the sciences of appetite. In part 2, I examine tensions over the proper role of science and medicine in addressing public and moral concerns raised by eating desires. In these years, I argue, the ambiguous nature of appetite became more problematic than before as researchers sought methods that promised new certainty while simultaneously striving to retain or establish authority over a phenomenon that lacked clear definition or boundaries.

Chapter 4 looks at two celebrated figures in the transition from the Enlightenment to the nineteenth century, the French physiologist Xavier Bichat and the English polymath Erasmus Darwin, who held strongly divergent views of the need and desire for food. Bichat associated ingestion and digestion with dark, visceral forces only minimally susceptible to understanding and management, whereas Darwin celebrated the beneficent, life-sustaining role appetite played in the organism. In ensuing decades such positions on the nature and impact of appetite would divide observers, even those who extolled objective scientific inquiry and purported to have no stake in moral debates over appetite and eating. Chapter 5 looks at the position of appetite in physiology as that discipline came increasingly to privilege animal experimentation and as leading figures called into question the expansive framework of the physical and moral that had exerted wide sway in the later Enlightenment. After 1800 physiology increasingly concentrated the study of ingestion and digestion on observable actions and reactions (pain, convulsions, weight loss) of animals undergoing vivisection or subjected to nutritional deprivation.[2] And although some physiologists studied appetite using experimental techniques such as vagotomy (cutting the nerves believed to be responsible for conveying the urge to eat), experimentalists often avoided the imponderables of appetite and gravitated to more tractable problems such as the sequence of acts involved in swallowing. Indeed, I argue that when experimentalists took up the problem of appetite, they mostly abandoned the laboratory and, drawing on the broad medical training that all physiologists of this era still received, culled evidence from medical case reports and from the observations of naturalists and travelers.

I look closely in part 2 at the use of laboratory animals in the study of appetite and eating, but I do not for the moment carry forward the story of how investigators looked at spontaneous eating in domesticated or wild animals. The early nineteenth century witnessed the decline of the encyclopedic natural history exemplified by Buffon, turning away from observation of living, eating animals to the study of dead ones viewed in dissecting chambers, laboratories, and museum collections.[3] Nor were appetite and eating central to early debates over the fixity or malleability of species. Richard W. Burkhardt Jr. has shown in respect to J.-B. de Lamarck (1744–1829) that although the celebrated French "transformist" is often said to have accorded the "*wishing* or *willing*" of animals a large role in evolutionary change, he in fact rejected appetites and desires as excessively voluntarist in favor of needs and habits associated with identifiable physical structures.[4] Given the limited interest in these years in observing eating in live animals, I defer discussion of such matters until I turn to the emergence of animal psychology in the later nineteenth century.[5]

After surveying the physiology of appetite, I turn in chapter 6 to the redoubled efforts of physicians in the first half of the century to understand and manage appetite disorders. Building on eighteenth-century nosologies, physicians of these years sought to clarify and reclassify the ills the Enlightenment had known as the vapors. In the process they created a new spectrum of psycho-gastric maladies—new or refashioned varieties of illness that blended mental or emotional distress with disturbed eating and troubles of the digestive organs. In the early nineteenth century doctors eagerly applied the new concept of neurosis to such ills, establishing "neuroses of the stomach" alongside diagnoses such as "gastralgia," "gastrodynia," and "dyspepsia." Yet while many physicians embraced the neurosis concept, others—especially those committed to the era's privileged method of pathological anatomy—either rejected the idea of neurosis or used it to label diffuse ills that lacked a clear organic foundation and thus seemed largely inaccessible to rigorous investigation. I argue, then, that in medicine, as in physiology, methodological advances made problems involving appetite, with its blend of psychic and somatic features, into dubious objects of inquiry. The one exception was provided by emergent "mental medicine," some of whose practitioners optimistically sought to trace mental aberrations (and the disordered appetite and eating often accompanying them) to physical lesions in the brain.[6] But even here, as appetite itself lacked any evident anatomical locale, psycho-gastric ills in which disturbed appetite was primary did not make such prospects likely.

All the same, in the study of appetite the emergence of mental medicine as a distinctive field was to prove a momentous development for inquiry into troubled appetite and eating. Mental medicine faced the immediate concern of ending practices of confining, restraining, and sometimes starving the mentally ill. The solution devised was the asylum, which, refurbished and reorganized, became an institution of huge importance to nineteenth-century society and, if largely unwittingly, a significant setting for observing disturbed appetite. Alienists struggled to provision the asylum and to treat patients whose disordered eating was in some cases life threatening. At the same time, psychiatric theorists strove to distinguish psychogenic disease that involved troubled appetite from the hodgepodge of blended mind-body ailments bequeathed by medical tradition or newly conceptualized by contemporary nosologists of psycho-gastric ills. Although some leading figures, especially Philippe Pinel, continued to uphold the vitalist tenet that mental alienation was "seated" in the abdominal viscera, others insisted that the brain was the center of mental maladies, including those marked by ills of appetite. Thus, some physicians began arguing that disturbed appetite, if not explained by evident somatic malfunction, was comprehensible only as a feature of mental illness, a step that furthered the psychologization of appetite and the sequestering of appetite disorders within the frame of psychiatric illness.

FOUR

Perils and Pleasures of Appetite at 1800: Xavier Bichat and Erasmus Darwin

The year 1802 saw the death of two celebrated figures who left a strong mark on the history of appetite. One of them, the iconoclastic physician Erasmus Darwin (1731–1802), was an old man who in the last decades of his life had drawn together the threads of a long career into a body of lasting work. The other, the precocious physiologist Xavier Bichat (1771–1802), died at the age of thirty from a wound sustained in the course of postmortem dissection, a method he had mobilized with special skill. In background and scientific temperament, these two men could scarcely have been more different. Yet in other ways they represented a unified moment in the history of European life science and, more specifically, of thinking about appetite. Both Bichat and Darwin were profoundly affected by the revolutionary turmoil of the 1790s. Despite his youth, Bichat enjoyed a prestigious position in the complex of institutions created by successive revolutionary governments to modernize France's medical and scientific establishment. One of the most acclaimed figures of the Paris Clinical School, Bichat was instrumental in unifying the sciences of anatomy and physiology in a new fashion. In one estimate, he effected nothing less than a "revolution" in physiology.[1] Nonetheless, Bichat was no friend of the French Revolution. Conservative in temper, he rejected one of its key principles by denying that human beings are "perfectible" and insisting that much of their physiological endowment was unchangeable. These were key tenets of his famed "two-lives" system, a physiological framework that said little explicitly about appetite but that greatly affected the subsequent thinking of biomedical investigators who did.

The English physician Erasmus Darwin stood in clear contrast to Bichat in scientific and political outlook. Darwin developed his thinking in association with a visionary cohort of English radicals who rejoiced at news of the French Revolution.[2] Although, like others in that company, Darwin grew disenchanted with the Revolution after the onset of the Terror, nothing shook his conviction that progress was possible.[3] Nonetheless, Darwin, disappointed with the outcome of revolutionary action and perhaps fearful of repression of revolutionary ideas and activities, moved away from politics to the strictly intellectual and scientific sphere as the domain best suited to achieve progress.[4] Indeed, Darwin was one of the earliest thinkers to embrace the idea of biological alterations in the human species as a method to foster social advancement, and he believed that an indispensable way to achieve such improvement was through changes in the food supply and in patterns of human eating. This idea rested on Darwin's elevated view of the appetites, which, in his thinking, were essential components of basic human endowment and tools of great power in the effort to remake individuals and build a better world. Darwin's view of appetite as a part of conscious, voluntary, improvable conduct formed a powerful contrast to Bichat's vision of organismic needs and desires that exercised a perilous and seemingly inalterable influence on human action.

Bichat's "Two Lives"

Until 1800 the dominant framework employed in the study of bodily functions was Galen's system based on the three central organs of brain, heart, and liver acting with the aid of three corresponding powers, faculties, and pneumas (fluids). In the centuries-long process by which Galenic physiology was overthrown, the most significant development was William Harvey's discovery of the circulation of the blood; it is at least arguable that the next most important move away from Galenism came with the work of Bichat.[5] In this connection two features of Bichat's work require emphasis: first, his creation of an anatomical-physiological framework that was dualist rather than tripartite in character, and, second, his demotion of the significance of organs in favor of "properties" inherent in tissues. Bichat's doctrine of the two lives posited a radical separation between the "animal life" of voluntary movement, reasoned thought, and cognition and the "organic life" of largely automatic, "lower" activities of the organism. While he accepted that change could be effected in the animal life through education and cultivation of good habits, he insisted that the organic life, which encompassed the work of the viscera and served as the seat of the passions, was largely fixed.[6] Animal life included those functions that connected the organism to the external world—inflow of sensory information, voluntary motor activity, and, most important, reasoned understanding and choice. Organic life, by contrast, entailed automatic activities—movement of the blood, beating of the heart, respiration, digestion—that were essential to life but proceeded without conscious management by the individual (*Researches*, 1–134).

Essential to this two-lives system was Bichat's identification of five "vital properties" that inhered in tissues distributed throughout the body.[7] Among other significant departures from earlier physiological thinking, this system rejected the claim made by Albrecht von Haller that sensibility was found only in nerves and structures supplied with nerves, while irritability was found only in muscles. Of greatest importance was Bichat's contention that sensibility was not limited to conscious feeling but, as the property of organic sensibility, was distributed throughout tissues of the body that enabled essential functions without conscious awareness. Although Bichat himself rejected the view that sensibility was confined to nervous structures, his scheme helped encourage the emergence of the dualistic system later seen as comprising the central nervous system of brain, cerebral nerves, and spinal cord, and the "vegetative" nervous system responsible for circulation, respiration, digestion, and related functions. His steps in this regard helped to focus attention, certainly not for the first time, but in a newly significant way, on what had long been regarded as the lower functions of the organism, the most "obscure" of which was digestion.[8]

Although Bichat's physiology is of great importance for the subsequent history of thinking about appetite, it is not easy to establish a clear picture of his conception of the workings of the urge to eat, ingestion, and digestion. Seeking to provide a "general" system, he did not organize his physiological inquiries in the usual fashion of discussing major functions in succession (circulation, digestion, and so on), but, instead, focused on the tissues throughout the body where the all-important vital properties resided.[9] It is possible, however, to reconstruct key features of Bichat's conception of the digestive process, which he discusses most thoroughly amid his examination of the nervous system. In his *Anatomie générale* of 1801 he stated the differences between the animal and the organic lives, the former having its "center" in the brain, cerebral nerves, and spinal cord, and the latter—the "nervous system of the organic life"—centered in the ganglia (*AG*, 1:115, 213). In this formulation it was the nerves of the organic life that influenced digestion and other involuntary functions (*AG*, 1:115). Yet elsewhere, Bichat offered a more complicated picture of the position of the stomach and digestion within the two-lives system when he described the stomach as a "composite organ" made up of many different tissues and sharing some connections to, and characteristics with, the organs of animal life. Thus, it was impossible to form "a precise and rigorous idea" of the "life proper to the stomach" (*AG*, 1:lxxxiv; *Researches*, 41). Moreover, although he held that in general the nerves of the two lives were separate from each other, differing in color, structure, direction, and other anatomical features, he also observed that they sometimes "communicated" directly, as seen with connections between the vagus nerves and the solar and cardiac plexuses (*AG*, 1:234–37). Such connections, he asserted, helped to explain how and why the brain, though generally not involved in organic life, did exercise control in an "indirect manner" over involuntary functions (*AG*, 1:158–59).[10]

Of central importance in this regard was the distinction Bichat drew between external and internal sensations. The nature of external sensations was clear: they depended on the body's sensory apparatus, the nerves of eyes, ears, nose, mouth, and the organs of touch. These structures were essential components of the animal life that connected the organism to its exterior environment. In contrast, the nature of internal sensations was obscure, as Bichat observed in pointing to the diffuse effects of wine and opium (*AG*, 1:170–71). Not surprisingly, then, when Bichat approached the internal sensation of hunger, he characterized it as especially puzzling because it depended on an unknown relation between the brain, as the center of the animal life, and the viscera, which principally served the organic life. When he asked, for example, whether muscles of the organic life suffered fatigue as did those of the animal life, Bichat argued that although "organic muscles" were less susceptible to fatigue than those involved in animal life, stomach fibers did experience fatigue and thus the "painful sentiment that . . . we call hunger" (*AG*, 3:355). In this fashion Bichat made explicit a connection between the brain as the center of animal life and the stomach, at least in its muscular structures.

Important as these nervous connections were to Bichat, his analysis centered not on structures but on vital properties. And here too the complexity of the stomach was evident: just as the stomach was anatomically composed of different tissues, its operations were governed by the diverse vital properties inherent in these tissues. Of singular importance was the muscular tunic, which, alone of the parts of the stomach, was endowed with the "sensible organic contractility" crucial to digestion (*AG*, 1:xliv). Yet the stomach also possessed the property of sensibility, as was true of all parts of the body served by the cervical nerves and of others in which sensibility operated independently of specific nervous structures. Although Bichat did not specify what role the sensibility of the stomach played in the phenomenon of hunger, he did suggest a key element of the problem when he asserted that hunger and thirst were sometimes "purely sympathetic" and felt even though the "cause that produces them in the natural order is not present in the stomach or intestines" (*AG*, 1:190). In such cases hunger (or thirst) resulted not from true visceral sensations but from "irregularities of perception," which "play the greatest role in sympathies of animal sensibility" (*AG*, 1:190). As perception was an operation of the brain, this meant that hunger could be a purely psychic phenomenon without any attendant physical promptings from the viscera.

Bichat saw the stomach and digestion, then, as standing somewhere between the animal and the organic life, and he regarded hunger as a sensation ordinarily caused by action of the stomach but sometimes as an error of perception. Although he said little explicitly about appetite, he did refer to "wants" with respect to external objects that could dominate the external life (*Researches*, 2). One occasion on which he used the term "appetite," moreover, had to do not with the urge to eat but with the "venereal appetite," which, unlike the "composite" phenomena of ingestion and digestion, functioned entirely outside the animal life, seemingly in a realm of its own (*AG*, 1:208–209, 3:339). Whether he regarded the appetite for food as standing apart in a similar way or as somehow analogous or joined to the sexual appetite is not clear, and the same is true for the relation between eating desires and the passions, which he definitively placed in the organic life. In keeping with well-established tradition, Bichat recognized that the passions powerfully affected the digestive system, but precisely how they were linked to eating desires he did not attempt to clarify beyond saying that perhaps the brain was united "by sympathetic relations" to the internal viscera or that the internal organs communicated directly through nerves to the voluntary muscles.

This, he said, was not known (*Researches*, 46–49, 54–55). In either case, to Bichat this power of the organic life (with the passions among its greatest instruments) to "snatch from the empire of the will" activities that "are naturally voluntary" was fraught with peril (*Researches*, 55). Human beings varied depending on the supremacy in their particular organization of the organic or animal life, and, although he observed in passing that the happiest man is he in whom the two lives are in equilibrium, the dramatic language he used to characterize the dangerous influence of the passions leaves little doubt of the superiority in his mind of individuals in whom dominance was exerted by brain and will rather than elements—passional, visceral, appetitive—of the organic life (*Researches*, 63).

It is curious, given Bichat's debt to vitalists of the Montpellier school—for whom desire was not just a psychological phenomenon but an essential stimulus of the independent "lives" of the individual organs—that he skirted the problem of the existence of innate desires and instead framed the issue in terms of degrees of sensibility. A crucial element of his thinking in this regard had to do with the importance of habit in all functions of the animal life (*Researches*, 34–41), an issue he approached in discussing what happens to the nervous system of the animal life in the process of growth and across the so-called ages of life. Childhood, he held, is the age of sensations. Everything is new for infants; attracts their eyes, ears, and sense of smell; and is experienced as "a source of pleasure." After childhood, he asserted, the capacity for both pain and pleasure progressively diminishes so that in old age "indifference" is the natural human state (*AG*, 1:206, 211). What this meant with regard to eating desires, Bichat did not clarify. Were these a part of animal life, the result of external sensations—the effect on eyes, nose, and tongue of objects whose ingestion caused pleasure—and thus something that diminished with advancing age? Or were eating desires a part of the organic life, a result of a primordial drive to take in substances that would preserve life in the face of the forces working for its dissolution? On this point Bichat gave few clues, although he did observe, again noting the ambiguous position of the eating function, that while the organic life overall was not dependent on habit, habit did perhaps play a role in the work of the intermediary functions of digestion and excretion.[11] Thus, Bichat's general position was that habit could obliterate the forces of pleasure and pain that exerted so great an influence in the animal life, but in respect to the automatic functions of the organic life it had no role. Where appetite and digestion fit in this thinking was, again, uncertain.

The questions about appetite, eating, and digestion left unanswered by Bichat's formulations were many. His system, whether conceived in terms of the properties imbuing tissues or of the nervous structures that many physiologists subsequently took as the foundation of his independent two lives, seemed to open a deep gap between functions that were conscious, willed, and voluntary, and those that were automatic, "physiological" in the truest sense, and unrelated to the "higher" functions. Whether and how this gap was bridged by the work of unifying structures or properties was to be a problem of enduring interest both to physiologists who endorsed Bichat's system and those who challenged it.[12]

Darwin's Celebration of "Natural" Appetite

A dramatically different approach to these matters was evident in the thinking of Darwin, whose goal was not to divide animal from organic, willed from unwilled, mental from physical, but to integrate all the functions of the human body—and indeed all the activities of animated and cosmic existence—into one unified framework.[13] In just about every respect—method and tone, goals and principles—Darwin's thinking differed radically from that of Bichat. Darwin was fervently interested in the vegetable world, and he investigated human physiology within a much more expansive context—a vegetable-animal-human continuum—than did Bichat. Indeed, Darwin's ambition was to encompass the whole of nature in one grand scheme.

In the last years of his life Darwin produced two major treatises that looked closely at problems of appetite, food, and eating.[14] His *Zoonomia, or The Laws of Organic Life,* a work in which the appetites figured centrally, was dedicated to all who studied the "operations of the mind as a science," a statement that indicated from the outset his intention to treat mental functions as part of nature. Darwin opened with a tribute to Newton and others who sought to discern the laws of physical nature. He then stated his own goal of extending this search to the laws of organic life with its two "essences or substances"—spirit, which initiated movement, and matter, that which moved (1:5). A committed vitalist, Darwin posited that all organic beings were imbued with a "living principle, or spirit of animation" (1:10) that, in humans and higher animals, was the source of four types of motion he termed the "irritative," "sensitive," "voluntary," and "associate."[15] In this scheme all movements or "contractions of fibrous parts of the body" developed in a sequence.

Resulting originally from irritations caused by external bodies, corporal movements were accompanied by sensations of pleasure and pain, which, in turn, created states of "desire" or "aversion" that then manifested in acts of volition. Finally, given sufficient repetition and intensity, some movements or contractions became linked to others by acts of "association." These four kinds of motion represented the "active state" of what Darwin called inherent "faculties" or "capacities" of "irritability," "sensibility," "voluntarity," and "associability" (1:32). Crucial to the unfolding of this sequence was the influence of habit, which Darwin saw as instrumental to acts of volition and judgment.[16] To Darwin, "irritations" originally caused by contact with external objects could, if repeated time and again, "become causable by the sensations" even if the irritation itself was no longer present. Similarly, once a habit was formed, the will, through an act of volition, could initiate a contraction in the absence of either irritation or sensation (1:34). Nonetheless, voluntary action—like all "motions" of the body—ultimately depended on the experience of pleasure and pain, because it was the recollection of pleasure or pain that caused the individual, acting out of the storehouse of feeling that Darwin called "desire or aversion" (1:35, 45), to seek or to avoid particular kinds of actions. Such voluntary choices could reflect the influence of single experiences or of "associate motions," "trains or tribes" of ideas associated with previous motions and called up by memory. An act of excess eating, for example, could create an aversion to food that in turn caused a voluntary act of avoidance.[17]

Although Darwin described in a general way the anatomical foundation of physiological activity—fibrils, muscles, nerves, organs—his objectives went well beyond those of anatomical science. Darwin sought to explicate in purely physical terms the interrelations of corporal and mental phenomena, or, as Edward S. Reed argues, to show that "the mind simply *is* the body and all its feelings."[18] Schooled in the psychological doctrines of the eighteenth century, especially the associationism of David Hartley, Darwin used as his most important device for this purpose the analysis of pleasure and pain.[19] While recognizing the priority of the "irritative" faculty, which existed not only in animals but in vegetables as well, he attended especially to the "organs of sense" and to the nature of the "sensorium," the clearinghouse for the "motions" conveyed by the sense organs that were invariably marked by the experience of pleasure and pain.

By the 1790s physiologists had extensively discussed the principle or property of sensibility in connection with bodily processes. It was Darwin's great achievement to attempt a deeper understanding of the relation between sensation and mental processes than most physiologists of his time—Bichat among them—cared to undertake, given the ample opportunity such inquiry supplied for clashes with religious orthodoxy on the autonomy of the intellect and free will. Darwin worked to distinguish among words—sensation, perception, idea—that he thought were employed too loosely or rendered useless for natural science by the connotations they carried from metaphysics. Having linked sensation to operations of the "organs of sense," and those in turn to the structure of the brain and nervous system (1:10–12), he defined the terms "idea" and "perception" in ways that gave them a distinctly materialist cast. Ideas were "notions of external things" synonymous with "sensual motions," whereas perception entailed "both the action of the organ of sense in consequence of the impact of external objects, and our attention to that action" (1:11–12). Likewise, Darwin linked attention, ordinarily seen as a purely mental operation, to sensibility: it was fixed by "the pain or pleasure that succeeds or accompanies" it.[20] Darwin's conception of appetite developed out of this understanding of sensation and the other kinds of motion peculiar to animated beings. His emphasis on pleasure and pain lent special importance to the initial stimulus of external objects, with the desire for food developing out of the pleasure first experienced in eating and its subsequent recollection. He wrote, for example, of the infant's pleasure in suckling: "When the babe, soon after it is born into this cold world, is applied to its mother's bosom; its sense of perceiving warmth is first agreeably affected; next its sense of smell is delighted with the odour of her milk; then its taste is gratified by the flavour of it: afterwards the appetites of hunger and of thirst afford pleasure by the possession of their objects, and by the subsequent digestion of the aliment" (1:146).

Thus, the appetite for food is aroused by the first experience of sensations caused by taking in objects (mother's milk) from the outside world rather than being somehow innate in the constitution. His view of the appetite for food becomes clearer if it is contrasted to other organismic "tastes" he saw as internal feelings lodged in the organs themselves. Like the Montpellier vitalists, he recognized in "every gland of the body . . . a kind of taste, by which it selects or forms each its peculiar fluid from the blood; and by which it is irritated into activity" (1:55). This line of thinking might have led him, as it did Montpellier vitalists, to an idea of innate desires or tastes in a larger sense. But when Darwin came to speak of the "appetites" he reverted to the pleasure-pain construct, seeing these bodily desires as created either by the pain occasioned by "the absence of . . . irritative motions" (the appetites of "hunger, thirst, lust, . . .

the desire of air") or the pleasure of an already directly experienced object such as "the smell of an hyacinth, or the taste of a pine-apple" (1:57–58). His emphasis was, then, not on urges immanent within the body but on those that developed in relation to the outside world, the hunger experienced when things desired could not be had ("when the objects of our pleasures or pains are at a distance" [1:58]) or the appetite stimulated by the memory of past pleasures.

Again, these comments seem to accent the role of consciously experienced pleasures and desires. Darwin was aware, of course, that good parts of the processes of ingestion and digestion proceeded unconsciously.[21] Of these, hunger was a special case: it arrived unbidden from a stomach "used to perpetual activity . . . [but acting] for a time with less energy" when deprived of the stimulus of food (1:71). But for Darwin this perpetual activity of the stomach (its "internal irritations") was problematic in that such motions suggested a realm of activity independent of the play of pleasure and dependent instead on "instinct." This Darwin could not accept: he rejected the term "instinct" for human beings, saying that although our "sensations and desires" are natural, a "part of our system," the word "instinct" refers only to the actions of animals. Indeed, he was not happy with the term instinct even for animals, observing that many animal actions characterized as instinctual—the fetus or chick swallowing, the puppy sticking out its tongue—were in fact "attended with consciousness" and established by the reiterated action of habit (1:137–39).

To Darwin, then, the appetite for food, though not instinctive, was "natural" and hence beneficent. Created by the experience of pleasure, it encouraged us to do what we need to live and enjoy life. Moreover, Darwin's commitment to progress encouraged him to see what was natural as capable of constant improvement. In other works, especially his agricultural treatise *Phytologia,* he urged that physical nature—the earth itself and the products it could be encouraged to produce through agriculture—was susceptible to endless improvement by those who sought to transform agriculture from the "art" of old to the "science" of the future.[22] Even more tellingly, everything in human nature, including the appetites, was susceptible of improvement. Perhaps reflecting his onetime revolutionary temper, when Darwin thought about improving human eating he did not rail against "diseased" or "indulgent" appetites but instead focused on those who suffered from a lack of adequate food and how they might be aided. Darwin depicted hunger as a pain from a "deficiency of stimulus" from food, just as cold is a pain caused by the deficiency of the stimulus of heat (1:75). The medical response to hunger was obvious: increasing stimulation of the hungry body by administering food. Thus, while most Enlightenment physicians attended to overeaters whose appetite needed curbing or "depraved" eaters whose cravings required reshaping, Darwin was concerned for the hungry. This was the great theme of *Phytologia,* which argued that humanity could overcome the horrors of dearth and usher in a new era of human felicity. Thus, in stark contrast to Bichat, Darwin looked to appetite as a source not of peril but pleasure, as the product not of urges fixed deep in an inalterable nature but of a potentially limitless field of beneficent experience.

What, finally, do Bichat and Darwin tell us about the status of appetite during the turn from the Enlightenment to the nineteenth century? Certainly Bichat was the more obviously influential of the two: his framework of the two lives established the paradigm for inquiry into the physiology of nerves until late in the century, not only in France but also in Britain, Germany, and elsewhere.[23] Darwin, however, came to be regarded by many in his native England as a dangerous materialist or even atheist.[24] Thus, when, in reaction to the turmoil of the revolutionary era, the human sciences in Britain moved in a conservative direction, Darwin's influence went underground. All the same, he has been judged the "most important European psychologist" of the early nineteenth century for his effort to ground psychology in the body and to overcome the dualist tradition that was fiercely reasserted after 1800 as a foundation of social order.[25]

In respect to appetite, Bichat and Darwin pushed investigators in often conflicting directions, partly thanks to the disparate methods they brought to bear. As one of the premier figures of the Paris Clinical School, Bichat favored the methods of pathological anatomy, but he also embraced experimentation, including vivisection, as an essential dimension of physiological inquiry.[26] In other ways, too, Bichat pushed physiology in modern directions. Although he accorded great importance to the role of the passions in the human constitution, he worked to excise from physiology many kinds of observation—introspection, reliance on travel literature, and the truisms of moralists—that had traditionally supplied knowledge of moral and emotional life. Instead, he blended clinical observation, postmortem investigation, and experimentation in a fashion that helped to establish physiology on a newly "positive" foundation. Following his example, many physiologists of succeeding decades moved to exclude from their frame of vision problems that did not seem susceptible to well-defined clinical and laboratory methods. In investigations of the drive to eat, this meant either a focus on elements of the problem that were manipulable at the dissecting table or in the laboratory, or, sometimes even in the work of those most committed to "positive" inquiry, tension and ambivalence about the kinds of claims physiology should assert.[27]

Darwin's work, by contrast, pushed unabashedly in the direction of an expansive psychophysiology based on eclectic methods, including approaches that investigators extolling rigor sought to banish. Very much a man of the eighteenth century, Darwin, while current with developments in anatomy and physiology (he greatly admired Spallanzani), remained in key respects a speculative thinker. In pursuing the sciences he knew best, including botany and medicine, he drew on methods—classification, domestic and field observation, the bedside work of the tradition-minded physician—that were increasingly subject to question by those whose hopes for scientific progress rested with pathological anatomy and laboratory experimentation. Indeed, Darwin was a strong critic of vivisection, the one method that was emerging as indispensable to the new physiology, decrying the futility of "cruel experiments on animals."[28] Both the expansive context of his work on appetite—situating its investigation amid the search for the "laws of organic life"—and his reliance on verbal and rhetorical display (poetic imagery, classificatory symmetries) helped to account for the fact that his influence was registered most fully by those who cultivated the hope for a unified science that could elucidate both physical and moral nature. If for some time his effort to create a "science of mind" that attended even to the lowly viscera roused disgust among students of lofty "mental" phenomena, his name nonetheless commanded respect among those who sought to pull all human capacities, high and low, within the purview of a wide-ranging "science of man."[29]

Aside from matters bearing on the methods and goals of science, Bichat and Darwin also pointed, on substantive issues, to future debates over appetite. Darwin, a cheerful optimist about human nature, celebrated the appetites, defending their essential role in the activities of life and imagining human improvement based on gratification of hunger and other needs. Far from seeing sure nutritional guidance in an instinctually fixed appetite, Darwin pointed to the role of pleasure and pain in "training" the appetite to new enjoyments and new forms of striving. Bichat, by contrast, helped to usher in a new era of pessimism about achieving and maintaining health. His two-lives system emphasized the automaticity of vegetative functions, but, in contrast to the ancients and to Enlightenment vitalists, he did not see beneficent direction at work in the organic life. Rather, characterizing it as a fixed physiological reality, he perceived conflict between reasoned and "vegetative" activity as an inescapable condition of human experience. Human beings had only a "determinate sum of forces or powers," and if energy was expended on one function, all others suffered ("when [the vital powers] . . . are increased in one part they are diminished in all the rest of the living economy" [*Researches*, 113]). Among other unhappy consequences, this physiological reality meant that nothing could be done to alter the character and temperament of individuals dominated by the passions or the influence of the internal organs; any such effort "would be as useless as the attempt of a physician to elevate or abase . . . the ordinary contractile power of the heart" (*Researches*, 125). As Bichat himself regarded digestion and forces at work in the stomach as a largely indecipherable composite of the animal and the organic, he left open the question of whether eating impulses were susceptible to management via training in good habits or voluntary choices. Yet overall he suggested that control in this realm was dubious given that "a thousand causes may at every moment" strengthen or enfeeble visceral functions (*Researches*, 26). Embracing Bichat's view of the inalterable character of the organic life, many physicians—those I will discuss later as "visceralists"—would regard maladies lodged in the depths of the organic life as singularly intractable. Ills of disordered appetite were to figure importantly among them.[30]

FIVE

The Physiology of Appetite to 1850

Scientific activity of all kinds revived and intensified in Europe when the turmoil of the revolutionary and Napoleonic eras ended in 1815. Yet the years of conflict left their mark: in succeeding decades much work in physiology reflected the political and social unease that civil upheaval left in its wake. With respect to the drive to eat, a major form of investigation in this era was the hunger study, which consisted of experiments on animals and sometimes humans to determine minimal dietary requirements. Although such experiments aimed to answer specific questions about nutritional needs, they also reflected a political environment in which conflict between the well fed and the hungry was a constant concern. This was evident, for example, in the work of the French commission that sought in the 1830s to determine whether gelatin from animal bone marrow was a suitable food for use in hospitals and asylums and also in the efforts of German scientists to produce ersatz meat products. Such projects simultaneously drew on and helped spur research on problems of appetite, ingestion, and digestion. Thus, the French commission took note of eating preferences in test animals, such as the dogs that at first would consume gelatin but eventually would "die of inanition rather than touch this substance."[1]

These developments in the sciences are illuminated by Michel Foucault's thinking about "biopower," the new stake public authorities had in the development and sustenance of a healthy population.[2] The most obvious stimulus to heightened concern with the biological well-being of the population of the nation-state was, of course, military—the necessity for a nation pretending to greatness to be able to field an army of fit soldiers. The Industrial Revolution also required its own kind of army, one of able workers, and from the viewpoint of sales, an ever-growing mass of consumers. If people were to purchase household goods, clothing, and other commodities that fueled industrial growth they had first to eat. What is more, in ways only now coming to be understood, "industrial foods" were a leading element in the Industrial Revolution and issues of their safety and palatability contributed to the ever stronger interest in the why and how of eating.[3]

Aside from public concerns, the physiology of the early nineteenth century was shaped by developments in allied sciences and, of course, by the field's own internal dynamics. In these years the scientific field that impinged most significantly on the physiology of appetite was chemistry, now transformed by the "chemical revolution" associated especially with A.-L. Lavoisier (1743–1794). At first focused on the inorganic realm, the new chemistry rapidly came to entail investigation into "animal" or "organic" chemistry, with digestion a major concern.[4] Not all physiologists welcomed the intervention of chemists and the relative roles of these sciences in explicating body processes became a matter of prolonged contention. At the same time, physiology registered the influence of other, rapidly developing life sciences. The early nineteenth century saw increased interest in the anatomy, morphology, and embryology of animals and plants, and the most ambitious goal of the time was to bring all organic beings—human, animal, and vegetable—into a unified theory of life.[5] This endeavor entailed both an effort to define the boundaries separating the organic realms, and, conversely, to identify the ways in which all living beings were alike. In this effort eating and digestion assumed particular importance. Bichat established as the dividing line between vegetables and animals the presence of the property he called "sensible organic contractility," essential in the digestive process of animals but lacking in vegetables.[6] Somewhat later the influential German physiologist Johannes Müller established as the crucial line of division the voluntary movements that animals, unlike plants, made in their search for food.[7]

Finally, it would be hard to overestimate the importance in these years of all issues surrounding the functioning of the nervous system, which now came to be seen as the most important unifying apparatus of organic function in the higher animals.[8] And just as life science at large was concerned with boundaries between human, animal, and vegetable, many who investigated nervous structures and action focused on phenomena that defined the boundaries between higher and lower animals. Among these, the two essential operations were feeling and movement. That muscular movement in some forms occurred without nervous involvement was by this point widely accepted; whatever it was called, Haller's irritability, the capacity for movement inherent in muscle tissue, had gained broad assent. But the extent to which sensation was dependent on nerves continued to be debated well into the nineteenth century. Bichat's contention that the property of sensibility could exist independent of nervous structures had strong defenders, but as time passed, many physiologists asserted that sensation was in fact wholly dependent on the nervous apparatus.[9] One of the great puzzles that continued to confront physiologists working in this vein was the problem of internal sensations such as the urge to eat. The nature of these *Gemeingefühle,* feelings registered generally throughout the body, difficult or impossible to localize, presented especially complex problems.[10]

Given these diverse contexts, the first half of the nineteenth century saw manifold approaches to issues surrounding appetite and eating. Nonetheless, within the physiological science that began to assume its modern shape, it was the panoply of experimental methods that came to have special standing. No figure was more instrumental in promoting experimentalism than the French physiologist François Magendie (1783–1855), whose work illustrates both the high expectations of experimental physiology and the difficulty of applying its procedures to an ineffable phenomenon such as appetite.

Need and Desire in the Physiology of François Magendie

In his famous manifesto of 1809, Magendie, a freshly minted graduate of the Paris Faculty of Medicine, urged that physiology would progress no further unless it sought explanations based on "facts" rather than "tedious reasonings" and "pure suppositions."[11] In a series of studies published in 1813, Magendie sought to apply these strictures in investigating aspects of ingestion and digestion.[12] Of these, his work on the causes of vomiting is of special interest, as it asked whether this singular activity of the alimentary canal operated under the control of the brain or the viscera.[13] In this work Magendie sought to disprove the long-standing view that vomiting resulted from stomach contractions (an idea challenged in the seventeenth century but forcefully reasserted by Albrecht von Haller). To this end, Magendie fed or injected live dogs with tartar emetic and then monitored the animals' reactions.[14] In his report Magendie described the methods he used to discern what was happening inside the body, as when he cut through abdominal tissue to feel directly, with his inserted finger, the state of the stomach. This technique showed that the stomach did not in fact contract but, flattened by pressure from the diaphragm and the abdominal muscles, remained passive. Seeking to determine which of these parts was the decisive actor, Magendie first severed the vagus nerves serving the diaphragm and then in a separate procedure detached the abdominal muscles to see the effect when these were disabled. Taken together, he urged, these experiments decisively demonstrated that vomiting, a digestive function long and erroneously attributed to the stomach, was in fact under the control of the cranial nerves and thus of the brain (*Vomissement*, 12–24).

This early work of Magendie gained much acclaim.[15] Then, in 1816, Magendie turned his attention to the famous hunger experiments in which he demonstrated the necessity of protein in the animal diet, investigations inspired by his interest in emergent "animal chemistry."[16] To ascertain whether animals could survive on "non-nitrogenous" foodstuffs, Magendie fed his dogs a series of foodstuffs in isolation: white sugar, olive oil, gum, and butter. The animals showed varied effects of these diets, including weakness, decline in muscle activity, and ulceration of the cornea. In the end they all died. Magendie concluded that animal bodies did not convert other foodstuffs to nitrogen or, as one theorist had proposed, take it in from atmospheric air, but had to ingest foods containing nitrogen. This work, "a classic in physiological literature," showed that in respect to hunger in deliberately manipulated animals, Magendie's chemically informed experimentalism was not only feasible but also productive of striking results.[17]

Along with other observations made in the course of his hunger experiments, Magendie took note of the state of his animals' appetite, reporting of the dog first fed only white sugar, for example, that even as his health declined he continued to be eager to eat but that in time, as his "forces diminished, . . . [he] lost his gaiety, and his appetite was not so lively" ("Mémoire sur les propriétés nutritives," 70–71). Although at this point Magendie drew no conclusion from this fact, it fit well with observations he would later make about the unreliability of appetite, which expressed the desire for pleasure in eating but did not reflect the true "physiological" state of the organism.

Magendie's experimental work positioned him to speculate more broadly about appetite, eating, and digestion in his influential textbook of 1816–1817.[18] Engaging with the work of physiologists from Haller to

Bichat, Magendie attempted to clarify the complex of issues surrounding sensation and sensibility, the difference between desire and need, and, regarding the urge to eat, the border between instinct and socially generated habits. "Hunger," he wrote, "or indeed appetite, which is only its first degree, must be distinguished from the sentiment that causes us to nourish ourselves with preference to one or another kind of food." Such "sentiments" differed essentially "from genuine hunger, which expresses the true needs of the economy," reflecting instead the influence "of civilization [and] habits" (*Précis*, 2:28–29). How difficult it was to clarify these differences was apparent in Magendie's uncertain terminology: designating appetite as the "first degree" of hunger, he left unnamed the "sentiments" governing food preferences. Nonetheless, the distinction he drew between legitimate and illegitimate eating choices was unmistakable. When he differentiated between need and desire, Magendie echoed views expressed by his Enlightenment predecessors. What was new was his determination to settle the question of whether sensations of appetite and hunger could be localized in a specific part of the nervous system. To this question he responded emphatically in the negative: "Hunger results, like all other internal sensations, from the action of the nervous system; it has no other seat than this system itself and no other cause than the general laws of organization." This truth was evident, he continued, given that hunger "sometimes continues although the stomach is filled with food," can fail "to develop although the stomach" is empty, and is "subject to habit, to the point of ending spontaneously when the habitual time for meals has passed" (*Précis*, 2:30). Thus, although hunger was an "internal sensation," seated in the nervous system, it could not be explained by reference to specific nervous structures.

In characterizing hunger in this way, Magendie staked out a strong position within the debate over the nature of internal sensations. As seen earlier, Cartesian psychophysiology attributed internal sensations to "movements of the nerves . . . that serve natural functions" while vitalists posited that feeling or sensation did not require specific material structures or movements but was experienced spontaneously as inherent "tastes" or "desires" active in the tissues or organs.[19] At the end of the century the significance of internal sensations was strongly reasserted by the influential physician P.-J.-G. Cabanis (1757–1808), an ardent advocate of a medical "science of man" integrating body, mind, and feeling. Cabanis challenged the distinction "analytic philosophers" made between "sensations" ("impressions that come from exterior objects") and vague "impressions" ("unperceived operations of sensibility") that arose from within the organism.[20] He emphasized the powerful influence of "dispositions of the internal organs," especially the abdominal viscera, which, when disturbed, produced "bizarre and extraordinary appetites . . . unheard of images (*images inconnues*) . . . [and] affections that overwhelm the will" (1:174). This way of thinking had culminated in the work of Bichat with his stress on the effects of freely functioning sensibility, including derangements of thought and will caused by sensations emergent from the hidden organic life.

Given this history, and Magendie's goal of refashioning physiology and revolutionizing its methods, it is more than a little surprising to see him defining hunger, of which appetite was the "first degree," as an internal sensation not attached to any particular nervous structure.[21] Generally speaking, Magendie was one of that cohort of French life scientists who in these years sought to overthrow the notion of diffuse sensibility and to insist that sensation required nervous action.[22] Indeed, it seems that it was Magendie's commitment to this restricted view of sensation and sensibility that, in his early work on digestion, governed his choice of certain experimental projects and his neglect of others. In the experiments on vomiting, for example, he referred to the internal sensation of nausea but made the perceptible action of vomiting the focus of his attention. This maneuver typified Magendie's efforts to deflect inquiry away from obscure events of the inner life to external manifestations of internal phenomena.[23]

These observations suggest why it was that Magendie did not try to specify a location for the urge to eat. Although this stance was related to his conviction that sensation necessarily entailed nervous involvement, he pursued the attack on diffuse sensibility only in relation to the identifiable structures of the external senses: an internal sensation like the drive to eat he attributed only in a general way to "action of the nervous system." Steering clear of the "vegetative" as opposed to the central nervous system (and thus the organic life in which Bichat had at least partially positioned ingestion and digestion), Magendie even questioned whether the most important structure of that system, the so-called great sympathetic, was actually nervous in character, as it differed from "nerves properly speaking" in "color, form, consistency," and basic properties. If the "great sympathetic" played a role in internal sensation, this could not be known, because, unlike the "fearful pains" caused by interference with nerves of the central nervous system, its structures could be "pricked, cut, even torn out" with no "sensible disturbance of function."[24] Despite his high ambitions for experimental physiology, this meant that, beyond a general recognition that the brain governed the nervous system at large, no specific nervous basis of appetite, hunger, and other internal states could be identified.[25]

Yet Magendie's thinking about appetite and hunger entailed many complexities beyond the problem of localizing internal sensations. Indeed, for all his strenuous endorsement of experimental methods, Magendie continued to function in important ways within the orbit of vitalist constructs of the physical and the moral. Nowhere was this more evident than in his characterization of hunger as an "instinct," one of those "penchants, inclinations, needs" by which animals are "forced to fulfill the intentions of nature" (*Précis*, 1:179). Developing this view, he urged that nature established hunger and thirst as "instinctive sentiments that alert us to the needs of the economy," not only a general need for nourishment but also the necessity "to make use of . . . or to avoid" particular substances. The same was true for feelings of "satiety or disgust" that indicated the need to stop ingesting and let digestion proceed (*Précis*, 2:23).

Thus, Magendie held that living beings eat as prompted to do so by instinct. Yet the nature of instinct itself was far from clear-cut in Magendie's construction. Even in animals, and certainly in human beings, instinct operated in relation to the environment, in the human case to society. Thus, for Magendie instincts were innate yet also subject to habit, custom, and the influence of civilization (*Précis*, 1:180). All animals shared the instincts of hunger and thirst, but human beings also felt the prod of an instinctual, but dangerous, desire for agreeable sensations (*Précis*, 1:181). In human beings, then, "natural needs" multiplied, becoming ever more varied. These "new needs, which one can call *social*," included the persistent desire to feel pleasure even as the force of sensation diminished and the "ennui that incessantly torments civilized . . . men" increased. In this fashion natural needs became unrecognizable and even so basic an instinct as "hunger was frequently replaced by a capricious taste" (*Précis*, 1:181–82).

In such passages Magendie sounded less like a modern experimentalist than a traditional moralist or, to be more precise, a physiologist bent on explicating physico-moral relations. Yet in the end his claims about appetite and eating were those of a physiologist of the future. To sum up his position on the urge to eat and other internal sensations: they had no seat, were part of the nervous system at large, and followed the general laws of nervous organization. This meant that to Magendie hunger and appetite—along with thirst, pain, nausea, the passions—were all ultimately under the control of the brain as the emperor of the nervous system. In stark contrast to Enlightenment-era vitalists, he urged that internal "impressions" were, somehow, "transmitted" to the brain, only there becoming conscious sensations governed by its functions of intellect, judgment, and will. It was mind lodged in brain that made rational judgments about the fleeting signals that reached it from the viscera and brought these in line by exercising conscious choice and the powers of will. None of this, evidently, was based on experimental procedures of the sort that brought Magendie scientific glory, but there was, all the same, a kind of kinship between what he sought from his experiments and his physico-moral speculations: bringing dark internal sensations into the full light of day; transforming what was hidden into what was manifest; and ultimately changing life phenomena that were often painful and hard to control into measured, considered actions. In this construction appetite arose as an impression from the viscera, but it was the brain that must rule it. The shift that this stance implied from the physiology to the psychology of appetite was to be accomplished after Magendie passed from the scene, but it was an ineluctable result of his drive to expunge from science feelings and desires unaccounted for by observable corporal structures.

Digestive Experiments: A Place for Appetite?

The 1820s to 1840s saw vigorous investigation of all features of digestion not only in physiology but in emergent organic chemistry as well, where investigations concentrated on the composition of nutrients and the bodily fluids involved in digestion. The English chemist William Prout (1785–1850) devised a lastingly influential classification of foods as saccharinous (carbohydrates), oleaginous (fats), and nitrogenous (proteins), and the chemically sophisticated physiologist Theodor Schwann (1810–1882) conducted experiments in search of the essential "digestive principle" that led to the identification of "pepsin."[26] F. L. Holmes has shown how chemical inquiries into digestion culminated in the 1840s with a sustained debate between German and French researchers over the question of whether animal bodies merely "assimilated" the constituents of foodstuffs required to meet nutritional needs or transformed foods by internal chemical processes. For some time this debate centered on fat, specifically whether fat had to be consumed directly in foodstuffs or resulted from the transformation of sugars, starches, and other foods.[27] This set of problems prompted years of feeding studies in which animals were fed particular kinds of foods and the results—gain or loss of fat, overall weight gain or loss, stages of decomposition of foodstuffs in the stomach—were observed. In this debate little attention was accorded to the possibility that digestive processes were affected by extrachemical factors, including the psychic influences widely credited by physicians of the era. In chemical investigations references to appetite were rare, appearing only in reference, for example, to the assumed enthusiasm of experimental animals for eating certain substances (a measure of a food's assimilability) or the seeming influence of appetite on yields of animal products. A case in point was the French chemist J.-B. Boussingault (1801–1887), who performed feeding experiments at a farm he maintained near Paris, and who commented at one point that optimal production of milk was achieved in cows that "showed little tendency to take on fat, while they kept their appetite." Yet though he recognized appetite as a beneficial influence, he did not ask how the state of the appetite could be known or maintained.[28]

Inquiry into the role of appetite in the digestive process remained largely, then, the purview of physiologists. Their work was spurred by an announcement in 1823 by the French Academy of Sciences of a prize question on "the phenomena that unfold in the digestive organs during the act of digestion."[29] In the end none of the entries won the prize, but those submitted by the French researchers François Leuret (1797–1851) and J.-L. Lassaigne (1800–1859) and by the German physiologist Friedrich Tiedemann (1781–1861) and chemist Leopold Gmelin (1788–1853) earned special recognition and a cash award.[30] Tiedemann and Gmelin heeded the academy's injunction to consider the question in relation to four classes of vertebrates—mammals, birds, fish, and reptiles.[31] Leuret and Lassaigne, by contrast, organized their material by function, beginning with "prehension of foods," mastication, and swallowing, and once they reached digestion proper, discussing specific organs and processes such as chymification. They looked chiefly at human digestion, mentioning other vertebrates only occasionally for comparison.[32] Leuret and Lassaigne, who frequently invoked the name of Magendie, opened with a manifesto on behalf of vivisection, stating that plain observation was insufficient and experiments with living animals essential (4–16). By contrast, the German investigators blended animal experimentation with chemical analysis and anatomical and morphological study. Gmelin was a skilled chemist, and he and Tiedemann drew on recent chemical findings (including Prout's identification of hydrochloric acid in the gastric juice) in undertaking new analyses of saliva, bile, gastric juice, and other bodily fluids and of the constituents of diverse foodstuffs.[33]

The Germans also differed from their French competitors in the way they handled the "first phase of digestion," the urge to eat. Bypassing the problem of appetite and hunger, they began instead with chemical analysis of saliva and subsequently referred only occasionally to the state of the appetite in their experimental animals.[34] This decision to omit discussion of the internal states or sensations prompting ingestion meant that they only briefly commented on the experimental procedure most commonly used to discern the nervous foundation of the drive to eat—vagotomy, or sectioning of the vagus nerves. This procedure had a long history, but in the 1820s and 1830s it was employed with new enthusiasm by experimenters who believed that, thanks to methodological refinements, vagotomy could finally resolve the question of control by the vagus (or "pneumogastric") nerves over the desire to eat and succeeding phases of digestion.[35] Tiedemann and Gmelin did perform vagotomies to see whether removing part of the pneumogastric nerves and the connected "ramifications" of the great sympathetic would slow or prevent digestion, but they remarked only tangentially on the desire of animals to eat after the operation (1:372–74).

Here was an area of investigation, the origins and nature of the urge to eat, where the French exceeded their German rivals (209–16). Dismissing previous hypotheses on the "proximate cause" of hunger, they stated that they would consider only claims that it was the "nerves of the eighth pair that were destined to transmit the sensation of hunger to the brain" (211). Their own experiments disproved the claim that cutting the vagus nerves caused animals to lose interest in eating; horses from which they removed several inches of nerve tissue ate "as before and with equal appetite, at least in appearance, to that which they had when they were in full health." They concluded further that the persistence of the horses' readiness to eat indicated that appetite "resides also in the intestines, which receive nerves only from the great sympathetic" (211–12). Contrary results achieved by other investigators did not undermine their own conclusions, they argued, but showed only that the manner in which the operation was conducted, especially the necessity not only to cut through the nerve tissue but also to remove a sizable portion of it, was decisive (127–40).

Although researchers all over Europe used vagotomy and expected much of it, the results were inconclusive. The French physiologist J.-L. Brachet (1789–1858) concluded in 1830 that cutting the vagus nerves caused digestive processes to cease, but others such as the Edinburgh anatomist John Reid found that, in some animals at least, hunger persisted after such interventions.[36] For their part, Tiedemann and Gmelin suggested that the greater part of the nervous influence exerted was probably the work not of the vagus nerves but of the great sympathetic and that it was impossible to isolate its nervous connections with any accuracy. In any event, whatever its other merits vagotomy did little to distinguish clearly between appetite and hunger or to define either phenomenon more precisely. Both the French and the German researchers used the term "appetite" to describe animals' readiness to eat in the aftermath of the operation: Leuret and Lassaigne said of their vagotomized horse that after the operation he ate six liters of oats "with appetite" (134) and Tiedemann and Gmelin reported on a vagotomized dog that ate cooked egg whites "with avidity" (1:374). Yet while they sought to learn whether animals would still show interest in eating after vagotomy was performed, neither team devised any means to determine whether an animal reviving from surgery ate from pleasurable appetite or from painful hunger. Rather, once they established that the animals would eat—with or without "appetite"—they quickly moved on, Leuret and Lassaigne to the task of collecting remains from the digestive tract to determine the rapidity of digestion (134), and Tiedemann and Gmelin to further chemical analysis of digestive juices. While such material processes could be illuminated by experiment, appetite, it seemed, could not.

And yet a physiologist like Tiedemann was not ready, even if experimental techniques were inapplicable to appetite, to exclude the subject from the purview of physiology. In the 1830s he turned his attention to problems of appetite and eating in a fashion sometimes strikingly at odds with that seen in the treatise he and Gmelin had presented to the French Academy of Sciences. One volume of Tiedemann's textbook on human physiology was devoted to the drive to eat and to the nature and components of different foodstuffs.[37] In this work he set forth a terminology and classification of what he saw as the three stages of a drive to eat (*Nahrungstrieb*): the first, a pleasurable state of desire; the second, a troublesome feeling of emptiness and craving for food; and the third, extreme hunger, "one of the most painful and torturing" feelings experienced by human beings (22–23). This was an important break from the often-stated view that appetite and hunger represented two phases, appetite the first, pleasurable phase of desiring to eat and hunger the subsequent, painful phase of needing to eat. Tiedemann's formulation was also crucial in establishing the language of "drive," which was to have a momentous subsequent history in both the physiology and psychology of appetite.[38]

Tiedemann's chapters on the *Nahrungstrieb* relied largely on observations of functional disturbances such as diminished blood supply and weakened muscles seen in experimental animals deprived of food (26–32). But in discussing what happened to starving human beings, he made use not of experiments but of medical reports and of accounts by historians, travelers, and naturalists of the sort on which medical opinion had long drawn. Tiedemann presented the urge to eat as the most powerful of all the drives. Updating tales recounted by Haller and others, he cited horrific stories of individuals tormented by hunger—soldiers retreating from Moscow, prisoners abandoned in dungeons, shipwrecked sailors (35–40). Along with these accounts he included tales of the celebrated "miraculous maids," girls or women who claimed to have lived months or years without nourishment (41–42), and of famed individuals who constituted a kind of bestiary of depraved or "vitiated" appetite, the voracious eaters who consumed vast quantities of food and "polyphagous" eaters who made a sport of consuming live animals whole (54–55). Thus, like Magendie, Tiedemann accepted that extremes of human appetite, though not observable in the laboratory, must figure in any account of the physiology of ingestion.

In these years the most celebrated technique for studying digestion came into use accidentally when the American army surgeon William Beaumont (1785–1853) encountered Alexis St. Martin, a Canadian fur trapper shot in the stomach and left with a gaping fistula.[39] The story of St. Martin's injury and the experiments Beaumont conducted on him between 1825 and 1833 has been told many times, often amid extravagant praise of Beaumont's achievements in illuminating digestive function.[40] Beaumont undertook both simple observation, peering into St. Martin's stomach to gauge its appearance and movements, and experiments that included introducing "through the perforation, into the stomach," different "articles of diet" suspended by silk string or in muslin bags and then withdrawn for examination (125). Beaumont's principal goal was to illuminate the process of digestion in the stomach by exploring the "powers" of the gastric juice.[41] Seeking to test the validity of claims about "artificial digestion," he placed foodstuffs in vials with St. Martin's gastric juices to see whether and how quickly they decomposed. Beaumont also investigated the digestibility of different types of foods and gauged the significance of such findings for dietetics.

When Beaumont took up the question of the nature of hunger, he revealed a partial debt to Albrecht von Haller. Although he objected to Haller's view that hunger resulted from "friction of the internal coats of the empty stomach," he endorsed the Swiss physiologist's general idea that hunger was a "painful sensation, referred to the region of the stomach" (53). Beaumont rejected other theories of the origin of hunger, reserving special scorn for Magendie's claim that hunger resulted from general action of the nervous system, a view that he found to have no "exact meaning" (56). Accepting that nervous action was involved in "warn[ing] the encephalon . . . of the wants of the tissues," he nonetheless

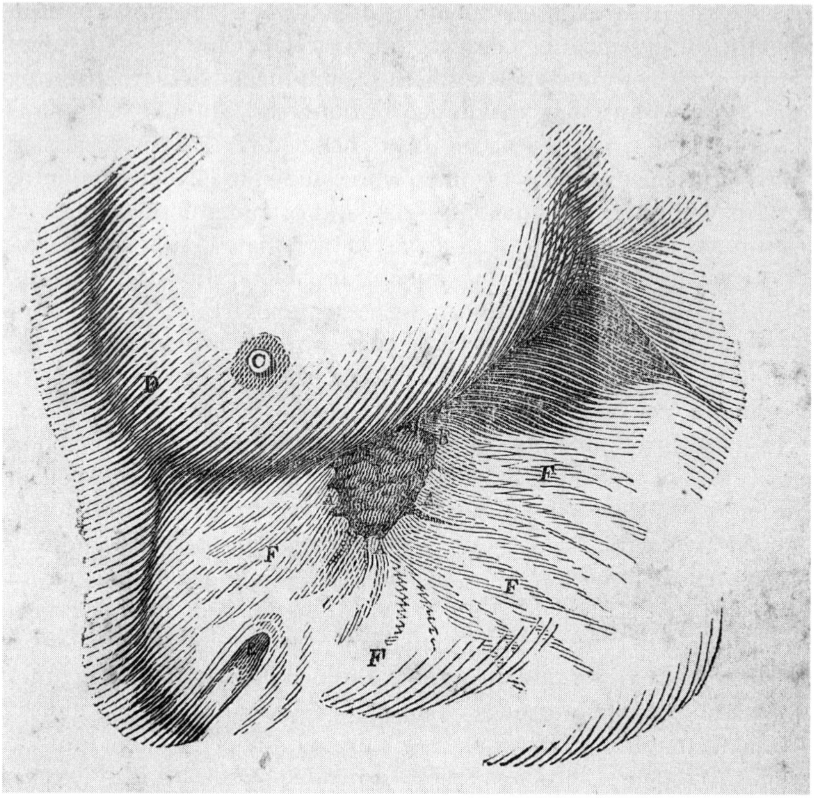

4 William Beaumont, *Experiments and Observations on the Gastric Juice, and the Physiology of Digestion* (Plattsburgh, NY: F. P. Allen, 1833), 25. Fistula created by a gunshot wound in the trapper Alexis St. Martin. Peering through the fistula, Beaumont observed stomachal activity that he compared to St. Martin's statements about feelings of appetite and hunger. US National Library of Medicine.

insisted that all sensations be referred to local parts (56). Beaumont's own theory of hunger attributed the phenomenon to "a *distention* of the gastric vessels, or that apparatus, whether vascular or glandular, which secretes the gastric juice" (57–58). This explanation encompassed appetite as well, since in Beaumont's opinion "appetite and hunger belong to the same class of sensations," differing "only in degree" in accord with "states of distention" (58). To test his view that hunger and appetite were sensations produced in the stomach, Beaumont had St. Martin fast until he was "quite hungry." He then introduced foods (beef, barley soup) through the "aperture" and found, on St. Martin's report, that these "satisfied the *appetite*" and that "he had no [further] desire to *eat*" (210). Beaumont believed that his experiments definitively proved that "the sense of hunger resides in the stomach, and is . . . well allayed by putting the food directly into the stomach" (208).

Despite this claimed success, Beaumont acknowledged that the subject of hunger was "involved in considerable doubt and obscurity" (57). Ultimately he urged that in this regard human beings enjoyed the benefits of providence in the form of an "irresistible sensation, which indicates the proper time for the reception of food" (57). "Man would be miserable and wretched indeed," he proclaimed, "if he depended solely on his own discretion and judgment to decide upon the quantity and quality of aliment necessary to supply the wastes, and administer to the growth, of the system" (57). Yet while urging that human beings enjoyed a God-given sense of what and how to eat, Beaumont adamantly denied that human eating was instinctive: animals, not human beings, were guided by instinct. Indeed, he held that overall the "mode of digestion in the human stomach" could not be "judge[d] . . . by studying animals" (93). This conviction rested partly on Beaumont's recognition of the differing mechanics of the digestive system in different animals, but it also reflected his conviction that in human beings ingestion and digestion were strongly affected by emotion. Speaking of the all-important gastric juice, he observed that "derangement of the digestive organs, slight febrile excitement, fright, or any sudden affection of the passions, cause material alterations in its appearance" (86) and noted that "fear and anger" could even "check its secretion" (87).

Beaumont's work was widely celebrated by physiologists, physicians, and dietary reformers. In North America and Britain, interest centered on his findings on the digestibility of foods and their application in dietetics, but in France and Germany physiologists paid close attention to his experimental procedure, which they praised even while finding much to criticize, especially "from the standpoint of quantitative and chemical analysis."[42] And although further experimentation on human patients with gastric fistulas got under way only later in the century, within a decade after the appearance of Beaumont's book, the French physiologist Nicolas Blondlot (1810–1877) undertook extensive experiments on dogs in which he created artificial fistulas.[43] In addition, in the late 1840s two German-trained researchers at the University of Dorpat, F. H. Bidder and Carl Schmidt, were inspired by Beaumont's work to devise an improved fistular apparatus for investigations with dogs and other species. Focused on chemical analysis of digestive fluids and gastric contents, Bidder and Schmidt only briefly noted a finding about the power of appetite to stimulate the flow of gastric juice ("psychic secretion") that was later heralded by Ivan Pavlov.[44]

By the 1840s experimental physiologists believed that they had made great advances over the past in good part by rejecting vitalist teachings and proceeding on the solid ground of experimentation. Beaumont derided his contemporaries for failing to see the "simplicity and *uniformity*" of digestive processes (37), and Blondlot celebrated the demise of speculative systems, each "destroyed in their turn" (3). Both men, moreover, saw the solidity of new findings as a happy result of joining physiology to chemistry. The fact of artificial digestion had become indubitable, which meant that vitalist claims that digestive processes depended on unique forces of life must be discarded. Nonetheless, these were years when many still believed that physiology had value chiefly as a handmaiden of medicine, and Beaumont and Blondlot continued, like others, to interpret their findings in the light of concerns for human health and disease.

In the final analysis, Beaumont sought to understand the character and work of the gastric juice because he hoped to see health improved through better eating habits. He severely criticized any form of excess and indeed seemed to reject any seeking after pleasure in eating and drinking. He took strong issue, for example, with those who urged that people could safely eat to the point of satiety, which to him stood "beyond the point of *healthful* indulgence, and is nature's earliest indication of an *abuse* and *overburthen* of her powers" (64). The standard he set, by contrast, was "*perfect satisfaction, ease and quiescence of body and mind,*" in short, the absence of any stimulus or excitement (64, emphasis in original). Similarly, regarding alcohol, Beaumont proclaimed that only strict abstinence accorded with the demands of health. Yet Beaumont was aware of the limitations of his experimental procedures; setting forth his table of the digestibility of "articles of diet," he acknowledged that the experiments required for solid knowledge of foods and their effects constituted "a Herculean task, which it would take years to accomplish" (45). What is more, Beaumont believed that the paltriness of his own powers reflected the incapacity of human beings, "miserable and wretched" without divine help (57), to discern what and how to eat. Thus, Beaumont denied that appetite itself could provide a reliable guide to eating appropriately. Even invalids and those who had already suffered from overindulgence kept on eating, appetite unimpaired (63, 109). Only a God-given inner "sense of perfect intelligence" (64) could show the way to healthful eating and digestion.

Beaumont's French admirer Blondlot had no recourse to divine guidance in considering the wisdom of eating choices, but he too, at the outset a confident herald of unchallengeable chemical knowledge, ultimately affirmed that in this regard an inexplicable "intimate sentiment," a product of the "admirable" guidance of nature, was at work (*Traité*, 385). Moreover, his own investigations led him to very different conclusions from Beaumont's on the role of the stomach versus that of the "general economy" in digestion. It was not states or movements of the stomach, he urged, that alerted the organism of the need to eat, but "internal perceptions" (384) that reflected the needs of the whole economy. Nor was the stomach responsible for "changes that take place in our digestive faculties" in the course of eating; rather, it was the "organism as a whole that rejects nutritive matters that it is not in a state to use or assimilate" (388). Why and how the organism rejected some ingesta while assimilating others remained a mystery. Thus, despite the advance of physiological science, an intimate and unknowable "sentiment" was the surest guide to healthful eating.

Appetite Local or Organismic? *Müller's Elements of Physiology*

Finally, to suggest the contested status of appetite by the 1840s, I turn to the textbook published by the one physiologist of this era to rival François Magendie in stature, Johannes Müller (1801–1858). Müller, professor of anatomy and physiology at the University of Berlin and editor of the leading German journal of physiology, published his *Handbuch der Physiologie des Menschen* in separate parts through the 1830s. Rapidly becoming authoritative throughout much of Europe, it was translated in the 1840s by the English physiologist William Baly, who added references to work done after its original publication, especially by British physiologists.[45] Baly's edition of Müller is thus a valuable guide to the state of physiological debate over appetite, eating, and digestion in Europe generally by the mid-1840s.

Müller was a vitalist but of a particular sort that flourished in Germany in the first half of the nineteenth century. With his predecessors, he held that there were clear distinctions between vital activity in living organisms and processes governed by physicochemical laws. Nonetheless, as Timothy Lenoir has shown, German physiologists of Müller's variety did not reject the view, as did vitalists elsewhere, that living organisms were bound by physical laws. Rather, in company with physiologists whom Lenoir labels "teleo-mechanists," Müller accepted that living beings conformed in many respects to such laws while also enjoying their own kind of lawful regularity. Given this theoretical orientation, Müller did not oppose experimentation on "life," but he was hostile to what he saw as the "mechanistic reductionism" of Magendie, charging the "French school" with ignoring the overall ends of vital organization and thus the teleological framework that alone could make sense of vital activity.[46]

Despite this critique, Müller made much use of the findings of Magendie and other experimentalists. Like Magendie he was closely attentive to the chemical analysis of foodstuffs, deriving from it his view of what constituted the chief "nutritive principles" (2:524). He also paid tribute to Magendie for the "important distinction" he had demonstrated between "azotized and unazotized aliments," reporting the findings of Magendie's hunger experiments and noting that these had been confirmed by the work of Tiedemann and Gmelin (2:525). Müller thus affirmed the value of chemical investigation, but he gauged chemical findings in reference to the results of other, more directly physiological studies of nutritional needs, such as experiments done in the past and by contemporaries to determine how long animals (amphibians, birds, seals, dogs) could survive without eating. Turning to human beings, Müller credited some evidence of extreme cases such as the putative capacity of the mentally deranged to endure long fasts while rejecting other stories of persons taking no food "for months and years" as simple "examples of deception."[47]

In drawing on the labors of chemists, Müller also praised the work that led to William Prout's tripartite classification of foodstuffs. He cited at length Prout's assertion, one extending far beyond the domain of chemistry, that when human beings combined foods, even "in the utmost refinements of . . . luxury," they acted on the human "instinct" to mix food elements that were essential to health.[48] Whether Müller endorsed Prout's view, he did not exactly say. Instead, moving on to his own observations about the nature of appetite and satiety, Müller linked questions about the instinctual character of eating, and its position among voluntary and involuntary movements, to the problematic of the nervous foundation of the prompt to eat. Characterizing hunger, appetite, and satiety as "sensations," he explicitly defined "appetite and satiety" as "in part taste itself, and in part sensations analogous to taste" (2:529). He also reviewed disputes over the results of experiments involving vagotomy. Noting claims by the French physiologist Brachet and others that appetite, hunger, and satiety were "local sensations" that "cease to be felt after division of the nervi vagi," Müller was tentative, concluding only that "it is possible that such is the case."[49]

For Müller the question of the relative roles of voluntary and involuntary action in ingestion and digestion formed part of the larger problem of the links between the central and the vegetative nervous systems. Although Müller flatly rejected some of Bichat's claims about the separation between animal and organic phenomena, he credited his French predecessor with having "exactly defined" the "functions of the two systems of nerves."[50] In respect to the nerves of the alimentary system, moreover, he endorsed a view closely resembling Bichat's when he stated that "the cerebro-spinal nervous system has but a limited influence," because only the "commencement and termination . . . of the canal have muscles which are subject to cerebro-spinal nerves and the will."[51] Like Erasmus Darwin, Müller framed the matter of the involvement of the will by pointing to the apparently instinctual sucking of newborns, both human and animal. Rejecting claims that the instinct to suckle constituted a Cartesian "innate idea," he argued instead for the action "in the sensorium of the infant" of an "irresistible impulse to the performance, when possible, of the motions of sucking" (2:545). Rather than lingering on this uncertain terrain, however, Müller turned to specific movements of the alimentary canal, assessing the roles of voluntary versus involuntary action in swallowing and esophageal contractions (2:545–50). He deferred the discussion of the larger issue of nervous control of digestion, returning to it amid his discussion of nervous structure and function.

Müller's name is principally associated with his doctrine of specific nerve "energies," the view that the nerves of sense are "capable of one determinate kind of sensation only," a theory Müller built from extensive observations, both everyday and experimental, about the effect of different stimuli on the sensory nerves.[52] His greatest interest was in vision, but he also inquired into the effect of diverse stimuli on the nerves of taste and smell, which he saw as closely associated with appetite and hunger. As noted earlier, he directly associated the sensation of appetite with taste. Hunger was another, more difficult matter: that hunger had a nervous foundation he asserted absolutely, as no sensation occurred without the work of a specific nervous structure. Addressing the problem of which nerves were responsible for conveying the sensation of hunger, Müller insisted on several points that had been much in dispute. First, he rejected Magendie's assertion that hunger was an internal sensation dependent on the general organization of the nervous system. Indeed Müller rejected the very concept of internal sensations, urging that there was no difference between sensations arising in response to "external agencies" and those "produced by internal causes." In this regard Müller broke with the tradition stretching back at least to Locke according to which the nerves of sense gave unique access to phenomena of the external world. Reflecting his schooling in German philosophy, Müller insisted that what sensations conveyed of the "qualities . . . of external nature" represented only those changes that external phenomena prompted "in the condition of the nerves" (3:1059). If sensation involved only states of our own body, clearly it made no sense to differentiate between those caused by external stimuli and those "constantly being produced by internal causes in all parts of our body endowed with sensitive nerves" (3:1060).

This stance on the nature of nervous action precluded, for Müller, the action of any diffusely distributed "property" of sensibility and committed him to the view that all sensations must be attributed to the action of sensitive nerves. Hence his interest in the work of Brachet and others who claimed to have shown experimentally that hunger disappeared once the vagus was cut. As I noted earlier, when Müller mentioned Brachet's findings in his discussion of digestion, he seemed to leave room for doubt, but in the part of his work devoted to nervous structures and activity he flatly stated that "Brachet has observed, that the sensation of hunger was no longer felt after the vagus was divided" (2:836–37). Müller suggested further that the sensitive nerves responsible for the sensation of hunger might also include the "many organic fibers derived from the sympathetic" (2:837). In sharp contrast to Magendie, who had questioned whether the great sympathetic was in fact "nervous" in character, Müller insisted that the ganglionic nerves were sensitive. Citing animal experiments that he and others had conducted using mechanical and chemical irritants, as well as clinical evidence of "painful sensations" patients felt "in diseased states of the parts supplied with sympathetic nerves only," Müller asserted that the celiac ganglion and other parts of the sympathetic system registered pain.[53]

The claim that sensations of appetite, hunger, and satiety must depend on specific nervous structures put Müller at the forefront of those who sought to "localize" all sensations. And yet, like Magendie, when Müller left the problem of the physical substrate of sensation and moved to the larger issue of the link between conscious desires that were part of "mind" and the bodily structures that made possible the realization of desires, he too acknowledged the vast distance between the known and the unknown. In the final division of his textbook, titled "Of the Mind," Müller asserted that "the peculiar quality of the mind is 'consciousness,' a something of which no further definition can be given" (4:1345) and stated that "the whole action of the mind consists of the conception of ideas and of emotions" (4:1353). Among these emotions he included those "desires" that reflected the general "organic *appetitus*" that impelled organic bodies to engage with the external world, strivings that were embodied not just in the brain as the center of mind but in the organism as a whole:

> In all organic bodies there is a tendency towards the exterior, and an effort to preserve their own integrity. Even plants are the subject of such a striving during their growth; and in the same sense has the striving even in man a much wider basis than the brain, and is an action manifesting itself throughout the whole organism, and determining the process of organization. . . . [T]his organic *appetitus* is reflected in the sensorium, and. . . . felt as an obscure power of the individual declaring itself in the manner of an instinct,—as a power which affirms itself constantly, and derives sustenance from all ideas connected with the self or individual being (4:1369).

Müller knew that physiology could not yet offer a general picture of the connection between such strivings of the whole organism and the action of brain and nerves. Referring to the claims of Franz-Josef Gall (1758–1828) that every propensity of the organism was governed by "a particular part of the brain," Müller observed that "there is a total absence of facts which might render it probable that organs are set apart for the different appetites."[54] Indeed, Müller denied that there were in fact different appetites, proposing instead that "only one identical appetite, one original striving for the maintenance of the integrity and extension of the power of self, exists." This "one striving or *appetitus* has different objects," he continued, and acts as directed by "those organs which afford specific modes of intercourse with the external world" (4:1369). Reasserting that such "organic influence is communicated to the body from the brain" along the course of the nerves, Müller recognized nonetheless that there was a mysterious concatenation of objective nervous effect and the peculiar "dispositions of . . . individuals"; the digestive organs might fail to respond to stimuli, for example, if "the appetite for food" was lacking (4:1370). Demonstrable physical stimuli acted differently, furthermore, on individuals whose "particular local idiosyncras[ies]" caused them to register effects preferentially in the liver, heart, intestinal canal, or "system of the spinal cord" (4:1370). Thus, once again, the great mystery of the psychophysical relation derailed efforts to explain the "strivings" of the organism by means of specific nervous or other organic structures. Müller was critical of Bichat and others who urged that organs other than the brain, specifically "the abdominal and thoracic viscera, have a certain connection with the mental functions" and serve as "the source of the passions" (2:816*). He acknowledged that the viscera were "affected during the prevalence of passions of the mind" but insisted that it was "by means of the change which takes place in the brain" that the passions exerted their effect. Yet he "confessed" that the nature of this link was "still involved in much obscurity" (2:816*).

In the first half of the nineteenth century physiologists achieved no consensus on the nature and functioning of appetite. Indeed, if its omission from Tiedemann and Gmelin's prize essay is a good measure, physiologists had not yet agreed that appetite was a problem falling within their purview. For those who did attend to the problem of appetite, the question of which methods should be employed remained unsettled. Leuret and Lassaigne were not alone in believing that vagotomy and related procedures targeting specific nervous structures would eventually yield convincing results. Yet others, including the commanding figures Magendie and Müller, regarded appetite as a phenomenon inaccessible by existing experimental methods. Magendie believed that hunger and appetite, though governed by the general laws of the nervous system, had no "seat," whereas Müller held that appetite expressed strivings of the whole organism for self-maintenance. Hence the tendency of these tribunes of the new physiology to revert, when speaking of appetite, to traditional forms of evidence—tales from travel literature, the history of shipwrecks, and reports from physicians and patients. How they judged the reliability of such accounts—how Müller determined that some accounts of prolonged fasting were "matters of deception" and others not, why Magendie accepted some reports from distant locales while rejecting others—they did not say. Yet it seems that their reliance on such evidence, which contrasted so strongly with "proofs" issuing from the laboratory, helps to explain why they were often the first to recognize the "obscurity" surrounding the why and how of eating. Only later in the century did such hesitancy give way and the urge to eat come to be regarded as a problem approachable with the methods of an increasingly self-assured physiological science. This conviction took hold only when physiologists overcame or set aside the objections posed by physicians who, while sometimes admiring the achievements of the laboratory, more often decried their futility in preventing or curing maladies of appetite. The next chapter explores medical approaches to appetite in the decades up to 1850, the last era of this history when physicians remained generally skeptical about the utility of laboratory methods and asserted that only by observing endlessly varying individuals could an elusive phenomenon like appetite be known.

SIX

Extremes and Perplexities of Appetite in Clinical Medicine

Historians have seen the early nineteenth century as a golden age of physiology, the era when its findings achieved a new degree of certainty and the discipline emerged as the leading field of life science.[1] Nonetheless, physiology continued to share common terrain with medicine, even while many physicians were either skeptical of or openly hostile to physiology in its experimental form. Medical criticism of experimental physiology was sometimes pragmatic, with commentators decrying the complexity and tedium of experimental procedures.[2] But it was often grounded in the vitalist view that vivisection shed little light on the normal functioning of healthy animals because "organized" beings were governed by unique laws of vitality. Some vitalists were especially hostile to the intervention of chemistry into physiology and to the idea that vital processes could be replicated outside the body, as with efforts to bring about artificial digestion in vitro.[3] Vitalist physiologists often joined hands with (or were themselves) practicing physicians who connected theoretical arguments about living nature to ideas about the healing art. Such physicians upheld the supremacy of the direct encounter with patients, often invoking the name of Hippocrates to defend medicine's peculiar methods and insights. Only medicine, these physicians asserted, could grasp the complex range of signs and symptoms, physical and moral, observed in sick patients. Manifestations of disease in individuals, they insisted, could not be illuminated by observing movements and processes of laboratory animals. It was precisely because medicine attended to the whole person that it could grasp phenomena that deterministic science, whatever its specific technique, could not. In the alimentary realm, this was especially true of appetite, the mysterious "first phase" of digestion that was so powerfully influenced by thought and emotion.[4]

In the early nineteenth century experimental physiologists generally did not reciprocate the hostility of medical traditionalists but, instead, accepted the need to join bedside and laboratory investigation. Certainly it is easy to find statements by physiological modernists such as Magendie deriding "fables" of medical tradition. Such assertions would become more strident after 1850, when, for example, the German experimentalist Emil Du Bois-Reymond (1818–1896) "denounced . . . 'the entire medical profession as obscurantist and bigoted.'"[5] Still, until late in the century all physiologists were trained in medicine and most made at least some use of clinical observations. Many physiologists concerned with digestion practiced medicine and took a special interest in aberrations of eating.[6] However, the reverse—training in and regard for the "auxiliary sciences"—was not true for most physicians of the early nineteenth century. Only gradually, and at a varying pace in different national settings, did training in the specialized life sciences come to be regarded as beneficial or essential to medical education.[7] This meant that up to midcentury (and in most contexts well beyond) there was a distinctly medical approach to problems like appetite that were seen to entail a poorly comprehended connection between mind and body.

Uses of Tradition: Medical Representations of Appetite and Hunger

The suspicion, and sometimes outright hostility, that physicians felt toward the new physiology at the opening of the century is evident in student theses on appetite and hunger prepared between 1800 and 1820 at the Paris Faculty of Medicine. The Paris faculty of these years is best known for its innovative stance, especially its embrace and promotion of pathological anatomy. But among its leading figures were physicians like Philippe Pinel, who owed a strong debt to Montpellier vitalism and advocated a traditionalist, "Hippocratic" conception of appetite and hunger. Because students usually simply restated claims made by their teachers, their theses provide a good index of tensions at the Paris school between old and new regarding appetite, foods, and eating.

One constant in theses completed at the Paris faculty was insistence on a clear distinction between appetite and hunger based on sensations of pleasure and pain, with some student authors pointing to appetite's dangerous "relation to pleasure" while others argued that appetite was beneficial in leading the eater to what was both pleasurable and nutritious.[8] These writings also frequently criticized the mechanisms physiologists had proposed to explain the origin of hunger, efforts they portrayed as "hazardous hypotheses," lacking demonstration, and having "no happy result."[9] Instead of focusing on causes, the theses concentrated on describing how appetite and hunger varied in accord with age, temperament, sex, season, climate, and the passions. Repeating dicta from classical and modern physicians, these student writers asserted, for example, that the appetite of children was intense but quickly satisfied; that people of the north, their appetites piqued by the cold, ate rougher foods than the indolent people of the south; and that mania or "sad passions" caused the desire for food to diminish.[10] These medical observations were often cast in vitalist language. Thus, one student explained the weak digestive power of melancholics by evoking the vitalist principle that the animal economy had only limited reserves of energy: in such individuals the vital forces were concentrated in the head, leaving the stomach to "badly perform its functions."[11] Similarly, another author held that children were marked by weak "digestive force" and that the diminished appetite of summer corresponded to the direction of the "vital forces" to the body's exterior during hot spells.[12]

In these writings vitalist arguments generally rested on evidence from introspection, historical anecdotes, travelers' accounts, and bedside observations, whereas newer modes of physiological inquiry were characterized as fruitless. This was especially true for appetite, which, one author asserted, belonged to "the empire of the intellectual faculties."[13] But hunger, too, though a corporal phenomenon more readily observable than appetite, had also escaped the grasp of physiologists; no method could penetrate the "hidden phenomena of our individual nature."[14] Citations to a wealth of authorities appeared in these writings but none to investigators from Spallanzani onward who had studied the urge to eat with the help of chemistry. Nor did pathological anatomy receive much notice; what postmortem results did appear were drawn not from recent investigations but from seventeenth- and eighteenth-century authors and

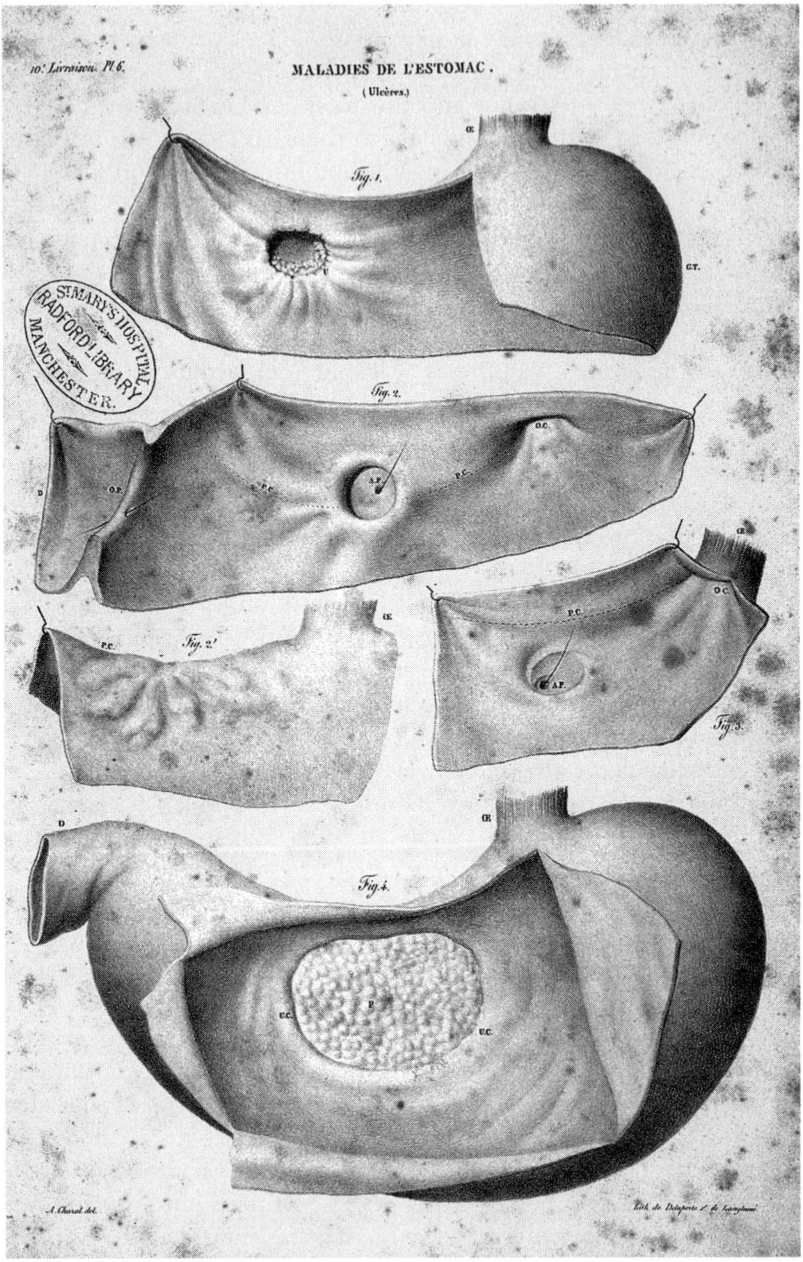

5 Jean Cruveilhier, *Anatomie pathologique du corps humain*, vol. 1 (Paris: J.-B. Baillière, 1829–1842), "Maladies of the Stomach." Pathological anatomists sought the origin of disease in lesions of the tissues such as the gastric ulcers depicted in this atlas of pathology. Wellcome Collection.

led only to general comments on the state of the stomach and intestines in individuals who had died of hunger. In lieu of inquiries using modern approaches, students cited Hippocrates as the premier authority.[15]

These student productions attended closely to extremes of human appetite and hunger, and here too they drew on medical observers back to the ancients. Narratives of prodigious fasting or eating often resembled traditional storytelling more than evidentiary accounts, marked as they were by repetition, minor divergences of salient detail, and stress on dramatic or horrific elements.[16] By the early nineteenth century a rich trove of eating tales—whether of excess, abstinence, or perversion—was available. Such stories were told everywhere in the medical literature. In Germany, the immensely successful journal edited by Christoph Wilhelm Hufeland (1762–1836), *Journal der practischen Heilkunde*, favored tales of fasting girls and women.[17] Hufeland himself recounted the case of sixteen-year-old Anna Kienker, who survived for eighteen months taking in only a few drops of warm water and who, when presented with food, showed no trace of a desire to eat. Kienker subsequently recanted, entering the ranks of despised "female deceivers," but such recognized deceptions did not interrupt the telling of these tales.[18] A few years later a hospital physician reported to Hufeland's journal the case of an aged woman who, ordinarily possessed of a "full appetite," took nothing but fresh water for five weeks and expressed "not the least desire for any other kind of drink or food"; in 1829 two other such cases were reported.[19] French and British accounts of "fasting maids" were also common. In 1809 an author reporting to the *Edinburgh Medical and Surgical Journal* gave credence to the "fasting woman at Titbury, who has at present lived above two years without food."[20] Textbooks and other compendiums similarly included, often in great detail, the histories of great and grotesque eaters.[21] Here then was a special role for medicine: observing and recounting from remarkable human experience the end possibilities of health and illness, life and death.

Gastralgia and Dyspepsia

Pathological extremes stood alongside more ordinary ills of appetite and eating in the classifications of disease that nineteenth-century doctors produced in seeking to improve on the efforts of Enlightenment nosologists. One of the most acclaimed productions of the Paris Clinical School was the updated classification of disease devised by Pinel, who signaled with his change of terminology from "nosology" to "nosography" the goal, widely shared by Paris clinicians, of offering precise disease descriptions before or in place of a search for causes. Pinel revised many labels and categories in light of shifts in medical method since the days of Sauvages and Cullen, especially the new attention to tissues and the intensified effort to discern in postmortem investigation the specific sites of disease. Such impulses led Paris clinicians to new insights into diseases of the heart and chest, but in other areas, including the spectrum of eating-related illness, their efforts led only to further disputes over the source, labeling, and treatment of such maladies as well as methods best mobilized to investigate them.[22]

Perplexities surrounding appetite, eating, and digestion are evident in work on diverse maladies recognized in these decades, both those linked to stomach and intestines (constipation, dyspepsia, enteralgia, enteritis, gastralgia, gastritis, gastrodynia, indigestion) and those connected to mind and emotion, especially hypochondria and other broad-gauge "nervous affections" that gradually displaced the vapors of old and invariably encompassed disturbed ingestion or digestion. Variants of all these English terms existed in other languages, and the alternative labels established in different national or linguistic traditions often involved minute variations in ideas of causes and symptoms. A blended psycho-gastric disease prominently discussed in France in these years was *gastralgie*, a diagnosis promoted by the Paris clinician J.-P.-T. Barras (1789–1851), who saw it as a leading form of what he and others, building on the newly minted concept of neurosis, termed "neuroses of the stomach."[23] Barras's point of departure was his hostility toward the medical system of F.-J.-V. Broussais, a Paris iconoclast who traced all illness to gastrointestinal inflammation and trumpeted bloodletting as the only efficacious remedy.[24] Barras denied Broussais's claim, arguing that a patient could suffer from gastralgia for a lifetime without ever manifesting signs of inflammation. Barras accepted that neuroses differed depending on their anatomical locale, and he devised for gastrointestinal forms a classification that resituated diseases from across the span of existing nosologies, including cardialgia, enteralgia, nervous vomiting, indigestion, anorexia, and a "gastro-enteralgic" form of hypochondria. In all of these, he urged, the state of the appetite was "an essential point in distinguishing neuroses from gastro-intestinal inflammations."[25]

A striking feature of Barras's treatise was a twenty-five-page, densely packed description of his own prolonged suffering from gastralgia. Describing four major attacks, he recorded ever more serious symptoms: fever, convulsive cough, hot skin, accelerated pulse, sweats, alternately increased and lost appetite, nausea, gas, constipation, and shifting pains. In the most severe bout, Barras suffered depression, fainting spells, a bizarre sensation that a "reptile" was moving about in his stomach (*sentiment de formication*) as well as suicidal thoughts, symptoms that reduced him to the condition of "the most vaporish little housewife" (4–30). Along with this account of his own illness, Barras appended thirty-four other cases of gastralgia, commenting on the sex and age of patients as well as the "passional upsets" and, in women's cases, reproductive phenomena that caused this kind of illness (38, 51, 98, 110).

In discussing the influence of gender, Barras indicated one of the ways that clinical observers approached ingestive and digestive problems that were inaccessible to laboratory investigators. Although in the later history of appetite, the sex of experimental animals was to be highly significant, at this point experimentalists had nothing to say about the "influence of sex" in the eating patterns of their subjects.[26] Here, then, was a matter of great importance to which only physicians could speak. Throughout his work Barras affirmed gendered patterns of appetite and eating that had wide currency among physicians: his anorexic prototype was a woman who had long had peculiar eating habits but lost all interest in eating when she entered menopause.[27] His characteristic bulimic was the famed Antonio Viterbi, a prisoner who had deliberately starved himself to death—experiencing "bulimia" or ravaging hunger along the way—to avoid the humiliation of public execution (248–61). In neither case, however, was the influence of gender absolute. Loss of appetite and unwillingness to eat might be more common in women than men, and reproductive influence was often supreme in women, but the disturbances of appetite Barras identified were not uniquely female or male in character or incidence. In the clinic, indeed, there were no absolutes: patterns were discernible and of great therapeutic interest, but ultimately innumerable variations marked health and disease. Nowhere was this more evident, wrote one physician at midcentury, than in "ills of everyday life" that were neglected by "patho-anatomists" in their single-minded regard for diseased tissues, but closely monitored by clinicians who attended to ordinary human suffering.[28]

In Anglo-American medicine, physicians continued to favor the disease label "dyspepsia" for troubles of appetite and related symptoms that marked diffuse ills of stomach and psyche. This trend is at least partly explained by the wide influence of William Cullen and the Edinburgh medical school throughout the English-speaking world. A collection of student dissertations held at the University of Pennsylvania indicates the great interest shown in dyspepsia in the United States in the early nineteenth century. One of the most prestigious medical schools in the young republic, the University of Pennsylvania drew students from every region of the country, in all of which dyspepsia was a widespread complaint.[29] Nonetheless, the causes, nature, and nomenclature of dyspepsia were all problematic: students often indicated alternative labels by first using dyspepsia in their titles but then shifting to different terms along the way, or, in some instances, directly addressing the problem of terminology.[30] Cullen's work was the usual point of departure for these students, who began by referring to "want of appetite" and other symptoms Cullen had set forth in defining dyspepsia.[31] All these dissertations offered lengthy descriptions of symptoms, the physical often mixed indiscriminately with distress identified with mind, the passions, temperament, or states of the "whole body."[32]

Yet already in descriptions of symptoms the crux of the diagnostic and classificatory puzzle of dyspepsia was evident: some of these works covered only somatic woes, whereas others included terrifying dreams, irregular and disturbed sleep, vertigo, hallucinations, or worries about business. In one instance the student simply incorporated as part of dyspepsia all the symptoms Cullen had supplied for the vapors—languor, listlessness, want of resolution, sadness, and "apprehension of great evil." In short, the problem of dyspepsia remained much as it had been in the late eighteenth century; opinion was strongly divided on whether it was a somatic ill, a "mental perturbation," or something in between.[33] The great hope of pathological anatomy, promoted in the United States as in Europe by at least some physicians, was that for all diseases the specific site, and with it the local or proximate cause, of disease could be identified. And although Pennsylvania student works do report efforts by American physicians to discern in autopsy the origin and nature of dyspepsia, the results were inconsistent, with one investigator describing the point of union of stomach and duodenum as the "most frequently found . . . seat of disease," and another pointing to the state of the "coats of the stomach."[34] Thus, the perceived causes of dyspepsia remained as varied as the symptoms, with etiologies ranging from specific disturbances of stomach or liver or bowels to general ways of living (indolence, sensual excess, urban "debaucheries and amusements") to "depressing passions," ingestion of noxious substances (coffee, tea, tobacco, opium), and even decayed teeth.[35] In this farrago, appetite continually reappeared in all its unhappy forms—lacking, deficient, voracious, perverted, or "wholly destroyed."[36] Usually, troubled appetite was a symptom of dyspepsia but sometimes—in the form of excessive desire for pleasure from food—appetite was itself the cause of the ailment. Treatments also varied, some urging that appetite be closely regulated, while others proposed that even noxious "cravings" must be satisfied; the old problem of distinguishing innate and beneficial guidance from the "false language of the stomach" remained.[37]

Distinguished as it was in North America, in the wider world of medicine of these years Pennsylvania was an outpost. Yet perplexities surrounding ills of appetite, eating, and digestion were equally evident in Edinburgh, often seen as the most forward looking of the British medical centers. Here too illnesses that entailed lost or deranged appetite were well known. Dyspepsia, hypochondriasis, and chlorosis were routinely reported both by physicians in private practice and those who saw lower-class patients in public dispensaries.[38] The fact that many dyspeptic patients came from the ranks of the needy accords with the claims stated in Anglophone sources of these years that dyspepsia afflicted individuals rich and poor and, even more significant, that "stinting one's self of food is quite as injurious as over-eating."[39] British authors recognized that dyspepsia and other "nervous affections" might well result from "gormandizing," but, as in the Pennsylvania dissertations, they adduced many different causes physical and moral. Frequently, moreover, the Edinburgh journal's editors or contributors commented that the very terms used to characterize these ills were unreliable and "specious."[40] They also despaired of help from postmortem dissection, noting that results had so far indicated that "dyspeptic symptoms" could arise from a diseased liver, gallbladder, or spleen. Moreover, this last comment appeared in a review of a book that referred not to dyspepsia but to "indigestion" and, in its subtitle, to "bilious and nervous affections of the head and other parts" as well as "disorders and diseases of the stomach and superior parts of the alimentary canal," a list that covered the gamut of possible etiologies.[41] The reviewers concurred with the author that ills of this kind included both stomach disorders that led to "disorders of the head" and the reverse, "sympathetic disturbance of the stomach from affections of the brain" (397). Ultimately they dismissed the issue of origin, noting that the medical practitioner could not worry over "nice distinctions" about the remote cause of disease but must, rather, focus on relieving symptoms by "the usual palliatives" (399). In sum, the Edinburgh physicians, like the Pennsylvania students, while agreeing on disturbed appetite and some other symptoms of a condition widely known as dyspepsia, could pinpoint neither the origin nor nature of this debilitating complaint.

Mental Medicine: From Epigastrium to Brain

The medical literature of dyspepsia suggests a common feature of maladies in which lost appetite was prominent: as puzzling mixes of somatic and psychic causes and symptoms, they had no agreed-on visible markers and as such could be known only, and then only uncertainly, through close observation of patients. While bedside observation was sometimes regarded as the province of traditionalists, many clinicians believed in progress and worked to refine the methods of listening and reporting on which observation of such phenomena depended. This was especially true within emergent "mental medicine," which saw the development of new styles of patient-physician dialogue and vastly richer case studies than those of earlier years.[42] In the France of Revolution and Empire, inspiration for such steps came chiefly from Pinel, who strove to chart the simultaneously psychic and somatic manifestations of disturbed appetite. Pinel is chiefly remembered for his determination to carve out a special domain of mental medicine, to shift interest in the etiology of mental illness from physical to moral causes, and to replace the old physical therapies, including dietetics, with "moral treatment."[43] Less well known is the fact that Pinel continued to endorse the vitalist tenet that the "original seat" of mental illness was in the "epigastric region," the "corporeal center" of sensibility, and from there "radiated" throughout the body.[44] Thus, Pinel, while promoting a clear division of mental from general medicine, also reaffirmed the holistic tradition in which health partook at once of mind and body, which meant that appetite was at once a psychic and somatic function. Accordingly, physicians on both sides of an emergent "visceralist-cerebralist" divide in respect to psycho-gastric ills invoked Pinel's name. Visceralists like Barras tended to stress stomachal origins of gastralgia and other ills encompassing disturbed appetite while practitioners of the new mental medicine moved to dislodge appetite from the body and to consider afflictions of appetite strictly as symptoms of disturbed mental function. This was to prove a fateful break in the medical history of appetite.

Pinel began investigating forms of mental alienation in the 1780s, and in his *Nosographie philosophique* of 1798 he described four "orders" of *névrose*. The first order encompassed the "vesanias," which Pinel saw linked in many cases to disordered appetite and eating. Surveying differing forms of insanity, Pinel noted approvingly George Cheyne's claim that madness was frequently tied either to the consumption of "rich foods" or, obversely, to fasting (*Nosographie philosophique*, 2:9–10). He established the rationale for such thinking when, in the first edition of his famed treatise on mental alienation, he linked mental medicine to philosophical sensationalism, drawing attention to painful and pleasurable sensations, hunger chief among them, that goaded human beings to fulfill the basic "ends" of existence—self-preservation, reproduction, and the protection of offspring.[45] Such theoretical pronouncements were matched by

Pinel's attention to practicalities of food provisioning in the refurbished asylum of the revolutionary era and the First Empire.[46] Given these dual impulses, Pinel frequently described eating and digestive symptoms exhibited by patients: the intestinal tightening that accompanied mania, the loss of appetite seen in melancholics, and the swings between "total abstinence" and "singular voracity" observed in hysterics. He also prescribed dietary measures in his plans of treatment. Troubled at the dangers posed by loss of appetite and refusal of food, he believed it essential to respond vigorously to such behavior, if necessary by forced feeding.[47]

Pinel's legacy to French medicine on the psychic entanglements of appetite and eating was mixed. Although overall he subordinated such matters to the moral dimension of psychic illness, when he concerned himself with bodily causes and manifestations, he credited traditional views that favored the epigastric region and stomach over the brain. From early in his career Pinel argued against conceptions of psychic illness that were exclusively brain centered, and as his formulations evolved in later editions of the *Nosographie philosophique,* he linked neuroses not to lesions in the brain or any specific organ but to disturbances of physiological activity—sensory, cerebral, locomotor, nutritive, or generative.[48] Pinel's developed view of the neuroses was that they originated in passional upset, manifested first in shifting psychosomatic symptoms such as loss of appetite, and often settled in one functional domain. By this means eating and digestion, though apparently demoted in importance by Pinel in favor of moral factors, could be, and by some of his self-declared disciples were, reinstated as essential elements of psychic illness. Thus, while Pinel was claimed by nineteenth-century alienists as the progenitor of a discretely mental medicine, he was also invoked by those who recast in the Pinelian language of neurosis long-standing teachings about the role of the "lower organs" in psychic disturbance.

Pinel's leadership of mental medicine passed in time to his most famous pupil, Jean-Étienne-Dominique Esquirol (1772–1840), who first directed a private *maison de santé* catering to well-heeled patients and later served as a staff physician in two large public asylums in Paris (the Salpêtrière and Charenton). Building on Pinel's work, Esquirol and his students were instrumental in developing mental medicine as a specialty both by undertaking important theoretical inquiry and by building an institutional base in the asylum. These asylum physicians, or alienists, developed a distinctive approach to mental illness that, despite important variations and disagreements, constituted a unified tradition. In regard to problems surrounding eating and digestion, Esquirol was that tradition's central point of reference.[49]

Like Pinel, Esquirol accorded a subordinate role to appetite and eating in the etiology of mental illness while nonetheless accepting the view that the seat of mental maladies "could be the digestive apparatus or the liver."[50] While he endorsed the principle that the passions often registered a powerful effect on the stomach or abdomen, he too emphasized the moral rather than physiological character of emotional disturbance. In the case of one hysteric, for example, Esquirol carefully noted the patient's eating patterns but attended chiefly to her "exalted imagination" and strange behavior. Similarly, in cases he used to illustrate how physical "crises" could resolve episodes of madness, Esquirol alluded to eating and digestive ills—loss of or heightened appetite, constipation, "gastric phlegmasia"—but he nonetheless emphasized the general march of disease and patients' overall response to treatment.[51]

In respect to treatment, Esquirol stated on occasion his intention to redress the balance upset by Pinel, who had undervalued physical therapies, and endorsed, along with hydrotherapy and other methods, dietetic remedies such as acidulated drinks to calm hysteria.[52] As did Pinel, Esquirol regarded failed appetite and refusal of food as especially dangerous and argued for the indispensability of forced feeding. To prove his case, he recalled the story of a patient who was called to military service during the Revolution but, to avoid the draft, gave "himself over to onanism so as to make himself ill and thus obtain an exemption." Previously healthy and cheerful, the man became melancholic and stopped eating for two, three, or five days at a time. Confined to Esquirol's care, he was treated with showers, massage, and hot flannels applied to the abdomen. When none of these succeeded, Esquirol prescribed the use of an elastic tube to introduce "liquid food into the stomach" via the nostrils and back of the mouth. After forced feeding, the patient's behavior "remained bizarre," but he began in time to eat on his own, and Esquirol was convinced that in cases of lost appetite this remedy was essential (*Maladies mentales*, 1:609–14).

Other patients whose eating pathologies Esquirol described included a young woman who starved herself to death after discovering her husband making love with her sister and a maniac who was "devoured by hunger," ate "everything that came to hand," and died when a sponge he consumed got caught in his esophagus (*Maladies mentales*, 1:535, 541). Embedded in Esquirol's reports was a consistent, if seldom focused, attention to how disordered eating and digestion varied in men and women. In an 1818 account of patients treated in private and public settings, fifteen males and fourteen females suffered symptoms including gastric fever, colic, the consumption of noxious substances, epigastric and stomach pain, vomiting, constipation, weight loss, and either refusal to eat or "voracity."[53] Five of those who lost appetite and weight were women and three were men; in contrast, the four patients with "voracious appetite" were all men. Reflecting on these observations, Esquirol noted that eating symptoms were frequently associated with sexual or reproductive ills: men suffered psychic and gastric symptoms after indulging in masturbation or sex with women, and women experienced psychic disturbance in connection with irregular menstruation, pregnancy, childbirth, nursing, or menopause. Despite this etiological difference, both men and women could experience voracious hunger or its obverse, lost or depressed appetite, although voracity was more common in men and loss of interest in food in women. Esquirol's observations in this regard were replicated in other such accounts into the 1860s.[54] As with Barras, this gender breakdown was not absolute; the accounts of alienists seldom proposed a rigorous gendering of the sort that characterized "uterinist" explanations of hysteria or that, later in the century, would come to characterize thinking about anorexia nervosa. Nonetheless, the influence of sex on disturbed appetite was repeatedly affirmed, and with it the exclusive capacity of clinical observers to recognize such constants of lived human experience.

While Esquirol's work remained important to alienists long after his death in 1840, as early as the 1820s critics of the Pinelian-Esquirolian approach initiated a drive to link psychiatric illness to organic malfunction of the brain and cerebro-nervous system. Perhaps the most single-minded of the French cerebralists was Étienne-Jean Georget (1795–1828), who served for a time as Esquirol's student intern. Georget held that lesions of the brain and cerebro-nervous system could be discerned in most, if not all, autopsies performed on patients who in life had suffered from mental alienation, and he argued strenuously against the role of the stomach in the origins of mental illness.[55] While acknowledging that "illness in virtually all our organs is accompanied by sympathetic lesions of the digestive canal . . . loss of appetite, thirst, disgust [for food], . . . diarrhea or constipation," he insisted that in mental alienation digestive upset was "not a cause, but an effect, of the [primary] cerebral affection" (*De la folie*, 173, 174). Thus, it was only as a complication of cerebral illness that Georget attended to deficient or voracious appetite and other disturbances of eating and digestion.[56]

Other prominent figures in the Paris Clinical School upheld similar views of the cerebral etiology of mental illness and of the subordinate role of disordered eating and digestion.[57] To midcentury and beyond, this cerebralist view gained broad acceptance among alienists, who either neglected or deliberately disavowed the role of eating and digestion in causing mental illness. From the 1840s onward, alienists did evince considerable interest in the strictly therapeutic issue of how to respond to lost appetite and refusal of food. Yet they were generally concerned only with the mechanics of eating, especially the complications of forced feeding, rather than the underlying disease process. This focus on mechanics was evident in an 1845 article by François Leuret, who detailed difficulties attendant on forced feeding and urged use of a feeding tube of his own design that could be kept in place for weeks or even months at a time. From this point the intricacies of forced feeding, especially the relative benefits and dangers of different apparatuses, drew regular attention from alienists.[58] But aside from these largely therapeutic interests, mental medicine generally came to reject the physical-moral framework of Pinel and Esquirol and to focus strictly on mental phenomena such as delusions and hallucinations while evincing little concern for feelings and states associated with ingestion and digestion. In this respect mental medicine of the early nineteenth century resembled the philosophical tradition that, as Carolyn Korsmeyer has shown, regarded matters of taste—the taste of food—as unworthy of attention.[59]

These patterns in French psychiatry that led to the neglect of appetite were evident elsewhere and in some cases were even more marked. In the first half of the nineteenth century British psychiatry was divided on questions of causation, some urging a rigorously somaticist origin of mental illness in lesions of the brain and central nervous system, while others, either because they rejected this kind of materialism or thought the search for lesions in autopsy to be chimerical, concentrated on the moral causes of mental disease. Reviewing opinion on the etiology of mental illness in 1824, the editors of the *Edinburgh Medical and Surgical Journal* observed that British physicians attributed more importance to brain lesions than did the French, citing the views of William Lawrence (1783–1867), for a time a surgeon at Bethlem hospital, who claimed that organic brain disturbance was always involved in mental disease. By contrast, those championing moral causation of mental illness, such as Alexander Morison (1779–1866), cited such factors as the ill effects of education that failed to bring "the inclinations and affections under the control of religious and moral principles."[60] In the British context as well, then, states of the appetite drew attention within mental medicine only as symptoms that accompanied illnesses seated in the brain or the soul. As in France, the greatest concern was the loss of appetite seen in mentally disturbed patients, and in Britain too, if a bit later in time, an extensive discussion unfolded of the therapeutic uses of forced feeding.[61] Again, as in France, this discussion centered on mechanics with little concern expressed for why the patient refused to eat. While physicians continued routinely to observe that disturbances of mind and passion caused the appetite to disappear, they accepted this association as one long recognized in medical tradition, and they did not seek an explanation, in body, mind, or soul.

Despite its ubiquity as a marker of health or sickness, appetite was not readily susceptible to investigation by the methods that came to represent medical progress after 1800. Once pathological anatomy became the favored procedure of medical science, such study could proceed in only one of two directions. Either appetite was directly linked to operations of organs or tissues that, when diseased, left visible lesions, in which case it was the organs and tissues that were of prime importance rather than appetite, or appetite was some puzzling mix of somatic and psychic experience, a phenomenon that left no visible sign and that could be known, if at all, only through methods unconnected to the dissecting table. In this way, bedside observation remained perforce the method of choice, the one way physicians had of confronting the desire to eat, or its lack, as a part of general health or sickness. Physicians did make progress in these years in refining the methods of listening and reporting on which observation of appetite depended, especially within the emergent field of mental medicine. For physicians like Pinel, still grounded in vitalist conceptions of mental illness as seated in the epigastrium, this development brought enhanced attention to patients' internal sensations as a source of their afflictions. Yet by 1850 mental medicine had largely rejected the vitalist-inspired physical-moral framework in which Pinel had been schooled and begun targeting strictly mental phenomena rather than internal states of indeterminate origin. Taken together, these directions in the medicine of appetite encouraged enhanced attention to the brain, as the presumed seat of mental maladies that disrupted the exercise of appetite and other functions, or, when this quest proved largely futile, to cerebral functions of reasoning, judgment, and will. This way of thinking about appetite, which was chiefly the work of French physicians, was to lead in the next half century to a protracted struggle over appetite itself, whether it was primarily visceral or cerebral in origin, and to competing approaches to maladies of appetite that varied markedly depending on whether appetite was adjudged an inner call or a matter of conscious, often disordered, choice.

PART THREE

Intelligent or "Blind and Unconscious"? Appetite, 1850–1900

It can in no way be assumed that there is a specific and predestined connection between desire for a food and the nutritional capacity of the sought-after substance. CARL LUDWIG, *TEXTBOOK OF HUMAN PHYSIOLOGY* (1861)

Introduction to Part Three

Historians speak of the Second Industrial Revolution to characterize momentous developments in Western economic and social life that unfolded in the second half of the nineteenth century. Industrialization had long been under way in Western Europe, but the later nineteenth century saw its spread throughout regions hitherto only partially affected or not at all, most dramatically with the rapid industrialization that occurred after 1871 in the newly established German Empire and from the 1890s in parts of Russia. As elsewhere, these shifts brought the mechanization of production, the extension of great connecting networks of transportation and communication, and the inexorable shift of population from country to city. For a long time historians who explored these transformations emphasized the role of iron and steel and of the extractive, electrical, and chemical industries. More recently, historians have attended to the significant role of industrial food: the coming of mechanized farming, the standardization of food "products," the spread of new modes of food distribution (chains, supermarkets), and related developments.[1]

The growth of industrial power went hand in hand with the consolidation and strengthening of the nation-state, which entailed in its turn a drive to undermine political and sociocultural forms of particularism. Stated in terms of biopower, this meant new efforts, often state-funded ones, to standardize the production and consumption of food: a scientifically devised diet ensured a stronger population and a stronger population enhanced national power. The embrace by public authorities of these ideals and goals did not of course ensure smooth processes of change. Rapid industrialization brought heightened conflict between the well fed and those who were denied the enjoyment of increased plenty. The early nineteenth-century conflict between the well-off and the "dangerous classes" developed after 1850 into a sustained confrontation between socioeconomic elites and working people in which demands for adequate nourishment were a central feature.[2] Industrialization and urbanization also conditioned women's accelerated push for an improved political, economic, and social position. The intensified contest over gender roles found expression not just in the spheres of work, wages, or the suffrage but also in intellectual and cultural efforts to define the capacities and "frailties" of women, the former often thought to be limited and the latter to stem from perverse or pathological appetites. In this domain as in others, scientific authority loomed ever larger.[3]

Perhaps most important for the history of the life sciences and medicine was the impact of industrial values of precision, standardization, and regulation, most notably in experimental physiology (explored in chapter 7). The physicochemical school of physiology spearheaded in Germany by Carl Ludwig was resolutely committed to the use of precision instruments and the standardization of experimental techniques and representation of results, but the simultaneous embrace in French physiology of Claude Bernard's conception of automatic "self-regulation" of the body's "internal milieu" also entailed a taming of organic processes once regarded as spontaneous and unpredictable.[4] Bernard's formulation spurred a search for the specific mechanisms involved in regulatory processes (nervous, chemical, and—after 1900—hormonal and "behavioral") that continues to the present day. Yet in the late nineteenth century these new models of precise and regular functioning of bodily apparatus did not go altogether unquestioned. A striking instance of continued fascination with physiological extremes was the phenomenon of the so-called hunger artist, whose prolonged fasts drew not only the rapt attention of a voyeuristic public but also the close scrutiny of scientific observers. As I show in chapter 8, the most notable of the scientific inquiries into such performances, conducted in 1888 by the Italian physiologist Luigi Luciani, revealed a tentative, late-century willingness to probe the power of "psychic factors" to influence apparently inalterable material determinants of ingestion and digestion.

The embrace of industrial values in the sciences developed alongside broader processes of "scientization": the growing importance of science in public life, the spread of positivism and a pragmatic materialism largely detached from philosophical underpinnings, and the ever-increasing reliance by public authorities on scientific expertise. Experimental physiology may have led the way, but life science at large reflected and promoted these shifts. With the publication of Charles Darwin's epochal *Origin of Species* (1859), the drive to elide distinctions between human and animal long upheld by theologians and philosophers gained force. Darwin's evolutionary theory placed human beings firmly within a "nature" no longer governed by providence but instead ruled blindly by material forces and competitions. Among its other momentous effects, Darwinism supplied a powerful new rationale—shared paths of evolutionary descent—for gauging human appetite in relation to that of animals.[5]

Along with its broad impact in biology, Darwinism also played a major role in the transformation of the social and human sciences, especially psychology (examined in chapter 8). After 1850 psychology gave up its near-exclusive emphasis on problems of mind and began addressing the relation between mental and bodily phenomena even in regard to the "lower" functions of ingestion and digestion. Registering the influence of Darwinism, self-styled physiological psychologists moved to compare in animals and humans survival-enhancing determinants of eating, whether appetite for beneficial substances or disgust toward what was deleterious. In "animal psychology," moreover, Darwinism provoked renewed debate over the relative role of instinct and intelligence in animal "choices," such as those made when animals discriminated between what was edible and inedible.[6] Whether eating was governed by an inner, "instinctual" prompt of appetite or must be guided by intelligent, trained choice thus became an abiding theme in the psychology of ingestion.

Where did physicians stand in all this? As I explore in chapter 9, the prestige of experimental physiology and Darwinian biology increased pressures on physicians to align their clinical observations with findings of the increasingly "positive" sciences that, so many argued, must provide a strengthened foundation for medicine itself.[7] More and more, physicians concerned with appetite and ingestion framed their claims in terms supplied especially by advancing nerve science. Long divided over whether troubled appetite and eating resulted principally from physical problems in the alimentary tract or from mental and emotional aberrations, clinicians separated into competing camps of "peripheralists" and "centralists," who traced the urge to eat and patterns of disturbed eating to fundamentally different nervous origins. Somaticist explanations of appetite disturbance, often linked to malfunctions of visceral nerves, competed with psychic ("central") etiologies for "anorexia nervosa," which from the 1870s emerged as the paradigmatic appetite disorder. Physicians who embraced the latter view continued the move toward a brain-centered understanding of the urge to eat that had already emerged in the physiology of François Magendie and in mental medicine, but both frameworks reflected the increased disposition of physicians to explain bedside observations in terms supplied by life science. The move to scientize medicine—to require of physicians extensive training in the sciences and to structure debates in accord with claims made by investigators outside the clinic—advanced at a varying pace in different national and socioeconomic contexts. But everywhere the domain of appetite was marked by the drive to replace inquiry into individual propensities with the search for scientifically demonstrable regularities.

SEVEN

The Drive to Eat in Nutritional Physiology

While many physicians moved after 1850 toward a psychological perspective on appetite, physiologists, especially in Germany, sought a more thoroughly materialist understanding of ingestion and digestion. This effort reflected the influence of disparate currents ranging from the increasingly broad sway of scientific materialism to specific advances in physiological chemistry and other fields that seemed to promise new clarity on problems long recognized as shrouded in obscurity. Although no physiologist of these decades denied the role in ingestion and digestion of what Carl Ludwig (1816–1895), architect of the physicochemical school of German physiology, termed "psychic factors," materialist investigators emphasized the role of cellular, metabolic, and nervous processes in governing all phases of the intake of food, digestion, and nutrition. In so doing, they argued for the indispensability of laboratory methods and moved increasingly to divorce physiological inquiry from clinical observation of human eating.[1] In these years physiologists made what many regarded as remarkable gains in knowledge of human nutritional needs. Yet despite the seeming promise of a strictly material approach, basic questions persisted about all phases of the alimentary process, including the origin and nature of the urge to eat: Did eating respond to local impulses or to diffuse bodily need? Did eating choices reflect genuine nutritional deficiencies or capricious desires for ingestive pleasures? Did the urge to eat have a material site, and if so, where should it be sought? Far from providing solutions to these problems, the work of these years increasingly fostered the emergence of competing camps of "peripheralists" and "cerebralists," a division that had important consequences not only for physiology itself but also, among clinicians who accepted the claims of scientific medicine, for the diagnosis and management of troubled appetite.

Jacob Moleschott and the Beginnings of Nutritional Science

A central figure in the drive for a rigorously materialist physiology of nutrition was the Dutch-born Jacob Moleschott (1822–1893). In 1842 Moleschott left Holland for Heidelberg to study with leading life scientists, including Friedrich Tiedemann, whose 1836 textbook on nutritional physiology had enjoyed great success in Germany and elsewhere.[2] Moleschott's teachers trained him in use of the microscope and introduced him to the cluster of problems surrounding the "metabolic power" at work within cells.[3] It was Tiedemann's invitation to Moleschott to help him revise his manual *Nahrungs-Bedürfniss* that led to Moleschott's own treatise of 1850 on that topic.[4]

In his dietetic treatise Moleschott elaborated on Tiedemann's teaching on nutritional needs and the human drive to eat, and he strove to support his mentor's ideas with new findings from physiological chemistry. Throughout the work he made frequent reference (positive and negative) to chemical research undertaken since the publication of Tiedemann's textbook, especially by Justus Liebig (1803–1873), whose acclaimed work on animal chemistry had appeared in 1842.[5] Yet Moleschott's book was more than a dietetic manual based on the new chemistry, incorporating as it did a vision of broad social reform based on nutrition. Moleschott worked for a time as a practicing physician, and throughout his life he was a political radical. His teachings on nutrition were always informed by a reformist purpose, that of ensuring that the basic nutritional requirements of all human beings were met and fostering human progress by improving the material substrate of thought and character.[6]

Moleschott's first task in *Die Physiologie der Nahrungsmittel* was to set forth the new concept of metabolism and to demonstrate that states of hunger and thirst developed as consequences of metabolic action.[7] To this end he explored the subjects of blood formation, nutrition, secretion and excretion, and metabolic deficiency. He then turned to the nature of appetite and hunger, using the terms *Eßlust* and *Appetit* interchangeably to designate the pleasurable first stage of the urge to eat, and *Hunger* for the painful condition that followed when the call of appetite was not answered. Dividing appetite from hunger by emphasizing the pleasures of the first and the pains of the latter was by this point commonplace, but Moleschott, mindful no doubt of the near starvation suffered throughout Germany in the "hungry '40s," infused with urgency what might have remained a dry matter of definition. As essential materials were lost through bodily excretions, he argued, the blood "demanded renewal," and although for "civilized men" this call remained a "pleasant . . . mild sensation," promptly answered, among those deprived of food frightful consequences ensued (77–78). Following Tiedemann, who had moved beyond the traditional dualism of appetite-pleasure and hunger-pain to a third stage of "torturing" hunger, Moleschott grimly depicted the mental and physiological state of "inanition" brought on by even a day or two of privation. Describing torments experienced throughout the body, he urged that it was such extremes of suffering that, throughout history, had impelled the hungry to seek relief by desperate means, devouring carrion and even human corpses.[8]

Despite Moleschott's philosophical commitment to materialism, his ideas about appetite were an uneasy blend of materialist physiology and insights gleaned from medicine and emergent anthropology. Denying that appetite required an elaborated psychic foundation, he observed that it returned unbidden throughout life, and not only in those with full mental equipment but also in nurslings, cretins, imbeciles, and the insane, all of whom experienced appetite though lacking "full or clear consciousness" (80–81). It could not be otherwise for any living being, given that appetite emerged as a consequence of inexorable metabolic processes. Indeed, there was a clear parallel between the rapid return of appetite and the lively metabolism of warm-blooded animals, growing children, and energetic youth, as well as those who felt the "invigorating affects of joy and hope" (82).

For Moleschott, attributing appetite to metabolic states did not imply uniformity or a rigid determinism because, he argued, metabolic processes differed widely depending on disparities of age, sex, climate, temperament, and custom. Thus, Moleschott recast the traditional influences on health recognized since antiquity in the new language of physiological chemistry, urging that metabolic activity explained the effects on appetite of the four temperaments—choleric, sanguine, melancholic, and phlegmatic (82). Given this emphasis, Moleschott's discussion of the influence of eating customs on appetite is of special interest. Endorsing the claim of Haller and other physiologists before him, he held that both the periodicity and intensity of appetite depended on customary mealtimes.

Although he acknowledged that pleasurable or stressful activity such as a child's play or a scholar's intense mental labor could interrupt or destroy appetite, overall the call to eat returned and grew stronger simply because it was mealtime. Leaving chemistry behind, he proposed instead a nervous explanation of eating customs by noting how repeated stimulation of the nerves involved in ingestion could result in eating habits that easily became "needs." So it was, for example, that the English and the Dutch, habituated to drinking strong tea before retiring, suffered sleepless nights if they drank coffee instead.[9]

Moleschott regarded illness as the most powerful cause of variability in appetite. On this subject he followed Tiedemann closely, considering both ills that were direct manifestations of disordered appetites—malacia, pica, bulimia (*Heißhunger*)—and others that were accompanied by disturbed appetite, including pregnancy, fevers, alcoholic "dyscrasia," and, especially, diseases of the brain and the passions (84–86). As was traditional, he defined malacia and pica as ills principally of young girls who, at the onset of menstruation, experienced irresistible inclinations to eat nonfood substances such as clay, lime, ash, coal, and even spiders (85). Like others impressed with the reportage of the great naturalist Alexander von Humboldt (1769–1859), Moleschott described in detail the "dirt-eaters" of the West Indies whose cravings brought on loss of appetite, heart palpitations, and cold and yellow skin, and whose intestinal canals were shown in autopsy to be filled with earth (85–86). He concluded with accounts, again borrowed from Tiedemann, of "professional eaters" ("Pamphagi" or "Allotriophagi"), muscular men who made a trade of gluttony, consuming huge masses of stones or living animals covered with skin and hair (88).

The significance that Moleschott lent to such observations becomes clear when his comments on appetite are compared to his treatment of hunger, where, instead of using clinical and anthropological evidence, he addressed the physiological basis of "the feelings that announce the need for food" (88). In this connection he offered a historical survey that began with Haller's theory of hunger as dependent on the stomachal folds grating against one another and continued with the work of those who had later affirmed that hunger was stimulated chemically, the result of swallowed saliva, secreted gastric juice, or bile (89–90). In place of these outmoded notions, Moleschott set forth what he saw as the definitive explanation of hunger: changes in nutrition altered the composition of the blood, which, in turn, affected the stomachal nerves that conveyed the feeling of hunger to the brain (91).

In an interesting reversal, however, Moleschott distanced himself from this straightforward materialist explanation of hunger, especially from what he saw as overemphasis on the role of the stomachal nerves. Instead, he argued, local feelings of hunger and general conditions of nutrition belonged to "one interconnected whole" (92). Only a view that recognized the interrelations of blood and nerves in the nutritional process, he urged, could explain phenomena associated with appetite and hunger that had long baffled observers. Among these was the often misjudged feeling of satiety. Eaters, he said, often confused the feeling of fullness in the stomach with actual satiety, which was not in fact achieved until genuine nutrition was effected in the blood. Acknowledging that hunger was stilled if the stomach was filled with oil or warm water or by the ingestion of narcotic substances, he nonetheless reasserted that hunger and satiety were the result of "altered nutrition of the stomachal nerves" (92–93). He did not suggest how this localist view that hunger depended on the state of the nerves of the stomach could be reconciled with the unitary view he also extolled.

Historians remember Moleschott's *Physiologie der Nahrungsmittel* principally as a contribution to the development of nutritional science, and indeed his remarks on the functioning of appetite and hunger led him to an extended discussion of the components and effects of diverse foodstuffs.[10] In this regard he drew on the work of chemists who had attempted to clarify such questions as the amounts of specific nutritional materials found in animal flesh (207). This part of Moleschott's work supported what one scholar has regarded as his most important contribution, the establishment of uniform standards of nutrition. Jane O'Hara-May has argued that his work contained the first "quantified dietary standard," one that was based on a uniform "unit of body weight" and that assumed "that for a normal, healthy adult output is equivalent to intake."[11]

Although this perspective on Moleschott helps to establish his place in the development of uniform nutritional standards, it overlooks the extent to which he was still committed, at least in some respects, to what O'Hara-May sees as the premodern notion that appetite alone could serve as an appropriate guide to food choices.[12] Certainly, Moleschott did not shy away from laying down nutritional rules, especially for children.[13] Yet he also argued that in the end there is only one guide to appropriate eating, especially how much to eat, and that was the "drive" to eat: "In general the drive to consume foods is the sole way correctly to determine amounts. The civilized man whose palate is not over-stimulated by delicacies and the misuse of spices can, with no disadvantage, eat and drink until his appetite and his desire for drink are satisfied. Ill consequences follow only when foods are consumed to the point of surfeit (*Uebersättigung*). 'Stop eating while the food still tastes' is the surest rule for guarding against overloading of the stomach" (560). This insistence on the individual's innate capacity to know how much to eat remained unchanged in the second (1859) edition of Moleschott's treatise, which in other ways indicated an effort to align his account of appetite and hunger more clearly with materialist science.[14] In the later edition, Moleschott made a number of changes that de-emphasized or recast the role of appetite, adding, for example, a statement that linked appetite to enhanced glandular secretion (179). In 1859 he also moved away from earlier characterizations of appetite that had perhaps come to seem too folkloric in character, eliminating his earlier reference in the 1850 edition to young girls eating spiders when afflicted with pica (85) and omitting all reference to the outlandish eaters he had portrayed as gulping down living animals—claws, feathers, and all (88).[15] At this later point, moreover, Moleschott tended to associate appetite with pathological deviations (adding, for example, a paragraph-long description of disturbed appetite in hysterical women) while using more clearly physiological language ("need for food," *Nahrungsbedürfniss*, and "drive to eat," *Nahrungstrieb*) for states of bodily depletion. Yet even as he made these changes, Moleschott did not discard his claim that appetite was a reliable guide to eating.

Despite the more materialist cast of Moleschott's 1859 edition, his work has been characterized overall as humanistic, and that word is apt both for his method and his goals. He recognized great variety in human eating habits while nonetheless pushing for standards that would ensure adequate nutrition for everyone; a healthful diet was essential, he urged, to a fulfilling and dignified human life.[16] Although his work was admired and cited for decades, in subsequent years German physiology attended ever more closely to the physicochemical processes examined by the materialist Moleschott while abandoning his humanistic concerns. Increasingly, physiologists strove to fulfill the overarching goal of establishing their discipline on a par with the exact sciences of physics and chemistry, and, as part of that endeavor, developing precise criteria for optimal physiological function. Kamminga stresses, as part of this new direction, the banishment from nutritional physiology of issues that intersected with—or, as in Moleschott's case, were inspired by—politics.[17] The drive toward a physiology that approached appetite, hunger, and eating as strictly scientific problems is exemplified by the career of Carl Ludwig, whose work at the physiological institute of Leipzig was pivotal in setting the direction of German physiology of the later nineteenth century.[18]

The Urge to Eat in Physicochemical Physiology

Ludwig took his medical degree in 1840 at the University of Marburg and in the 1840s became interested in developing more exact instruments for physiological investigation. In 1846 he invented the revolving drum recorder (kymograph) to measure variations in fluid pressure, the first of the new instruments that allowed physiologists to claim "objective, quantitative analysis" of physiological events.[19] Between 1855 and 1865, after failing to land a post in Prussia because of his liberal and anticlerical views, he held a position at the military academy in Vienna. In 1865, he took up a faculty position in Leipzig, moving in 1869 to the physiological institute, which in the later nineteenth century became the model for similar institutions around the world. Ludwig directed the institute for almost thirty years, supervising students from not only Germany but also England, the United States, Scandinavia, and Russia.[20]

Between 1852 and 1856, Ludwig published what was to be the most influential physiology textbook of the period, *Lehrbuch der Physiologie des Menschen*.[21] In the second volume of this work he took up problems in the physiology of nutrition, including the working of the desire to eat (*Nahrungsbegehren*) and the origins of hunger. Like so many of his contemporaries, Ludwig contributed to a physiology that, as William Coleman puts it, endorsed the "hegemony of the blood."[22] Thus, in looking at nutrition, Ludwig began with the formation, composition, and movement of the blood, focusing on the "nutritional need" (*Nahrungsbedürfnis*) that prompted renewal of the blood by the supply of foods made available through digestion. The need for food was first signaled, Ludwig observed, by "a nagging or pressing sensation in the region of the stomach" that, when it "persisted for some time . . . produces an unpleasant, painful impression and a pronounced desire for solid food" (2:583).

One of Ludwig's earliest achievements had been his demonstration in 1851 that secretion of saliva was dependent on nervous connections, so it is unsurprising that in considering hunger, he turned immediately to the role of the nerves.[23] Here he distinguished between what he called "low-grade hunger," the product purely of the action of the stomachal nerves, and higher-intensity hunger, which also involved the nerves of the intestines and perhaps others spread throughout the body (2:583–84). Despite this stress on the role of the nerves, Ludwig denied that feelings of hunger could be attributed purely to action of the vagus nerves and that cutting the vagi interrupted transmission of the hunger sensation. He acknowledged, rather, that psychic factors alone could cause an animal to desire to eat regardless of the state of this purportedly essential nervous connection. In any event, for Ludwig the crucial issue in regard to nutrition was not the feeling of hunger but whether the products of eating made their way to the large intestine, there forming the all-important nutritive blood. Only this process could produce the sensation Ludwig presented as the obverse of hunger, satiety. Yet here too the eater's feelings were not easily explained, and they could be wholly mistaken: the eater might perceive as satiety a purely mechanical sensation of compression or see his hunger stilled by ingesting nonfood substances such as narcotics (2:585–86). In contrast, eating nutritious substances might not assuage hunger, which could persist if the products of digestion did not, for whatever reason, proceed to the intestines. In none of these cases were the eater's feelings a reliable guide to the process of nutrition; rather, physiological need, created by unrestored losses in the composition of the blood, required assimilation of the specific foodstuffs now known to be essential to the maintenance of life (2:587–90).

Still, despite this emphasis on physiological need, at key points Ludwig mentioned psychic factors as well as the role of culture in eating, as when he observed that animals in which the vagus nerves had been cut might eat purely because of "psychic causes" (2:584). Like his predecessors, he held that hunger reflected not only need but also the role of "psychic stimuli" such as the arrival of mealtime (2:585). The desire for food was affected by memories and sometimes obstructed by the "psychic condition" of "disgust" (2:587). He also observed that pleasurable stimuli (spices, additives) that worked on the sense of taste were essential to the digestive process (2:590). Thus, the culinary art merited attention: Ludwig recognized that human beings were nourished by an extraordinary variety of foods that reflected the influence of region, "degree of civilization," and race. Indeed, he argued that any properly "physiological view" of ingesta must take the preparation of food into account, because most foods were eaten only after cooking (2:590–91).

Although Ludwig acknowledged the role of psyche and culture in ingestion and digestion, he invariably drew his discussion back to the procedures and insights of materialist science. Thus, in discussing the desire for food, he mentioned the role of memory only alongside such factors as the temperature of the body and the resistance the teeth encountered in chewing. Spices he mentioned not because they made food taste good but because they helped to stimulate salivary, stomachal, and intestinal glands to action. Cooking mattered not as an element of cultural or social life but because it rendered foods digestible (2:587, 590, 591). Thus for Ludwig, the prompt to eat—whatever the role of unreliable sensations of desire, hunger, or satiety—was ultimately grounded in physiological need, and it was the duty of science to illuminate that need and show how it could best be met. Turning to the problem of nutritional equivalencies, Ludwig focused on the debate pursued since the work of Liebig on whether some foodstuffs served best or exclusively for the production of heat, whereas others produced muscle power or the "plastic" capacity to build tissues.[24] Here he adduced evidence from recent feeding studies, posing such questions as whether muscle exertion was required for the decomposition of protein, and offering, as the way to find the answer, the results of experiments in which dogs were held immobile in crates (2:600–601).

When Ludwig shifted to problems involved in digestion, he also occasionally mentioned psychic elements. Again he emphasized physical and chemical changes—the mechanical work of mouth, throat, stomach, and intestines as well as the "chemical work of the digestive juices" and the formation of digestive products (2:604–52)—while referring to the work of the psyche in passing. In looking at the movements of mouth and tongue, for example, he asked which actions were subject to the will and which were not. In analyzing vomiting, he referred to the "powerful passions or feelings of disgust" that could, along with purely physical events, cause this phenomenon (2:605, 621). But generally he stressed such topics as the capacity of the body to store nutrients, the chemical composition of outflows, and the effect of hunger on particular tissues and organs. Presenting the results of recent feeding studies, he drew special attention to those obtained by the Munich physiologists T. L. W. Bischoff (1807–1882) and Carl Voit (1831–1908) when they fed varied diets to a single dog maintained over a long period.[25] Human beings appeared only at the end of this discussion, where Ludwig reported observations from around the world of the physiological effects of particular diets on different human types from Eskimos to prisoners in Bombay (2:713–14). He ended by taking up questions about growth: its relation to diet and differing patterns observable by sex, social class, region, and, "the most important variable," the race of the individual (2:717). Why race was the crucial variable, or indeed why any of these topics mattered, Ludwig did not explain; on questions of this kind he provided a recitation of facts and nothing further.

Thus, Ludwig's textbook indicates that by this point experimental physiologists of his complexion were principally interested in the question of what nutrients were required to sustain animal and human life.

Although he recognized that what both humans and animals ate was related to desires and feelings, he discarded such factors as a guide to what constituted appropriate eating. Acknowledging the role of smell and taste in the food choices of "higher animals," he nonetheless discounted the significance of food desires: "It can in no way be assumed that there is a specific and predestined connection between desire for a food and the nutritional capacity of the sought-after substance" (2:587). Strikingly, in this instance, he did not feel compelled to supply numerous experimental results to support his view but referred instead to a single instance of a dog that rejected food saturated with water even though the food retained its nutritional capacity but eagerly consumed sawdust flavored with tasty broth (2:587). The implication was clear: unlike hard physiological realities, eating desires could be characterized only by anecdote.

One can venture several lines of interpretation in considering why Ludwig accorded so little attention to psychic factors and rejected appetite as a guide for consumption. One is that he continued to reflect Bichat's conception of the independence of animal and organic functions. Bichat's name never appears in the *Lehrbuch*, but his conception of the two lives undergirded the work of two German researchers Ludwig did cite, F. H. Bidder (1810–1894) and A. W. Volkmann (1800–1877), who had recast Bichat's conception in terms of the independent functioning of the central and vegetative nervous systems.[26] Ludwig's adherence to updated versions of the two-lives system meant that, while he recognized the undeniable influence of psychic factors on appetite and hunger, he saw this influence as subordinate to nutritional needs and, in any event, not susceptible to rigorous investigation. Ludwig's overall approach to hunger lodges him among peripheralists, physiologists who saw hunger as seated ordinarily in the stomach and paid little heed to more general corporal or psychic phenomena connected to it.[27] This view of hunger was particularly strong in Germany, where many investigators continued to work, wittingly or not, within the tradition established by Haller. The contest between those favoring a peripheral interpretation of hunger and opponents who argued for central or general views became increasingly important in the later nineteenth century (see the subsequent discussion of Ludwig's contemporary Moritz Schiff). For the moment it is important to stress how thoroughly Ludwig's support for a viscerally based view of hunger and his emphasis on objectively demonstrable physiological need encouraged him to exclude investigation of the role of the psyche, and hence of appetite.

At a more theoretical level, it seems likely that Ludwig avoided questions of psychophysical interrelations because of his disdain for any interpretation of biological phenomena that implied an inherent wisdom of the body or innate capacity, of either the organism as a whole or individual organs or systems, to exercise beneficent choice. This kind of thinking was associated with teleology and, more damning yet, with vitalism, theoretical frameworks that Ludwig sought throughout his career to banish from physiology.[28] In any event, it is clear that the kind of science Ludwig promoted suggested that healthy eating was not something that could be determined by individual choice but, instead, must depend on scientific authority. This was evident in the distinction he drew between "natural" and other foods, the first supplying simple and essential constituents of nutrition whose value was affirmed by science, whereas the latter consisted of concoctions devised by the culinary art, always a potential source of deception. Finally, it is worth noting that insofar as he recognized variability in digestive power from one individual to another, he emphasized differences not in what they desired or chose to eat but in what he referred to as "unfathomable peculiarities" in the act of digestion as it unfolded in the intestinal canal (2:591).

Ludwig's institute at Leipzig was the most important center of physiology in Germany into the 1890s. Using the physicochemical approach that Ludwig encouraged, a number of his students took up problems connected to digestion and nutrition, including the formation of urea, lipid catabolism, and the formation of lactic acid in relation to muscular chemistry.[29] Elsewhere in Germany, especially at the institute headed by Voit in Munich, the chemical approach to nutrition dominated. One of Voit's students was Max Rubner (1854–1932), who pioneered the study of the caloric values of foods. Perhaps no assertion of the era better illustrates the disregard for eating desires than Rubner's doctrine of the "isodynamics" of foods, termed now (and still debated) as the principle that "a calorie is a calorie."[30] Throughout his career Rubner held that for the mass of people nutritional needs were uniform and that individual deviations caused by "excessive or weak appetite" should be discounted.[31]

To be sure, physicochemical frameworks did not exclude all other ways of thinking about appetite and hunger within German physiology. Ludwig's long-term rival E. F. W. Pflüger (1829–1910), for five decades professor of physiology at the University of Bonn, made extensive use of ingestive phenomena in arguing for the existence of deep-seated "needs" whose satisfaction prompted "instinctive" actions that restored the organism to a state of well-being.[32] For Pflüger, beneficial eating choices were guided by instinct, which was in turn mediated by the play of sensations of pleasure and pain. This "teleological mechanics" of ingestion was but one instance of a general "causal law" governing needs and satisfaction of needs: "The cause of every need of a living being is at the same time the cause of the satisfaction of the need" (76). Pflüger's idea of need was virtually indistinguishable from his concept of drive, and indeed, he conceptualized living beings as bundles of drives—to consume, to grow, to change, to reproduce—for each of which there must exist a specialized mechanism that served to preserve both individual and species.[33] Nonetheless, in respect to ingestion and digestion, Pflüger's thinking could lead to a straightforward emphasis on mechanics that aligned him with the kinds of investigation promoted by Ludwig. Taking up the issue of digestive secretion, for example, Pflüger insisted (against much unacknowledged experimental evidence) that glandular secretions (salivary, stomachal, intestinal) never occurred during hunger, when the stimulus of food was not present; such a "task performed for no purpose" would violate Aristotle's rule that "God and nature do nothing in vain" (80). Thus, in considering the urge to eat, Pflüger began with deep-seated drives that worked in concert with psychic processes, but he ended by looking to physical processes like glandular secretion for confirmation of his teleological assumptions.

In contrast, it was the problem of secretion that prompted the Breslau physiologist Rudolf Heidenhain (1834–1897), another opponent of physicochemical "oversimplification" of "vital phenomena," to admit at least the possibility of a strictly psychic or nervous prompt to digestive activity, one suggested by the fact that "the mere sight of food in hungry individuals suffices to get secretion underway."[34] Heidenhain noted that German investigators had repeatedly seen this phenomenon since the first report given of it by Bidder and Schmidt in 1852, and, in his own discussion of the "influence of the nervous system on the formation of gastric secretions" (116–21), he showed himself open to the possibility of such psychophysiological reactions in a way excluded when Ludwig set aside "psychic factors." Indeed, Heidenhain showed the way to a new methodological exactitude in observing secretory processes with his use of a *Blindsack*, or stomach pouch detached from the larger stomach, which became the model for the apparatus later developed by his student Ivan Pavlov to gauge the relation between appetite and gastric secretion.[35]

Nonetheless, the physicochemical approach to nutritional physiology that dominated German physiology afforded little room to deep-seated organismic "needs" or to "psychic factors" involved in the urge to eat or in digestive processes. Appetite was seldom mentioned, and hunger was commonly attributed to the work of peripheral nerves. In France, in contrast, an alternate framework emphasized the general needs of living organisms and their satisfaction by the complex interworking of a multiplicity of autonomous "regulatory mechanisms."[36] This conception built on the thinking of François Magendie, who had argued in the 1830s that the urge to eat developed in response not to specific bodily mechanisms but to a general need for bodily repair registered throughout the whole nervous system.[37] In the hands of Magendie's most illustrious pupil, Claude Bernard (1813–1878), this view of the general origin and impact of nutritional need came to form part of Bernard's broad theory of organismic defense and self-regulation.[38]

Claude Bernard: Nutrients and the "Internal Milieu"

Bernard first developed his idea of the internal milieu in the 1850s, as he came to view vital phenomena functioning in relation to the demands of the organism's external environment. In his thinking the internal environment of cells, tissues, and organs was created by the blood (and subsidiary fluids generated from it) that provided protection from factors that disturbed or threatened their function from the outside. Bernard was especially interested in heat, or how the body (human and that of higher animals) could maintain an internal temperature required for survival despite changes in external temperature. He also explored how a steady supply of essential nutrients was necessary to maintain the composition of the blood in a state required for survival despite vagaries of alimentation.[39] One of Bernard's most celebrated achievements was his demonstration of the glycogenic function of the liver, a fact he established in the face of long-standing views that sugar had to be taken into the body from food.[40] When he first began to conceive of this function as an element of an overall bodily system of self-maintenance, Bernard was troubled by the idea that the liver not only generated sugar but also varied its production in accord with bodily needs. Frederic L. Holmes explains that Bernard rejected this notion because it implied the exercise of a kind of "intelligence" on the part of the organ, a vitalistic view he could not accept. In time, however, Bernard came to regard as irrefutable the evidence that the liver, like other organs, did indeed vary its "response" to "perceived" needs.[41] Convinced of the validity of his broad conception, he argued that the task that lay ahead was to identify and illuminate the working of the perhaps innumerable interrelated mechanisms responsible for maintaining a balanced internal environment.

Bernard's doctrine of the internal milieu was a conception he arrived at after long years of painstaking experimental investigation of specific bodily processes, including those involved in gastric digestion. In the course of this research, he mobilized an array of experimental techniques, in many instances modifying commonly used procedures or devising new ones as he pursued fresh lines of inquiry. One technique that he made much use of, for example, was vagotomy, which he employed repeatedly to determine the effects of the operation on digestive and other bodily activities. So far as I can tell, Bernard never performed vagotomy specifically to illuminate an animal's urge to eat, but he did take up issues of appetite, hunger, and satiety in the course of work he did on other problems. Thus, in research he pursued on reflex movements governing the cardiac and other sphincters, he denied claims of prior authors that hunger and thirst were governed by action of the pneumogastric nerves. He specifically refuted arguments that vagotomy destroyed the feeling of satiety, a claim some experimenters had made to explain why, when animals revived from the operation and regained their appetite, they sometimes ate avidly but then regurgitated and ate again with no stopping. Bernard had accepted this interpretation himself, he said, until a "fortuitous circumstance" indicated the true explanation. Pursuing "a very different objective," he operated on a dog that had undergone vagotomy and in which he had created a stomach fistula. Even when this dog "ate well," he observed, nothing entered its stomach. He then realized that the food had lodged in its "paralyzed esophagus without crossing the sphincter" into the stomach. At this point the "phenomena observed became marvelously clear": the esophagus was filled with ingested material, the animal had to regurgitate, and, as no food had reached the stomach, the animal continued to be hungry and to eat again and again. What was at issue, then, was not any sensation of satiety, normal or disturbed, but a persistent contraction of the sphincter caused by disturbance of the nervous network in the region, in short, a purely local and mechanical effect that required no explanation based on a larger apparatus of "feeling."[42]

Similarly, it was in investigating a problem far afield from the nature of appetite that Bernard commented on its role in stimulating the production of saliva. The context in this case was the problem of how to secure adequate amounts of saliva and other fluids for use in investigating their role in digestion. Commenting on the role of reflex action in salivation, he noted that "thought alone, and even more so the sight of certain substances, suffices to produce an abundant secretion of saliva." This was "a phenomenon known to everyone" and "popularly expressed by saying that one or another thing causes the mouth to water." Thus, copious salivation was "considered the expression of a powerful desire." Bernard noted that he and others had used "this purely moral influence" to produce large amounts of saliva in experimental animals. In a horse kept fasting for twenty-four hours and then shown a bale of hay, its "appetite was so excited that it produced a flow of saliva as abundant and continuous as the flow from a faucet."[43] Such examples show that Bernard's interest in the presence or absence of the urge to eat was purely incidental and that he could offer two very different explanations—one local and mechanical, the other dependent on a general (and unexplored) "moral influence"—to account for its origin or effects.

Nonetheless, Bernard's overall approach to vital phenomena encouraged a way of thinking about appetite that differed fundamentally from that of researchers who looked strictly to local processes and structures for explanations of the urge to eat. His concept of the internal environment directed attention not only to the general condition of the organism but also to the interrelated work of complex balancing mechanisms. Ultimately, Bernard's significance to the study of appetite lay in the fact that he proposed a way of approaching problems of "intelligent" organismic action that nonetheless broke with vitalism and privileged experimental investigation of concrete mechanisms. Bernard distanced himself from philosophically grounded conceptions of organismic ends and insisted that study of the internal milieu must proceed by investigation of specific physicochemical processes that, acting together, assured bodily equilibrium. How exactly this conception was to be applied to regulation of the intake of food Bernard himself did not explore. In his posthumously published work on vital phenomena common to animals and plants, Bernard summarized the conditions that had to remain constant for life to be maintained, including water, heat, oxygen, and nutritive reserves.[44] In regard to nutrients he again emphasized not the direct action of the external milieu—in this case the foods taken in by the organism—but the way in which inner processes handled what came in from the outside. What mattered, in short, was not the choices of ingesta made by the eater but the processes of assimilation and the "mechanisms" that assured the uptake of appropriate materials "from . . . variable aliments" and that "regulated the proportion [of such materials] that entered the blood" (1:121). "One *lives*," he concluded, "*not from foods taken in at the moment but from those that one ate previously*, [that are] modified, and created, in a fashion, by assimilation" (1:122, emphasis in original). It was to be some time before other researchers began searching for the "regulatory mechanisms" involved in desiring, seeking out, and ingesting foodstuffs. Indeed, it was not until the 1920s and 1930s, when a full-blown revival of Bernard's thinking got under way, that the broad implications of the doctrine that came to be known as "homeostasis," for appetite as for a broad range of issues, became clear.[45] Yet once Bernard's conception made headway, one of the most powerful ideas governing inquiry into the nature and working of appetite was that the intake of food depends, at base, on the body's need for particular nutrients ingested in particular quantities and that responses to that need are regulated by innate mechanisms that ensure a state of internal equilibrium.

Hunger and Appetite in Moritz Schiff's Digestive Physiology

Well before these later developments of Bernard's legacy, this emphasis on general states of bodily need in prompting ingestion marked the work of an important contemporary of Bernard, Moritz Schiff (1823–1896). Schiff was German born and mostly German trained; he studied in Heidelberg with Tiedemann and in Berlin with Müller. But his training also included work in the 1840s with Magendie and François-Achille Longet (1811–1871).[46] A military surgeon caught up in the Revolution of 1848, Schiff was one of the physicians and scientists whose careers in Germany were hindered by their liberal politics and activism. Failing to obtain a position in Germany, he took up posts in Bern and Florence before becoming professor of physiology in Geneva.[47] Some of his earliest work was done on the inhibitory action of the vagus nerves on the heart, and for this purpose he did repeated experiments with vagotomy. Although this research focused on cardiac activity, he became interested generally in the neurophysiology of the vagus nerves and in digestion.

In 1867 Schiff published a two-volume work on digestive physiology in which he reported many experimental results but also offered general insights on hunger, thirst, and appetite.[48] Although Schiff was sometimes highly critical of Bernard, his overall perspective shows the influence of the latter's conception of the internal milieu. Thus, he conceived of hunger as a "symptom" of the impoverished state of the blood caused by "regular exercise of our organs" (1:30) and criticized those who saw hunger as a local phenomenon resulting directly from the state of the stomach. Although he acknowledged that hunger was sometimes felt in the stomach, he denied that this local feeling implied a local source; sensations could originate elsewhere and be felt in the stomach "by reflex action" (1:31). He also argued that hunger was felt in bodily locations other than the stomach, a point he sought to prove using the unusual method of asking ordinary people, non-scientists, where in the body they felt hunger.[49] Personally interviewing soldiers at a local garrison, Schiff found that many of the men experienced hunger in the neck or chest, and sometimes in no determined locale at all, but rarely in the stomach. Schiff also discounted the view that hunger resulted from an empty stomach, using William Beaumont's observations of the trapper Alexis St. Martin to argue, instead, that "the stomach empties completely several hours before hunger returns" (1:32). For Schiff, any assumed correlation between eating and the fullness or emptiness of the stomach was disproved by experiments done with animals—guinea pigs, rabbits, marmots—which showed that animals routinely ate even when their stomachs were full.

Schiff concluded from such facts that hunger was not peripheral but "essentially general in origin and independent of the nerve endings of the stomach."[50] Commenting on the results of vagotomy experiments, Schiff ignored disagreements about the outcomes of such trials, urging that they clearly proved that hunger was not dependent on the pneumogastric nerves, the only "distinctively sensitive" nerves in the stomach. Although it was true that temporary anorexia ensued when these nerves were cut, after a time the sensation of hunger was felt as before (1:36–37). Schiff argued, further, that hunger subsided only when the purpose of eating had been served, when "the materials destined to repair the losses [suffered by] the blood" had been elaborated and absorbed (1:39, 43, 46). An individual could eat his fill but if, for whatever reason, absorption was blocked, hunger persisted. Only when the "general need" of the body was met did hunger disappear (2:380).

Although Schiff's perspective resembled that of Moleschott and Ludwig in emphasizing that hunger was truly satisfied only once digestion and absorption had been successfully accomplished, he devoted greater attention than either of these predecessors to the feelings that prompted or accompanied ingestion. In his discussion of the urge to eat, Schiff used the terms "hunger" and "appetite" interchangeably, as when he referred to dogs that were vagotomized but then showed "the most indubitable signs of hunger" or the full "reestablishment of the appetite" (1:37). Yet he acknowledged that this usage was inexact and noted important differences between hunger and appetite:

> So far we have used interchangeably the expressions *hunger* and *appetite,* which, in ordinary language, designate only the nuances of the same sensation. If we seek to define more rigorously what we mean by hunger and appetite, a distinction that, however scholastic it appears, has nonetheless a certain practical interest, we would say that hunger is the sensation of *needing* to eat, while appetite is only the *agreeable* sensation

> that . . . revives in us the *desire* to eat. In the immense majority of cases, these two sensations form . . . only one, and one feels appetite *because* one feels hunger. However, in the terms of our definition, hunger and appetite are not identical and do not necessarily coincide. Indeed, the first can exist without the second and vice versa. We can desire a food that our body does not really need out of simple gourmandise. (1:50–51)

Thus, not only did Schiff recognize the error in conflating the "sensations" of appetite and hunger, but also, more significantly, he argued for the therapeutic utility of this seemingly scholastic distinction. He noted that in many illnesses patients suffering "impoverishment of the blood" knew that they should eat but "found the idea of taking in food repugnant" (1:51). How was one to explain this absence of the desire to eat, he asked, when the patient clearly needed food? Schiff explained this phenomenon by stressing the crucial role played by pleasure in eating and the dependence of appetite on gustatory sensations that conveyed the "agreeable effect of foods." Seen in this light, loss of appetite resulted from "local troubles that alter the special sensibility [of taste]" or "pathological sensations" engendered as food moved through the digestive pathways (1:51). This was a view of appetite that at one and the same time emphasized its importance and, in contrast to hunger certainly, de-emphasized its physiological standing: appetite was crucial in providing a reason to eat, the pleasure provided by food, but at the same time it was dependent, in ways that hunger was not, on strictly local conditions such as a "bitter or repulsive taste on the tongue" (1:51–52). Still, Schiff concluded that distinguishing failure of appetite from failure of digestive and absorptive processes was essential in treating cases of anorexia (1:52).

As was perhaps inevitable, given that Schiff provided no practical means to evaluate signs of appetite versus those of hunger, he quickly reverted to using the terms synonymously, as when he asked whether animals injected with pepsin seemed to experience enhanced appetite or hunger (1:54–55). The therapeutic uncertainties that resulted from conflating the two became apparent when Schiff, concluding his discussion, asked what should be done with patients who did not want to eat. Commenting on the error of physicians who failed to feed patients stricken with typhoid fever, thus adding to the "disorders caused by the fever . . . the murderous effects of inanition," Schiff observed that he had seen beneficial effects of feeding in the absence of appetite both in postoperative experimental animals and in human patients (1:56, 57). Thus, although Schiff was more attentive to appetite and its pleasures than Ludwig was, he eventually arrived at the same conclusion: that appetite was not, at least not always, a good guide to the question of when and how to eat. Nor, despite his recognition of its importance, did he provide any guidance on how to distinguish between hunger and appetite except insofar as human patients themselves could state their feelings about the prospect of eating.

Moritz Schiff's work posed a strong challenge to all theories of the urge to eat that privileged the empty, grinding, or contracting stomach. Yet for many experimentalists, especially those working in the Hallerian tradition, identifying a definite bodily cause or point of origin for hunger was the question of greatest interest in seeking to explain the urge to eat. Despite criticism by Schiff and others, peripheralists continued to attribute hunger to the painful condition of the empty stomach.[51] As with the earlier work of Beaumont and Blondlot, this view was tested in experimental animals with artificially created fistulas and, in a celebrated French case, in a human patient with an occluded esophagus who was fed through a fistula created surgically.[52] But ingenious new methods appeared too, as when one Berlin physician inserted a tube into a patient's stomach to gauge the volume of its contents and then correlated that measure with the individual's reported feelings of hunger.[53] Thus, peripheralists tried to pinpoint the timing and character of the stomachal activity that produced the hunger sensation, although, as later investigators would complain, they largely ignored the underlying problem of how gastric movements could produce a sensation of this kind.[54] Contrary to the views of peripheralists, those who supported a central origin of appetite and hunger held that all internal excitations were transmitted to the brain—precisely how, they acknowledged they could not say—and that it was only in the brain that such stimuli served as goads to actions such as the decision to eat. This commitment to the primacy of the central nervous system encouraged investigators, with the French in the lead, to attend with increasing interest to the psychology of eating. By the 1870s, issues connected to ingestion were addressed, if sometimes reluctantly, within emergent physiological psychology, whose proponents moved beyond the focus of their field on "higher" mental functions to posit important claims about eating desires and behavior. The efforts made within this new field to join psyche and soma in the study of appetite are treated in the next chapter.

EIGHT

The Psychology of Ingestion: Appetite in Physiological and Animal Psychology

In the first half of the nineteenth century, physiology and psychology occupied largely separate domains. Phrenologists, to be sure, had sought to tie psychic activity to specific sites in the brain, but the term "psychology" generally indicated commitment to an autonomous science of mind. Thus, before 1850 psychology was closely tied to theology or social philosophy, even when, as was the case in Britain, steps were taken toward a psychophysiology that strove to understand mind as a part of nature.[1] To the extent that experimental physiologists laid claim to psychological processes, their efforts—Johannes Müller's work on specific nerve energies, Pierre Flourens's (1794–1867) experiments on the gait of decerebellate pigeons—were carefully delimited.[2] Yet after midcentury, thanks to a range of influences, not least the impact of Darwinian evolution, the boundaries between physiology and psychology became increasingly porous. The drive to integrate mental processes into a determinist and naturalistic framework had gained strength in the 1860s and 1870s, eventuating in the creation of new institutions dedicated to what contemporaries called "physiological," "experimental," or sometimes simply "the new" psychology. A founding text of the discipline was Wilhelm Wundt's *Grundzüge der physiologischen Psychologie* of 1874. The creation of Wundt's laboratory and its outfitting with experimental equipment is seen as a crucial benchmark in the rise of physiological psychology.[3] Yet momentous as these German developments were to prove for the general history of the field, in the realm of appetite and hunger it was French researchers, whose work remained closely tied to physiology and medicine, who took the lead.[4] The impulse to psychologize eating proclivities, already evident in the physiology of François Magendie and in French mental medicine, informed not only investigations in physiological psychology proper but also one of the era's most telling "interdisciplinary" studies of ingestion, the Italian physiologist Luigi Luciani's probing of the psychophysiology of the so-called hunger artist. Finally, French students of "animal psychology," especially entomologists, also looked closely at ingestion, taking insects as the paradigmatic model of an instinctual eating that putatively illuminated the experience of human eaters, especially females.

The "New" Psychology in France

A key figure in the emergence of physiological psychology in France was Charles Richet (1850–1935), who assumed the chair of physiology at the Paris Faculty of Medicine in 1887. Richet was a man of eclectic interests, ranging from gastric secretion to hysteria to anaphylaxis (work that gained him the Nobel Prize in 1913) to paranormal phenomena.[5] He was unusual in the French scientific community for pursuing both the MD and a doctoral degree in science. At the invitation of the renowned surgeon A.-A.-S. Verneuil, with whom he interned in the 1870s, Richet devoted some of his early work to the nature of gastric secretions. Indeed, Richet became the first investigator since Beaumont to observe closely a patient with a gastric fistula. In this instance the fistula was surgically created by Verneuil to allow the young man, whose esophagus had been occluded when he accidentally drank potassium hydroxide, to be fed through the opening. Richet observed this patient ("Marcelin R . . .") for months, taking samples of his gastric juice (unadulterated by saliva thanks to his esophageal occlusion) and noting his states of thirst and hunger.[6] He took particular interest in the gastric secretions accompanying the patient's artificial feeding, the mechanical movements involved in the emptying of the stomach, and the time required for digestion of assorted foods and drinks (milk, fats, and alcohol). Although not focused on the nature or origin of hunger, Richet used this case to argue in passing against mechanical explanations of hunger, reporting that Marcelin could not feel whether his stomach was full or empty and that "hunger manifest[ed] itself with him in a kind of weakness" (162).

In a move his biographer presents as an anticipation of Pavlov's work on "psychic secretion," Richet also recounted that his patient began secreting gastric juice when given sweets or lemon to chew on, before any food substances were introduced into his stomach via the fistula.[7] This finding that gastric secretion could be set in motion by seeing or chewing pleasant substances was not new; German researchers had reported it in 1852, and Heidenhain commented on it in his 1883 textbook.[8] And while some physiologists took Richet's evidence as strong confirmation that the simple appearance of foods could stimulate gastric secretion, not everyone was convinced. Commenting on Richet's "striking" report, Heidenhain noted that this effect might result purely from a vasomotor reflex that led only indirectly to secretion.[9] In any event, what is clear is that Richet did not draw special attention to this unexplained stimulus of gastric secretion, a fact he himself ruefully recounted in his memoirs when acknowledging Pavlov's superior insight.[10]

In subsequent years Richet became increasingly interested in psychic and nervous phenomena, including hypnosis, telepathy, and clairvoyance, and he became a leader in inquiry into the occult and spiritism.[11] The extent to which his psychological studies came to influence his view of eating and digestion is indicated in essays he wrote for the multivolume *Dictionnaire de physiologie*, which appeared under his direction beginning in 1895.[12] The most interesting of these, "Dégoût," sheds light both on this "obstructive and inhibitory" phenomenon and on the "dynamogenic" phenomenon of "appetition."[13] Initially Richet presented disgust as a strictly physical condition characterized by obstructed swallowing, nausea, and vomiting (755). But, taking up the problem of how disgust for certain foods developed, he also considered the influence of heredity, illness, and, especially, habit and training.[14] Although he affirmed the power of education in forming habits, Richet also saw disgust as an instinctual mechanism that alerted humans and animals to the dangers of poisons, parasites, and putrefying substances, and instinct itself as an expression of the "law of finality" (759). Urging that the sense of taste was the principal "agent" of disgust, he nonetheless held that the "imagination" was powerfully involved: "If someone alerts us, rightly or wrongly, that food has been sullied, we cannot eat it despite all our efforts" (760). He concluded that "nature" inspires disgust for what is harmful and "appetition" for what is nourishing (760).

Despite Richet's apparent interest in appetition, the *Dictionnaire de physiologie* included no article on appetite as such.[15] As had been true since the eighteenth century, experiences of pain or displeasure seem, once again, to have appeared easier to grasp and represent than those associated with pleasure. Nonetheless, Richet's analysis of disgust did indicate the extent to which French researchers attended to involvement of the central nervous system in eating desires, in so doing continuing the physiological tradition from Magendie to Schiff in which ingestive choices reflected at least partly what Magendie had seen as "mental events." Although Richet saw reflexive reactions (animals' retching to rid themselves of noxious substances) at work in eating "instincts," he also noted the role of mental habits formed by pleasures and pains that could be purely imaginative in character (756–57).

Another attempt to link the physiology and psychology of appetite appeared in the work of Henry Beaunis (1830–1921), a graduate of the Montpellier medical school and physiology professor at Strasbourg and Nancy before becoming, in 1889, the first head of the new Institute of Physiological Psychology in Paris.[16] Beaunis explored problems surrounding appetite and hunger in an 1889 treatise on internal sensations in which he attempted to distinguish between needs and what he called functional sensations.[17] The status of hunger as a need was unquestioned: signaled by "more or less vague localized sensations" in the muscles of mastication and the digestive organs as well as a "general sensation" resulting from a "multiplicity of . . . obscure . . . sensations . . . from diverse regions of the organism" (33), hunger developed, if unsatisfied, into unspeakable pain (25). Appetite, in contrast, was a form of pleasure associated with taste and gourmandise. It was not just "light hunger"; indeed, one could experience it without being hungry. Focused on certain foods rather than others, appetite was roused by "condiments [and] spices," and concerned with quality rather than quantity (24–25n1).

Medically trained, Beaunis employed a variety of methods in studying the urge to eat. He examined his own feelings of appetite and hunger, measured how much weight he had to lose for hunger to set in, and culled newspaper accounts of cannibalism among shipwreck survivors. He also performed some sixty vagotomies, confirming that in time vagotomized animals resumed eating, and thus the urge to eat did not depend on the action of intact vagus nerves (24–35). Ultimately, however, Beaunis acknowledged that no method had yet been devised to illuminate the nature of internal sensations, especially the fundamental question of the difference between pleasure and pain. Throughout history "physicians, physiologists, [and] philosophers" had studied pain "in all its faces" (237). Pleasure was another matter: anyone could follow his instincts to find it, yet its "expressive manifestations" were less distinct and precise than those of pain (237). It remained for the psychology and physiology of the future to devise "instruments" and "procedures" to gain access to the "intimate nature" of pleasure (239). Although Beaunis was dissatisfied with what physiological psychology so far had to offer, his work demonstrated a readiness to explore even the lowly pleasures and pains of ingestion, whose discussion, he knew, was likely to "excite a certain repugnance" (24). In the event, however, his contribution failed to spark new interest among his contemporaries in appetite and hunger; direction of the institute passed within a few years to Alfred Binet (1857–1911), and it became known chiefly as a center for studying ways to measure intelligence and related subjects.[18]

Such facts serve as a reminder that the phrase "new psychology" had competing meanings in the later nineteenth century. Although the term could indicate a new drive to overcome the subjective character of psychology and to ground the discipline in experimental procedures akin to those used by physiologists—in short, making psychology dependent on physiology—it could also mean recasting phenomena long regarded as primarily somatic in nature as purely psychological realities. Thus, researchers such as the psychiatrist Hippolyte Bernheim (1840–1919), known for his "suggestionist" interpretation of hysteria, challenged physiological determinism and emphasized the influence of psychology on bodily states rather than the reverse.[19] Bernheim proposed that the desire for food was under psychic control in ways that defied the seeming certainties of "needs" established by the body's internal milieu. Broaching the perennial question of how long a person could survive without food, Bernheim proclaimed that it was not the exhaustion of nutritional reserves that killed an individual deprived of food but the psychic disposition of the individual, the desperate feeling of hunger.[20] The doubts that this perspective raised about physiological laws are strikingly evident in one of the stranger investigations of hunger and appetite of the era, a result of collaboration between the Italian physiologist Luigi Luciani and a celebrated hunger artist.

Psychophysiology of the Hunger Artist

The phenomenon of the hunger artist, largely concentrated in the 1880s and 1890s, involved individuals who sought money and fame by voluntarily starving themselves while being watched by scientific observers and members of the public. Like the séance and other occult rituals, the performance of the hunger artist was a curious blend of the carnivalesque and the positivistic spirit of science. In some instances the performer himself was a scientist or physician, such as Dr. Henry S. Tanner,

6 Photograph of the "hunger artist" Giovanni Succi. Observers noted Succi's "good appetite" in the days before his fast in Milan in 1888. Universitätsbibliothek Johann Christian Senckenberg, Frankfurt am Main.

who underwent a forty-day fast in New York in 1880.[21] But mostly the events were initiated by "artists," who approached "men of science" to lend authority to the performers' claims of physiological heroism. Such was the case with Giovanni Succi, who, in 1888, asked Luciani to supervise a prolonged fast he intended to undertake to counter suggestions that he was a charlatan.[22]

At the moment Succi approached him, Luciani was a professor of physiology in Florence, where he conducted numerous animal experiments.[23] His medical degree was from the University of Bologna (1862), but he saw the work he did at Ludwig's institute in Leipzig in 1872–1873 as his most important scientific training. In his investigations of the "periodic cardiac rhythm" (now known as Luciani's phenomenon), he made use of the graphic method favored at the institute.[24] He later turned again to questions of automaticity when he studied respiration. The most celebrated work of his career was research he did in the 1870s and 1880s on the functions of the cerebellum.[25]

Luciani explained his favorable response to Succi's request by observing that the phenomena of hunger and nutrition affected every part of physiological science.[26] After agreeing to become involved, Luciani turned to the Accademia Medico-Physica of Florence to help him organize a commission to oversee Succi's fast. Conditions were duly established: Succi would live in a place selected by the commission and would consume only specified liquids (2–4). Succi began fasting on March 1 and stopped on Easter, March 31, 1888. The account that Luciani published of the fast the following year included a foreword by the by then aged Moleschott, who hailed the work for its outstanding contribution to knowledge of inanition (vii).

By 1888, Succi had already conducted a number of prolonged fasts. According to the account he gave Luciani, he had discovered his ability to resist the effects of hunger while living in Africa, where he suffered a liver disease that caused him to stop eating and found that his strength and sense of well-being increased rather than decreasing. Succi returned to Rome intending to demonstrate his remarkable capacities to the public, but he was betrayed by companions and incarcerated in an asylum. He then returned to Africa, where he performed self-experiments that confirmed his powers of resistance to both poisons and hunger (17–19). In the following years Succi staged public fasts in different locales, including Paris, where he met with skepticism on the part of physicians from the Academy of Medicine but obtained the help of a commercial sponsor (19, 27–28). In Florence, Succi sought out Luciani hoping to counter the fame of a Paris rival who had sustained a fifty-day fast and become the favorite of the public.[27]

In his account of the fast, Luciani provided copious detail about Succi's biography and his physical and mental characteristics, recording many observations of his bodily features using state-of-the-art instruments including a dynamometer to gauge hand strength and a hematometer to measure the content of hemoglobin in the blood.[28] Overall, he judged Succi to be a mature man in excellent health. To confirm Succi's statements about his attitudes toward food, Luciani questioned him extensively about his eating and drinking habits and made inquiries of people around him, who affirmed that in the days before the fast Succi enjoyed a "good appetite" (8). Noting that Succi generally prepared for a fast by eating a great deal of food, Luciani placed him within a classification he had developed of types of people who ordinarily ate normally but sometimes ate to excess. Unlike those who remained largely unchanged or others who became dyspeptic, Succi fit the category of individuals who needed little food and, when consuming too much, built up strong bodily reserves (9).

Luciani inquired at some length into Succi's mental state, commenting that when it became known in Florence that the man was to undertake a thirty-day fast, people assumed he was either deranged or a master of deception (12). Luciani was not so sure, and he interrogated Succi closely to learn his mental history and current state of mind. He discovered that during Succi's sojourn in Africa, he had taken an interest in spiritism, coming to think of himself as an indomitable "lion-spirit" or, alternately, as a Christ figure who dealt in immortality (17–18, 25). To gauge Succi's credibility, Luciani sought documents about his history and learned that he had twice been incarcerated in insane asylums and diagnosed with a variety of mental ills (16, 31). Luciani showed a remarkable willingness to enter into Succi's thinking, incorporating in his own account long passages from an interview the latter had granted the physician who oversaw his fast in Cairo. He also doubted reports of the faster's insanity. While he recognized that Succi's writings made him appear "crack-brained" (24), he concluded that the man seemed to be of sound mind.

The great bulk of Luciani's study of Succi and his fast consisted of close monitoring of his major physiological functions, in this respect capping decades of physiological investigation into the effects of prolonged hunger and inanition. Luciani considered in turn his subject's body temperature, respiration, circulation, heart and muscle irritability, secretions, sensory capacities, and the general state of his nervous system. He reported many results in numerical tables, and, wherever possible, compiled graphic records of Succi's bodily condition. One of Luciani's chief goals was to show how different bodily tissues were affected by what became, amid a prolonged fast, a "struggle for life" between strong tissues and weak (78). Summarizing previous research, he added his own observations on the relative condition of Succi's heart, liver, spleen, stomach, and blood throughout the fast (77–104).

Luciani was especially interested in the matter of Succi's muscle force, partly because in releases to the press the faster had claimed that he grew stronger rather than weaker as his fast unfolded. Luciani tested Succi's right and left hands before the fast began and daily thereafter, observing that the right hand grew stronger up to the twentieth day but then declined, whereas the left hand grew stronger until the fourteenth day but then weakened. These observations supplied one of the most important contexts in which Luciani considered the extent to which Succi could affect the results of tests he underwent by the power of "autosuggestion." Despite his great confidence in his apparatus, Luciani came to accept that Succi could bend the results of certain tests by an exertion of "subjective will" and especially by autosuggestion; indeed, such an individual, so strongly convinced of his own powers and a believer in spiritism, became "a slave to his own illusions" (56). To gauge the effects of autosuggestion and the exercise of will, in one set of experiments Luciani seated Succi so that he was unable to see the writing lever of the dynamograph (58–59). In the end Luciani concluded that use of this instrument settled the matter of Succi's muscle force: contrary to the hunger artist's claim, his strength—in both the right and the left hand—diminished as his fast drew to an end. Nonetheless, Luciani did not deny that Succi's muscle force increased for a time during the fast. Struggling to explain this "paradox," Luciani concluded that following the "best physiological principles" these observations should be attributed to the power of autosuggestion.[29]

Aside from autosuggestion, Luciani also explored the overall role of psychological factors in Succi's capacity to resist hunger. Summing up what was learned from the fast, Luciani declared the results unprecedented: for the first time a faster had lived for thirty days ingesting no solid nourishment without suffering significant damage to his bodily functions or overall health and sense of well-being.[30] At the end Succi was far from the accepted physiological criterion for the final stage of inanition, a weight loss of 40 percent. As photographs showed, he had lost considerable weight, and overall his physiological functions were not optimal. But in general, the experiment demonstrated that Succi did enjoy, just as he claimed, remarkable capacities of resistance to the effects of hunger (210). To explain these results, Luciani combed the "rich" literature of the past with its accounts of fasting ascetics and famished victims of accidents and natural disasters (70–74). But ultimately he asserted it was only modern physiology, with its "objective proofs" (211), that could illuminate the results of Succi's fast and, in so doing, establish a new scientific understanding of hunger. Noting that Succi had begun the fast with extensive reserves, Luciani argued that these sustained him for a long time thanks to his age and maturity, which precluded the great physiological demands of growth, as well as his innately slow metabolism and sluggish respiration (9–10). More generally, Succi's experience proved that a healthy man could survive a long fast with few ill effects if he took certain precautionary measures, pursued the fast under given conditions, and—a claim of a wholly different order—was possessed of particular character traits ("Charaktereigenschaften") (76). In this fashion Luciani arrived, after all his painstaking tests and recording of results, at the conclusion that resistance to hunger was in good part a matter of "character" and "individuality" (229).

Luigi Luciani enjoys high standing in the annals of physiology, partly for the "significant research" he did on hunger.[31] Generally, these assessments make no mention of Luciani's interest in the power of psychological factors in the physiology of hunger, but for my purposes the most interesting question raised by his study of Succi is what connection he saw between psyche and soma in the hunger artist's feelings about eating. Put plainly, when Succi fasted, did he feel the promptings of appetite or hunger, or not? In this regard it is illuminating to compare Luciani's own reticence on this matter to comments by his German translator M. O. Fraenkel. Discussing the difficulties he faced in translating Italian, French, and other expressions observers used to characterize Succi's experience, Fraenkel noted that he had rejected language that carried overtones from the age of religious asceticism, especially the term "fasting," and used, instead, the more exact terms "hunger" and "hunger feeling." These were more appropriate, he asserted, since the German "'Hungern' of course includes the concept of the desire for food."[32] Whether Luciani thought the same is unclear. Like many of his predecessors, the Italian tended to use the terms "hunger" and "appetite" interchangeably, differentiating between the two only when it suited him. His observation that in the days before the fast Succi enjoyed a "good appetite" served, seemingly, to show that Succi was not a chronic anorexic (8). But Luciani was chiefly interested in the painful "Hungergefühl" that Succi experienced at the outset. Indeed this painful feeling became phase one in a new three-part periodization of hunger that Luciani offered as an overall conclusion of the study: an initial painful feeling of hunger, a second phase of "physiological hunger" in which vital functions continued but reserves diminished, and the final stage of "Hungertod" (225, 229–30, 233). In this process the crucial factor was not the conscious feeling of the faster but what Luciani called "more or less strong disturbance" of the "regulatory apparatuses" and the corresponding state of the blood.[33] So long as the blood continued to be well nourished, the essential functions of the organism continued (91–93).

Yet later Luciani shifted the blood out of this primary position, saying that, in the final analysis, the body's overall condition was governed by the nervous system through which signals were conveyed of the state of the blood and other vital tissues and organs. Nerves had figured little in earlier sections of *Das Hungern*, but by the end the state of the nervous system appeared as the key factor in hunger and inanition, explaining why it was that well before physiological reserves were depleted death could occur, thanks to the "disorganization . . . of the regulatory system" for which nervous structures were responsible (234, 237). These concluding remarks on the nervous system, which functioned not just by means of this or that part but always as a whole, "in its integrity" (237), suggest how it was that Luciani could establish as essential to the organism's resistance to destruction not only material metabolic processes but also the influence of character and individuality. Summing up Succi's case, he vehemently denied, as some had suggested, that the hunger artist could go without food because he was mentally disturbed.[34] How absurd, he argued, to claim that a thirty-day fast would be a matter of indifference to someone afflicted with insanity. "Character" and "individuality" made a difference, yes, but these—along with powers of autosuggestion—eventually confronted inescapable physiological limits. Claims about the purported ease with which desperate individuals moved beyond hunger feeling to hunger death belonged not to the sphere of physiology but to that of "poets and novelists" (229). The same was true, it would seem, for ineffable appetite, which scarcely found a place in Luciani's physiological science. Thus, in the end, while crediting Succi's character and psyche for augmenting his powers of resistance to hunger, Luciani affirmed the supremacy of bodily regulatory mechanisms, whose relation to the psyche he did not seek to explain. All the same, it must be said that with works such as Luciani's *Das Hungern* the history of appetite reaches a turning point. The fact that physiologists of Luciani's time had come to believe they must reckon with showmen like Succi and give credence to powers such as autosuggestion indicates that the psychic factors long pushed aside as elements of ingestion and digestion could no longer be ignored. In keeping with physiological tradition, Luciani's focus was on human functions. But in these years even work on the lowliest animal creatures suggested that eating was not a strictly material process but one grounded in animal "psychology," however primitive in character.

Feeding in Insects and Other Animals

In the second half of the nineteenth century the life sciences were dominated by questions of origins and development given new prominence by the Darwinian "revolution." Darwinian theory, with its central problematic of species change, intensified interest in every facet of animal organization and activity on which the power of natural selection could be exerted. The Darwinian impulse fixed new attention on all forms of animal behavior, and eating habits, whether explicitly linked to appetite or not, found a place among them. At the same time, competing evolutionary traditions continued to exert strong influence, especially Lamarckian theory with its stress on the formation and hereditary transmission of changed habits. Earlier in the century Lamarckian transformism had encouraged little work on appetite or on the activity of eating, as, with few exceptions, investigators had concentrated on problems of structure rather than function.[35] After the appearance of *Origin of Species*, this emphasis shifted as neo-Lamarckians who faced the Darwinian challenge began pursuing close studies of the behavior of particular species.[36] And even naturalists who held to biblical and other views of the fixity of animal forms were stimulated to a new interest in animal behavior by the ambition of refuting Darwinism.

A notable consequence of the Darwinian revolution was the enhanced attention accorded to animal psychology, a subdiscipline that gained traction after 1850. A host of pertinent titles appeared between the 1860s and 1890s, some of them the work of Darwin's disciples such as George Romanes (1848–1894), who championed animal intelligence, and others the labor of anti-Darwinians such as the French entomologist J.-H. Fabre (1823–1915), who celebrated the wisdom of the Creator in endowing animals with unerring instincts. Within animal psychology sharp debate unfolded over the relative importance of instinct and intelligence in directing animal behavior. Although this theme had long been important to moralists and sometimes to physiologists, Darwinism gave new significance to the search for the dividing point, if any could be said to exist, between the animal and human, the instinctual and the intelligent in behavior.[37] Darwin himself regarded the question of instinct as vexed: in *Origin of Species* he described instinctual actions in a general way as those performed on the basis of no prior experience, but he declined to attempt more than a working definition of the concept.[38]

It was within this frame that investigators inquired into the instinctual or intelligent foundation of animal eating choices and habits, doing so in some instances with the help of innovative methods for which a new vocabulary of psychological experimentation ("deprivation studies," "isolation studies") was gradually to emerge. In his study of evolutionary interpretations of animal behavior, Robert J. Richards points, for example, to the work of Douglas Spalding (1841–1877), a student of the Scottish psychologist Alexander Bain (1818–1903) and a talented scientific amateur.[39] In the 1860s, Spalding drew on the work of naturalists back to Spallanzani to study animal eating. Observing the barnyard activity of newborn chicks, he asked whether eating was accomplished automatically under the guidance of instinct or prompted by some kind of training or learning.[40] Practicing an early form of deprivation studies, Spalding "hooded the chicks to prevent them from learning correct responses" and found that when the hoods were removed, the chicks were still able to measure distance accurately and to distinguish appropriate foodstuffs from inedible objects on the ground.[41]

Such findings were of great interest to a much more renowned figure in emergent animal behavior studies, C. Lloyd Morgan (1852–1936), the South African–born author of influential work on animal intelligence. A determined Darwinian, in the early 1890s Morgan performed experiments similar to those of Spalding. As Richards states, Morgan concluded that "the instinct to pick up and consume objects of a certain size came built in, but the animals had to improve their aim with practice and learn what objects were edible."[42] Educated in the British idealist tradition, Morgan was principally interested in the nature of mind. In the course of a long career, he argued strenuously that mental processes were subject to the same evolutionary forces that shaped physical characteristics. Like many Darwinians he was hostile to what he saw as facile anthropomorphizing of animals and overvaluation of animal intelligence. Yet he also found objectionable the attribution to animals of foolproof instincts. Instead, Morgan sought to devise an evolutionary theory of instinct in which behavior that began with adaptive individual choices improved the survival prospects of the animal and, in accord with natural selection, became part of the hereditary equipment of its species. Morgan regarded the animal's "decision" to consume certain foods rather than others as largely a matter of trial and error and thus as part of an incremental process of learning and adaptation.[43]

While Morgan's work was embedded in long-standing scientific and philosophical debates, a great deal of work on feeding habits was carried out by investigators who observed animals purely for pleasure or for pragmatic, local, often economic interests. Such was the case, for example, with reports on the feeding habits of birds that appeared in the United States in these years in such journals as *American Naturalist* and the *Auk*.[44] These reports often included discussion of the merits of competing methods for studying the feeding of birds—examination of stomach contents, retrieval of food pellets—but they generally did not ask why animals ate as they did. Only when graduate education in ornithology was established did such problems come to be framed in theoretical terms.[45] Thus, despite Morgan's example, much Anglo-American inquiry into animal appetites and feeding habits tended to remain largely descriptive while the impassioned debates over animal instinct and intelligence unfolded on the ground of loftier problems of mind.[46]

One of the most ideologically freighted approaches to the psychology of animal appetite and eating appears in the work of late nineteenth-century entomologists, especially in France, where insect studies sparked lively public interest.[47] Such investigations are of great importance to this study for two reasons. First is the fact that entomologists were exceptional among early students of animal behavior in the sustained attention they accorded to appetite and eating. While students of other animals struggled to devise reliable methods of tracking ingestion, or, for differing reasons, were preoccupied with other issues—courtship and mating, nesting, animal language—those who observed insects often paid close attention to eating patterns.[48] Second, the very fact that insects stood so low on the chain of being suggested that they were governed wholly or predominantly by instinct and that their behavior thus provided clues to what was natural and unnatural in appetite and eating habits.[49]

As far back as Buffon's *Histoire naturelle*, French investigators who sought to understand eating patterns in different animal species had framed the question as a choice between instinct and learning. Buffon himself, though adopting overall the view that animal eating was instinctive, gave some role to learning when, for example, he alluded to the training in what to eat that barnyard chicks got from their mothers.[50] In France as elsewhere, Darwinism spurred study of the origins and nature of instinct, even though French reactions to Darwin were often hostile or skeptical. In any event, as French investigators often asserted that insect behavior was a miracle of instinct, insects became central to the debate over the relative importance of innate and learned behavior. An interesting dimension of entomological inquiry was the attention given to the role of sex differences in eating behavior. Female insects drew special attention because their behavior raised questions not only about instinctual eating but also about much broader concerns attached to the problem of maternal instinct. Indeed, this early entomological literature was heavily laden with anthropomorphic language that embodied both implicit and explicit judgments about female behavior that often said more about human females than the insects under the entomologist's gaze.[51]

A touchstone of animal behavior studies in France was the 1877 treatise *Des sociétés animales* by the social theorist Alfred Espinas (1844–1922).[52] In this work Espinas set himself the rather grandiose goal of defining the origins and nature of society, and the foundation of his thinking was the assertion that a continuum linked the social organization of the lowest organisms to the most advanced human societies. One of his basic premises was that social organization originates in the need of organisms to accomplish the two primordial functions of nutrition and reproduction; eating habits, therefore, formed a large part of Espinas's inquiry. Combing the works of naturalists, Espinas looked closely at work on insects, arguing that insect social units that tightly bound mother and young constituted the first "families" and that the creative power responsible for this original insect family was the inexplicable force of "maternal love." Utterly selfless, the "mother" wasp, ant, or bee expended her vital force in the task of nourishing her progeny, creating a community whose members, like her, lost any sense of their own "fears and . . . desires."[53] Yet ultimately, Espinas judged this insect family an inferior form because it lacked the adhesion and devotion of the male. Here too feeding habits were determinative, as they expressed the equally primordial, in this instance destructive, urge of the female to kill the insect father before the birth of the young (189), a phenomenon vividly exhibited in the move of the "voracious" female spider to devour her mate (139).

In considering the "societies" of the lower organisms, Espinas accorded great importance to the desire and search for food, "cares . . . that provoke, with the differentiation and coordination of organs, the true perfecting of organisms" (112–13). Overall, Espinas stressed the importance of physiological need in supplying this impulse, yet he also noted that innate eating desires might exceed the bounds of needs, even to the point of endangering the life of members of the group. Such was the case with individuals in one species (in this case, infusoria rather than insects) that ate much and grew to a large size, "an appreciable advantage" in a world "in which voracious appetite condemns the smallest to being devoured" (84). A key element of Espinas's analysis, moreover, involved competition between the needs of nourishment and reproduction. Societies based on reproduction stood higher on the "social scale" than those based on nutrition; only in the former did affective ties and "a consensus of emotions and representations" develop, conditions evidently superior to bonds formed in the interest of securing nourishment (124, 170–72). Although in some respects this theme linked Espinas's discussion to the long-standing physiological antinomy between functions of nutrition and those of "relation" (to the outside world), he emphasized its implications for the instinct-intelligence dispute. Like Morgan, Espinas framed the question of the role of experience in developmental terms: "education" became more important the higher the species stood on the "social scale" (245). Thus, one measure of the superiority of birds to insects was the fact that bird "parents" taught their young what to eat, passing on "intelligent" ways of getting food, while insects ate automatically (277–78).

An acclaimed figure who went beyond Espinas's style of armchair theorizing to minute observation of insect behavior was J.-H. Fabre (1823–1915). Of humble birth, Fabre worked for a time as a science teacher in a lycée, but in the 1880s he retired to a village in Provence, where he lived as a recluse, creating a makeshift laboratory and pursuing idiosyncratic experiments.[54] His investigations frequently touched on maternal instinct, a topic that took him, like Espinas, to the subject of the conflict between the desire to eat and the desire to nourish progeny. Among Fabre's most famous insect portraits was that of a type of scavenger beetle called the Spanish Copris (*Copris hispanus*). According to Fabre, the mother Copris was the very exemplar of "maternal self-denial": "So intense is her affection that she loses all desire and need of food."[55] Single-mindedly devoted to amassing provisions for her young, she takes "good care not to touch the food prepared for her family. She will go hungry rather than let her grubs suffer" (79). Here was a refinement of Espinas's self-abnegating insect mother: Fabre's mother beetle did not just neglect her own need to eat; rather, her need to eat disappeared amid her labor to provide for her young. Like Espinas, however, Fabre also described fearsome eating habits among female insects. The praying mantis, he wrote, was a "savage ogress . . . a cannibal . . . [who] makes a habit of devouring her mate, whom she seizes by the neck and then swallows by little mouthfuls" (31). Thus, for Fabre the female insect exhibited wild extremes of self-denial and greedy murderousness. Fabre construed all such patterns of insect behavior in terms of fixed instincts, devised by the Creator and essentially unchangeable. He marveled at the ingenuity and miraculous efficacy of insect instincts while at the same time acknowledging that their behavior could be "stupid" if they were called on to perform tasks outside their usual range.[56] For Fabre, a vociferous critic of evolutionism, such instincts, whether admirable or lamentable, defined unchanging types on a seemingly infinite spectrum that exemplified the mystery and fullness of creation.[57]

The maternal self-abnegation evident in insect feeding was also a favored theme of Edmond Perrier (1844–1921), who became director of the National Museum of Natural History in Paris in 1900 and published an oft-reproduced book on what biology had to teach about the nature of women, especially "feminine psychology."[58] In this regard no creature was more illuminating than the insect: "The most characteristic of the instincts, those of the insects, are effectively, in the great majority, of feminine origin. They are formed of limitless devotion, infinite foresight" (364). An expert on wasps, Perrier praised the selfless devotion the mother wasp showed in supplying food and shelter to her larvae, working "without respite" for the whole of her "short life."[59] Perrier also admired the skill the mother wasp exhibited in bringing to the nest not decomposing bodies of dead prey but living organisms she paralyzed with a single strike. Yet Perrier mocked those who perceived intelligence at work in such actions: "If [the mothers] acted with discernment, we would have to suppose that they understood the dangers of cadaveric decomposition . . . [and] had precise ideas of insect anatomy. . . . This is too much for the small head of a wasp that did not attend medical school. One thus concludes: the head of the wasp is not inhabited by a veritable intelligence but by something that is at once more and less than intelligence, something that is blind and unconscious; this something is Instinct" (563). For Perrier, maternal insect behavior indicated unerring choice both in the selection of prey and in the feeding of young, so that, in this instance, instinctual eating and mothering went hand in hand. Perrier argued a Lamarckian case for the development of eating and feeding instincts when he urged that "the instincts of many animals would remain very imperfect without the aid of education," a pattern seen with young birds who were guided by their parents in "their first quest for nourishment." Transmitted through heredity, the "personal experience of the ancestors" became in time an instinctual heritage (565).

To see what was at stake in these constructions of insect appetites and eating, it helps to look ahead to developments after 1900. Up to 1914 entomologists usually mixed praise and blame in depicting the eating and feeding habits of insect mothers, but this tendency changed during and after the First World War, when researchers more often portrayed insect mothers as ruthless in satisfying their own eating desires. In a 1921 work on insect habits, for example, E.-L. Bouvier (1856–1944) of the natural history museum depicted insect mothers who had refined methods of mutilating their victims, "killing by gourmandise" and saving the tasty parts for themselves.[60] Similarly, Bouvier's student Émile Roubaud (1892–1962) made nurturing a matter of self-interest when he argued that female wasps nourished larvae not out of devotion but from the desire to get at a tasty exudate produced by their progeny.[61] Reading such texts, one cannot escape the sense that the sometimes-bitter debates over gender during the war and postwar years seem to have altered thinking about insect as well as human females.[62]

Although French entomologists seldom discussed eating habits in relation to any explicit view of how appetites formed, their work marked an important contribution to the view that instinctual eating desires differed essentially in the sexes. Crucial to Fabre's notion of female eating was the conflict between the mother insect's instinct for her own self-preservation and her drive to protect her progeny. Ultimately, the latter outweighed the mother insect's concern for self, effacing even her raw physiological need for sustenance. Here, then, was a representation of appetite that asserted the absolute primacy, in the female, of the drive to reproduce over the drive to sustain the life of the individual by eating. Evolutionists who otherwise disdained Fabre's idea of fixed instincts argued along the same lines in discussing female eating. Alfred Giard (1846–1908), who assumed the chair in organic evolution at the Paris Faculty of Sciences in 1888, denied that maternal love was a fixed feature of animal life and urged that actions appearing to result from motherly solicitude were actually those of mothers who sought their own advantage, only accidentally conferring benefit on the species.[63] Yet Giard too was struck by prolonged fasting in animal mothers of diverse species. Reporting on observations of a female python held at the museum, he noted that in the course of a long incubation "the female did not want to eat" (20–21). Thus, both fixists and evolutionists affirmed—but only for the female—Espinas's principle that in notable cases the urge to reproduce trumped the urge to eat.[64]

French entomologists saw their work as resting on a newly firm scientific basis, but as this discussion suggests, their representations of female insect eaters reflected long-standing stereotypes about human females, especially of the two kinds of mother—devoted and self-abnegating versus neglectful and self-absorbed—but also of "voracious" females who "devour" their mates or greedily hoard all the best bits.[65] Indeed, with some frequency these authors drew out the significance of their investigations for knowledge of the eating needs and desires of human females. A case in point is Giard's theory of parasitism, developed not in regard to insects but to marine organisms called "bopyrians." This work was devoted in part to mothers whose maternal instinct was diverted from nurturing of their own young to the care of invasive parasites.[66] Along the way Giard developed an analogy between parasites and the human embryo that led to startling conclusions about the nutritional needs of pregnant (human) mothers: the state of "dénutrition" experienced by the pregnant woman, he stated, was a natural phenomenon, one that was inevitable in her condition and that exerted a powerful "reflex nervous" effect throughout her system (372). Whatever the value of such observations for the understanding of animal parasitism, Giard's assertion that a pregnant woman was "inevitably" undernourished naturalized aberrations of appetite and eating that for centuries physicians perceived in women at critical points of their reproductive life. Such observations mattered, then, well beyond the sphere of entomology. By 1900 physicians like scientists, often claimed that their insights into appetite and ingestion derived not from tradition but from newly "positive" inquiry; in this context, insect studies lent valuable support to the project of discerning what was "natural" and what was not in eating behavior.[67]

Until the latter half of the nineteenth century ingestion was not a fit topic for students of psychology, a term that referred to the mind and its "noble" activities. Even Gall's phrenology aimed chiefly to localize moral and intellectual traits and said little of the appetite for food. When, for example, Gall broached the issue of the "carnivorous instinct," he moved immediately to the human "penchant au meurtre" and the wicked deeds of famed murderers.[68] Thus, ingestion was a matter for physicians and physiologists, who, as Beaunis noted, were steeled to consider facts that those untrained in the ways of science would naturally find "repugnant." After midcentury, this neglect of the "lower functions" began to give way in physiological and animal psychology and in outlier investigations such as Luciani's work with the hunger artist. To be sure, these first ventures were tentative: in the late nineteenth century "psychological" investigation of appetite and eating rested on a hodgepodge of methods—clinical, historical, introspective, experimental—and it lacked any agreed-on focus. Candidates for the psychological determinants of appetite and eating abounded—imagination, internal sensation, instinct, autosuggestion. All the same, however tentative its methods and ruling concepts, a psychology of ingestion had come into being. What headway it would make remained to be seen.

NINE

Peripheral or Central? Disordered Eating in Clinical Medicine

Physicians of the Enlightenment and early nineteenth century recognized disturbed appetite as a ubiquitous sign of illness but differed widely on whether states of appetite, whether lost, excessive, or "perverted," constituted "primary affections" or were purely secondary to more general pathological states. In the later nineteenth century the question of whether disturbances of appetite were "primary" or "secondary," idiopathic or symptomatic, continued to be a leading question for physicians who attempted to sort out the nature and causes of an array of ills involving appetite, eating, and gastric function. That disordered appetite was a secondary feature of such well-defined diseases as tuberculosis and cancer was taken for granted.[1] Yet defining the role of appetite in diffuse psycho-gastric ills remained as baffling as in earlier years despite the proliferation of diagnostic labels and therapeutic approaches. Ills of appetite continued to be subsumed within diagnoses of dyspepsia and broad-gauge neuroses of the stomach, and from the 1860s onward, these were encompassed by the new label "neurasthenia," which rapidly gained favor in the specialty of nervous disorders. Diagnostic overlap and confusion were strikingly exemplified by the view of the Jena neurologist and psychiatrist Otto Binswanger, who held that "nervous anorexia" was "a form of nervous dyspepsia linked to neurasthenia."[2]

Another new diagnosis bearing on appetite that took hold in the 1880s was "autointoxication," the idea that the organism at large might be poisoned by noxious intestinal residues, leading to both physical and mental dysfunction. Vigorously advanced by the French internist Charles Bouchard (1837–1915), autointoxication became a leading diagnosis in Germany, England, and the United States.[3] Bouchard's theory of self-poisoning, based on the new principles of bacteriology, asserted that the longer that digestive residues remained in the body, the more they produced toxins that were absorbed into the bloodstream, producing a range of (sometimes fatal) symptoms of general poisoning. Although Bouchard was chiefly concerned with the effects of autointoxication, he also considered its etiology, and here the state of the appetite was a leading factor. He blamed "excessive appetite" and "irregularity in our meals" for preventing the "very fine mechanical division of food" required for proper digestion and thus for inducing autointoxication (174). Bouchard also noted the frequently beneficial effects of diminished appetite, especially in preventing the dangerous practice of introducing "an additional meal into the stomach when the previous one has not been digested" (184). As James C. Whorton observes, autointoxication became one of the key "catchall" diagnoses of the era, "the category into which cases of headache, indigestion, impotence, nervousness, insomnia, or any number of other functional disorders of indeterminate origin could be dumped."[4] Disturbed appetite also belongs on this list, although whether its significance was greater as symptom or cause Bouchard did not make clear.

Although it is difficult to quantify such a matter, it seems that until the turn of the century "gastric" or broadly somatic explanations of troubled appetite and eating continued to dominate medical thinking. Yet the most momentous development of these years was the move by psychiatrists and neurologists to assert the primacy of psychic factors in the etiology of ills of appetite. This shift was clearest in descriptions of the syndrome first called "anorexia nervosa" or "hysterical anorexia," now commonly regarded as the precursor of the modern "eating disorder." Obesity, known by assorted labels and attributed to varied causes, also drew greater attention in the second half of the nineteenth century than it had in the first, and a prolonged debate over psychic and somatic origins of the condition got under way.[5] In this process anorexia was gradually transformed from symptom into disease, and obesity set on the long road toward its current status as a distinct pathology.[6] The self-conscious innovators who began to conceptualize forms of troubled eating as independent diseases or as distinct "symptom complexes" focused directly on the involvement of appetite in—or its irrelevance to—these ills, bringing about major shifts in thinking about the urge to eat.[7]

First Descriptions of Anorexia Nervosa

It might seem self-evident that appetite would figure centrally in discussions of a condition whose name—"anorexia"—literally means "lack of appetite." But in fact a striking feature of this history is the lack of consensus on the state of the appetite in anorexics. Appetite held a prominent place in the classic first descriptions of hysterical anorexia or anorexia nervosa that appeared simultaneously in France and England in 1873, one by the neurologically trained French physician Charles Lasègue (1816–1883) and the other by the English physician and sometime professor of physiology William Withey Gull (1816–1890).[8] In their descriptions, diminished appetite stood among the first symptoms listed. Lasègue was more attentive to the role of appetite, which he regarded as the element of digestion that was "the best analyzed by patients and the least easily studied by physicians" (386). He lamented the "poverty of the vocabulary" available to describe the "degrees or varieties of *inappétence*" and devised a tripartite classification of changes in appetite that included, as the first, loss of appetite without repugnance for food; second, "a more or less vivid repulsion for certain foods"; and, last, a state in which "any alimentary substance whatever provokes disgust."[9] He also briefly referred to "false appetites" ("demanding, imperious") that were better treated with a few drops of laudanum than ingestion of foodstuffs.[10] Though less concerned with nuances of appetite, Gull also regarded anorexia nervosa as entailing disturbance of appetite: he referred to his patients' "complete anorexia for animal food, and almost complete anorexia for everything else," noted that occasionally the appetite would become "voracious" for a few days, and questioned which factors were responsible for the "want of appetite" (22, 23, 25) characteristic of the condition.

Both physicians regarded these peculiar forms of anorexia as afflicting young women and girls principally; Lasègue based his views on eight cases, all of them young females between the ages of eighteen and thirty-two, and while Gull stated in passing that he had "occasionally seen it in males," he observed that this was a condition "mostly of the female sex, and chiefly between the ages of 16 and 23" (22). The three cases he presented, poignantly illustrating the look of the patients "before" and "after," were all of young women. Having noted the original loss of appetite, Gull proceeded to describe the ensuing effects, including weight loss, shrunken abdomen, restlessness, and, ultimately, "the depression of all the vital functions" (23). Lasègue's gendered description characterized the typical behavior of a patient in the context of the family—her refusal of food, the pleading of the distressed parents, her gradual decline to a bedridden state and the attendant disruption of family life—and the patient's apparent psychological state, ranging at the outset from a kind of exhilaration ("une façon d'alacrité," 391) to indifference to, finally, the sadness and anxiety of the advanced stages of the disease (399).

Both Gull and Lasègue struggled to understand the origin of this new condition, anticipating many of the explanations that would be offered by students of anorexia over the next hundred years. Attuned to physiological inquiry into the basis of appetite and hunger, Gull suggested that anorexia nervosa resulted not from "any gastric disorder to which the want of appetite could be referred" but from a "morbid mental state" that pointed to a "central . . . not peripheral" origin (25). Although he assigned one case specifically to "a failure of the powers of the gastric branches of the pneumogastric nerve" (26), Gull was inclined to attribute the illness to the special tendency of young women to "mental perversity," including willfulness and obstinacy. Lasègue's view was similar; as did Gull, he noted that there was no evident visceral disorder in these patients, although he did observe that most of them suffered from "chloro-anemic" conditions (392). Emphasizing that the failure of the patient to eat was at the outset voluntary, Lasègue nonetheless noted that the "decision" not to eat usually followed some kind of emotional upset and could not be mastered simply by an exercise of will (394). Again like Gull, Lasègue tied hysterical anorexia to a "mental perversion," a disturbance of the central nervous system. In this respect, he urged, deranged appetite resembled other symptoms of hysteria—persistent cough, pain in the larynx, paralysis—that occurred in the absence of any discernible organic cause. But unlike Gull, Lasègue vigorously rejected gastric remedies; patients got no benefit from these medicines but, instead, recovered only when freed of their "subdelirious preoccupation," an occurrence that might result from docile acceptance of the physician's guidance but that often occurred spontaneously when the patient underwent some "profound moral perturbation" (400).

These descriptions of anorexia in young girls were widely influential. Physicians in France, Britain, and the United States accepted the new terminology and characterized the condition in a fashion similar to that of Gull and Lasègue, including close attention to the state of appetite observed in anorexics. At the end of the 1870s, the American neurologist William Alexander Hammond (1828–1900), for example, began his discussion of what were popularly known in the United States as "fasting girls" by focusing on the patient's loss of any desire for food, which he attributed to an initial experience of "domestic trouble" mixed with a failure of will often seen in young women who did not subject emotion to reason. Overall, he ascribed female pathologies of will, including readiness to ignore the prompt of appetite, to an underlying peculiarity of female nervous organization.[11] In France the clinician Lucien Deniau emphasized the state of the appetite in a work of the early 1880s when he distinguished between two types of anorexia, gastric and hysterical, according to the source of the appetite loss, one resulting from a disturbance of the stomach and the other linked to hysteria. The gastric anorexic had no desire to eat but did not feel repelled by food, he claimed, while the hysterical anorexic felt deep repugnance to food.[12] Similarly, the neurologist Jean-Martin Charcot (1825–1893), responsible for training a generation of French neurologist-psychiatrists at the Salpêtrière, quoted a young patient who spoke directly of her total loss of appetite—indeed, her "horror of eating"—and remarked on the impossibility of restimulating the appetite of such patients by ordinary means. Charcot traced anorexia to an underlying "nervous predisposition" that made patients vulnerable to psychic shocks that healthy individuals would absorb without great difficulty.[13]

Already by the 1890s, however, some French physicians began to question earlier claims about the loss of appetite as an integral element of the syndrome Lasègue had described, a challenge that led some to dispute the disease label itself. The sought-after clinician Paul Sollier (1861–1933), who treated patients at a private sanatorium outside Paris, argued in 1891 that some anorexic patients experienced no loss of appetite and that the essential feature of the disease was that the patients would not eat, for whatever reason. He then devised a four-part classification of causes for refusal of food that encompassed "moral" reasons such as life weariness or the desire to be thin; strictly somatic problems such as oversensitive mucus in the stomach; sensory distortions such as those causing misjudgment of the amount of food on one's plate; and, only fourth, loss of appetite, or anorexia.[14] Other French authorities also denied or downplayed the importance of the state of the appetite. The neurologists Édouard Brissaud (1852–1909) and Achille Souques (1860–1944) argued in the mid-1890s that the misleading term "anorexia" had to be banished in favor of some label (they proposed *délire de maigreur*) that accented the psychic origin of the disease. Despite this apparently determined psychogenic approach, however, these authors also considered the possibility of a hereditary predisposition to the disease and even, looking to purely somatic origins, traced in two exemplary observations an initial physical trauma in the form of a blow to the hip and subsequent nervous hyperalgesia in the hip region.[15]

Yet another option was soon provided by physicians who came to regard loss of the desire to eat as secondary to some kind of psychic "idée fixe" or "obsession." Pierre Janet (1859–1947) proposed, interpreting the famous case of "Nadia," that she "perfectly retained the feeling of hunger" but refused to eat as "the consequence of an idea, a delirium (*délire*)," her morbid fear of getting fat.[16] Janet also reported on cases of "psychasthenic" patients who suffered "an exaggerated appetite and a constant need for nourishment" or, alternatively, "have no appetite and are, rather, disgusted by all foods" (1:410). These ills Janet traced to "gastric upsets having as their point of departure a nervous trouble" (1:412). Here again it was not loss of appetite or hunger that mattered most but an overall "weakness of nervous functions."[17]

Janet's works indicate the direction in which the increasingly influential psychiatric community was to take the study of appetite, a full-scale cerebralization of appetite in which its somatic grounding was underplayed while the psychic dimension became paramount. This story will be traced later, but for now it is important to note that a countercurrent to the cerebralization of appetite and the search for psychiatric causes of anorexia ran strong throughout this whole period and helped bring about a dramatic return to somatic explanations early in the twentieth century.[18] As Tilmann Habermas has shown, German-language physicians, including psychiatrists, showed little interest in the anorexia nervosa diagnosis in the nineteenth century, focusing instead on "metabolic changes" responsible for loss of appetite or on "gastric neuroses."[19] Physicians in Berlin, Vienna, and Budapest were often skeptical of the observations of anorexic patients reported by French and English physicians; when confronted by similar symptoms, they saw loss of appetite as the central feature of a special variety of gastric neurosis caused by such conditions as "hyperesthesia" of the stomachal nerves.[20] From this perspective, which was grounded in peripheralist explanations of appetite and hunger, the syndrome elsewhere identified as anorexia nervosa was an organic ill seated in the stomach and caused by disturbances of the stomachal nerves. Although this was to change somewhat after 1900 with the work of such psychiatrists as Otto Binswanger (1852–1929) and Louis Schnyder (1868–1927), somaticist interpretations of food refusal predominated overall, and it was to be a German physician, Morris

Simmonds (1855–1925), who, with papers published between 1914 and 1918, transformed anorexia nervosa into the glandular disturbance that bore his name.[21] Thus, while appetite came to be marginalized by physicians, especially French, who increasingly emphasized psychic phenomena such as morbid fears of weight gain that left the desire for food intact but caused sufferers to refuse food, German-language physicians continued to regard the state of the appetite itself as the key element of a complex psychosomatic disease picture, a disposition that helps to explain why it was these physicians who first undertook full-blown investigations of appetite after 1900.[22]

Obesity in Fin-de-Siècle Medicine

While the debate about the role of appetite in anorexia was unfolding, a parallel discussion took place in respect to obesity. As discussed in an earlier chapter, the problem of fat had commanded new interest in the early nineteenth century when the question of how it formed in the body became the subject of a protracted debate within physiological chemistry.[23] After 1850, medical men took determined charge of the problem of obesity, producing an extensive literature.[24] One helpful means of getting at the medical literature of obesity is to examine the summary articles offered in medical encyclopedias of this era, especially France's *Dictionnaire encyclopédique des sciences médicales* and, in the German-speaking lands, the *Real-Encyclopädie der gesammten Heilkunde*. The lead articles on obesity in these two compendiums indicate that, as was the case with anorexia, physicians struggled to define the place of appetite in the nature and causation of the condition.

The author of the *Dictionnaire encyclopédique des sciences médicales* article on obesity was Émile Demange (1846–1904), who had trained in Paris with Bouchard, the architect of autointoxication theory.[25] Demange defined obesity as "a pathological state characterized by more or less general hypertrophy of the adipose tissue," a state that, in varying degrees, "augments the volume and weight of the body, deforming its exterior and causing functional disorders of the organs surcharged with fat" (5). He also discussed synonyms frequently used for the condition—the German *Fettsucht*, the English "corpulence," and, a French term that signaled the essential role of the blood in obesity, *adipose* or *lipotamose universel* ("the expression is precise for we know today that the blood contains an exaggerated quantity of fat," 5). Demange first traced the history of obesity, asserting that it had been seen throughout the ages, certainly from the moment when "men began to enjoy the pleasures of the table" (6), and identifying celebrated physicians who had contributed most to understanding the problem—Hippocrates and Galen among the ancients and Haller and Bichat among the moderns. Bichat's views were especially important because he had associated obesity with the "cellular tissue" that was long regarded as a primary reservoir of fat. Demange nonetheless noted that Bichat's long-influential view had been overturned by histologists who had come to recognize that "[fat] constitutes a tissue of its own, the adipose tissue" (6). Considering next the anatomical viewpoint, Demange described the particular places in the body where fat accumulated, recognizing that it played a protective role for the body but observing that, when found in excess, it "altered the blood itself," causing the "hematopoietic and nutritive troubles" that figured among its most serious consequences (10). The most important symptom of obese persons was their "exterior deformation," but they also suffered difficulties in walking, sleeping, breathing, engaging in sexual relations, and, so some physicians averred, thinking. In short, "their life passes in a kind of vegetative state" (12).

Appetite appeared in Demange's discussion of obesity both when he described the condition and, principally, when he surveyed its causes. These causes he divided into "predisposing" and "occasional," with the former entailing special stress on the role of heredity. Offering figures amassed by other physicians, especially Bouchard, Demange observed that in one study of 124 cases, 61 were hereditary, equally divided among men and women. Telling evidence for obesity as a familial disposition was available particularly in cases of infant obesity, including one little child whose mother was "enormous in pregnancy"; who was himself obese at birth; and who, by age three, had an "enormous appetite" and grew so large he could not be lifted (17). Hereditary obesity differed from that brought about by occasional causes, including periods of convalescence, such as that of a girl who, recovering from a long illness, became obese in the space of two months: "She ate with great appetite almost all day long, and I fear that her career ended in sudden death" (21–22).

Thus, appetite appeared as a significant element in the "somber tableau" (15) of obesity. Still, what is striking about this compendium of medical knowledge on obesity is the extent to which the matter of why people became huge eaters, took up "noxious" eating habits, or placed themselves ("voluntarily or not") in a condition similar to that of animals fattened for slaughter (20) was not of great interest to Demange. Overall, he gave much greater attention to the physical process by which fat formed and to the incidence of obesity by age, sex, climate, region, and hereditary predisposition than he did to questions about why patients ate as they did. Attending closely to the contributions of "modern chemistry" (24), Demange also considered the question of whether eating fat made an individual obese or whether, instead, a range of food substances could contribute to fat formation. He took special note of the celebrated feeding experiment conducted by the Germans Carl Voit and Max Pettenkofer (1818–1901) on a dog first fed degreased meat, then a mixed diet of partially fatty meat and nitrogenous substances, and finally a diet that included these substances along with sugar and other carbohydrates as well.[26]

Chemical findings also had clear therapeutic significance for physicians called on to devise some remedy for obesity. Indeed, what fat people should eat in order to lose weight was perhaps the most impassioned question addressed in the obesity literature. Already a subject of great interest in earlier decades, this problem became a matter of even livelier debate from the 1860s onward, thanks to publications by an Englishman named William Banting (1797–1878). In a series of works, Banting outlined the particulars of a diet he adopted on the advice of his physician and that, after twenty years of failed attempts to shed weight, had finally helped him slim down and recover his health. The Banting diet entailed cutting the quantity of food, but its most important feature was the rigorous exclusion of all fatty, sugary, and farinaceous substances.[27] The "Banting cure" enjoyed a tremendous vogue for a time, but it was also severely criticized by physicians who urged that the most important teaching of modern physiology and physiological chemistry was that fat was essential to the healthy body. The German physician Wilhelm Ebstein (1836–1912) argued vehemently that human beings could not be fed on one class of food alone and that foods containing "no nutritive ingredients except albumen, as for instance flesh destitute of fat, are not proper food for man."[28] This struggle over the particulars of therapeutic diets, so familiar to us from the apparently endless arguments of our own time about what to eat to lose weight, was hotly joined throughout the late decades of the nineteenth century.[29]

But what about the motivation of the "huge eaters" who, as Demange also recognized, filled the ranks of the obese? Demange did state on occasion that obesity was essentially a matter of overeating and that, to be cured, the obese must learn to overcome their "detestable habit" of gluttony (31). How they were to do so was not, however, a problem to which he devoted much attention. Closing his discussion of the most beneficial therapeutics, he held that diet was the most important element of treatment and that reducing the amount of food—taking the eater's consumption below what French chemists had established as the "maintenance ration" (30)—was the only way to cure obesity. He considered the range of medicines—purgatives, alkalines, mineral waters—that physicians used to help the reducing process along, but he stressed that eating less was the key. He recognized that for people who were accustomed to eating a great deal, eating less was difficult: after some improvement had been seen, the patient must be allowed some relief. The physician directing the cure must take care to avoid "monotony" so that the patient did not become discouraged (30). Demange also observed that spas were useful in achieving the isolation of the patient from the social surroundings in which he or she customarily overate: "A stay at a thermal station is often the only means of forcing the patient to break with a habitual bad regime" (33). So, psychological factors did enter in. But Demange was chiefly interested in the material reality of fat, how it formed and, in keeping with the primary concerns of Bouchard, how it set off pathological changes in the blood. Following his teacher, Demange concluded that obesity was a "malady of the blood," a physical illness to be placed alongside diabetes in the larger class of "dystrophies" (27).

Interestingly, Demange concluded his look at obesity with a section devoted to *maigreur*, asserting that in many respects thinness was a pathology "analogous to obesity" in its causes and therapeutic indications (33). Like obesity, *maigreur* was found in many different degrees. When it became extreme, it indicated "a poor state of nutrition" and presaged grave constitutional maladies such as tuberculosis and cancer (33). Also like obesity, "constitutional thinness" was often hereditary; it was frequently found in conjunction with other maladies, especially within the "vast class of dyspeptics" (34). Yet in some ways the most striking feature of this purported "analogy" between *maigreur* and obesity was the elision of the most dramatic form of *maigreur*—anorexia, in any of the forms described by French physicians (hysterical, gastric, or mental). Despite Demange's claims that obesity and thinness were analogous, in fact the medical literature of the period provided radically different views of the conditions. The anorexia literature had little to say about chemical processes of "dénutrition" that might be at work, certainly nothing comparable to the extended discussion of the body chemistry underlying obesity. Instead, anorexia was chiefly ascribed to moral or psychological failings of the patients themselves, teenage girls who, whether from "coquetry" or rebelliousness, refused to conform to normal standards for the gratification of appetite. Certainly, obese patients—men, women, people of all ages—were not blameless, although, according to

Demange, at least half of them became obese because of a hereditary predisposition over which they had little control. Nonetheless, the difference between anorexia and obesity in terms of the moral culpability of the sufferer was pronounced. And although obesity was more common in women, it was not a rigorously gendered disease: both men and women could become obese. It was also an ill that disproportionately affected those in middle age. That these victims, unlike those of anorexia, suffered ills that their physicians often attributed to physical predispositions and causes beyond their control is scarcely surprising. Fat people were to be counted among the worthies of society: one could overeat only if one had the means to do so, and this meant that those most affected were men who controlled scarce economic resources or women who depended on such men. By contrast, undereating to the point of emaciation was largely a matter of will. Failure to manage the appetite in accord with appropriate standards of health and behavior was characteristic especially of girls and young women whose weak or rebellious will was a truth confirmed not only by medicine but also by every authoritative discourse of cultural life.[30]

The summary of German-language literature on obesity published in the Viennese *Real-Encyclopädie der gesammten Heilkunde* showed even less interest in the vexed psychology of the condition than did Demange. This overview was the work of the Austrian gynecologist and balneologist Enoch Heinrich Kisch (1841–1918).[31] Like Demange, Kisch defined obesity as the accumulation of fat beyond the normal, surveyed the places in the body where it ordinarily developed, and discussed at length causes and treatments of the disorder. Kisch offered extensive discussions of how, physiologically, obesity developed. Like Demange, he pointed to the work of Voit and Pettenkofer, lauding their experimental demonstration that a diet containing average amounts of meat along with a great deal of fat caused the largest accumulation of fat in the body (554). He also noted the etiology of obesity proposed by the pathologist Julius Cohnheim (1839–84), who had argued that obesity resulted from abnormally low levels of fat combustion that occurred when oxygen failed to bind with fat in the cell.[32] A somewhat surprising difference between these French and German contributions is the greater attention that Kisch devoted to the loss of beauty—which he, following the Greeks, defined as grace and symmetry of form—in the obese. Like Demange, he held that obesity was more common in women than men, and he lamented the distortion of the female body, especially of the breasts, seen in fat women (556). He attributed women's obesity in part to their "sedentary ways" and preference for sweet and fatty foods, but he suggested, finally, that physiological defects, especially a "nutritive antagonism" waged between reproductive and nutritional processes, were perhaps of greatest importance.[33]

Despite Kisch's more extended attention to the claims of experimentalists, ultimately, he, like Demange, adopted a clinical standpoint in respect to obesity. In attempting to distinguish normal from abnormal in fat accumulation, both men cited standard authorities on appropriate ratios of fat to overall body weight but finally concluded that it was impossible to establish with quantitative rigor the borderline between normal and pathological levels of fat. Demange asserted that only the presence of "functional disturbance" could indicate pathological levels of obesity and Kisch said much the same in urging that only the "subjective element of pain" could be relied on.[34] Hence the long recitals, in both cases, of the myriad forms of damage done by excess fat to major organs and essential functions; it was these pains, lesions, and dysfunctions, not physiological norms, that justified the rigorous therapeutic measures recommended by both physicians.[35]

The therapeutic regimes proposed by Demange and Kisch covered the gamut—dietetic, medicinal, gymnastic, and, extensively in Kisch's case, balneological. Both physicians evaluated competing diets, and Kisch, who worked summers as a supervising physician at the spa at Marienbad, presented a detailed regime of his own devising, based at least partly on the dangers of fat consumption pointed to by Voit. Unlike his French contemporary, he made no mention of the psychological difficulties of following a severe eating regime, although he briefly considered Ebstein's claim that the obese should consume fat, as it was the food that most quickly led to a feeling of satiety.[36] His only explicit mention of patients' appetite came when he surveyed a debate over whether fat-reducing cures produced a loss not only of appetite but also of protein supplies in the body; diminished appetite he accepted as an essential goal whereas diminished protein levels were not. His final judgment on fat-reducing cures rested on such material effects: the best fat-reducing method, he concluded, was that in which fat was decreased while muscle power and protein supply remained unaffected (569). In sum, these encyclopedic surveys suggested that even when physicians of the late nineteenth century considered overeating and other forms of patient behavior as the cause of obesity, they scarcely mentioned the problem of why people ate as they did and thus had little to say about the relation between states of the appetite and fatness.

Dyspepsia, Neurasthenia, Neuroses of the Stomach

The same neglect of the why of eating characterized much medical discussion of the diffuse nervous ills in which disturbed eating, if not disturbed appetite, was prominent. From the 1850s onward, many physicians attributed dyspepsia not to the unfortunate dietary choices of patients but to a chemical disturbance, "a pepsin insufficiency in the stomach," and prescribed commercially prepared pepsins as a remedy.[37] The literature of neurasthenia was also largely silent on the question of what caused patients to eat in certain ways. S. Weir Mitchell (1829–1914), widely acclaimed for his innovative treatment of neurasthenia, noted that women who suffered from "nervous exhaustion" amid a general decline of health often ate little, but he did not ask whether they stopped eating because of diminished appetite or because of some other condition attendant on nervous disturbance. Overall, Mitchell stressed such physical ills as the loss of fat stores and impoverishment of the blood, along with detailing his proffered cure, a combination of seclusion, rest, massage, electricity, and foods selected by the physician. What patients themselves desired or chose to eat was scarcely at issue.[38]

Unlike dyspepsia and neurasthenia, the diagnosis of gastric or "visceral" neurosis did involve close attention to appetite. Indeed, aside from anorexia nervosa, gastric neurosis, a diagnosis especially favored in England and Germany, was linked to disordered appetite more than any other such condition. The English physician called to give the 1884 Gulstonian Lectures, T. Clifford Allbutt (1836–1925), attributed visceral neurosis to "the greediness or recklessness" that made the stomach a "helpless receptacle" and to implacable "cravings" and "attacks of 'want' " seen with these ills.[39] Nonetheless, he took care to contrast these neuroses, which resulted directly from disturbances of the nervous system and were "as real as the doctor's own toothache" (21, 47) to the "irregular or depraved" appetites characteristic of hysterics, individuals who enjoyed "limited reason" and acted on "foolish impulse." Thus, in visceral neurosis the state of the appetite was a by-product of genuine disease caused by nervous malfunction, whereas in hysteria a "depraved" appetite reflected underlying mental and moral flaws.

The most elaborate analysis of the place of appetite in gastric neurosis was the work of German-language clinicians, such as Theodor Rosenheim (1860–1939), who wrote on the topic for the *Real-Encyclopädie der gesammten Heilkunde*.[40] Rosenheim classified neuroses of the stomach in three divisions, those affecting motility, sensibility, and secretion, each subdivided in turn into "irritative" and "depressive" disturbances of the function in question. Disordered appetites figured prominently in Rosenheim's second class, neuroses of sensibility, which included, among others, gastralgia, bulimia, and anorexia nervosa. In Rosenheim's construction, hyperesthesia of the stomach membrane often led to a loss of appetite or, alternatively, to an intensified *Hungergefühl* (407). Bulimia was an "anomaly of the specific sensibility of the stomach by which feelings of hunger and satiety are transmitted" (413), whereas anorexia nervosa resulted from "anesthesia of the nervous apparatuses that transmit the hunger feeling," causing not just loss of appetite but also "loathing of food" (415). Rosenheim took care to distinguish appetite from hunger, presenting the former as a desire for specific foods but the latter as an "absolute" expression of need (415). Although Rosenheim's attention to disordered appetites was more elaborate than Allbutt's, he shared with his English contemporary the conviction that the ultimate source of gastric neurosis—and the resulting, symptomatic states of the appetite—was nervous dysfunction. Both acknowledged, moreover, that, aside from the involvement of the vagus nerves, the specific nerves in question were unknown. Allbutt readily referred to "vagus neuroses" (37), but he dismissed as "mere gymnastics" (76) explanations that targeted other nervous pathways.[41] Yet while the specific nerves involved remained unclear, here at least was a definite physical explanation of strange appetites and ways of eating: cravings, bulimia, and anorexia were only symptoms of underlying nervous dysfunction seated in the viscera. Complexities of psyche and behavior as probed by Gull and Lasègue were brushed aside.

In the later nineteenth century both scientists and physicians questioned the reliability of appetite as a guide to healthy eating. Just as digestion had long been regarded as the most obscure of the major bodily functions, appetite was often characterized by scientific and medical investigators as essentially unknowable, a blend of the physical and the psychic that could not be subjected to rigorous methods of investigation. Given that this period was the high-water mark of confidence in experimental methods, it is scarcely surprising that researchers were confounded by a phenomenon such as appetite that they could neither define nor measure. Physicians like Lasègue observed that the problem of appetite was medical in character, meaning that it could be known or gauged only when observed by physicians who knew the signs of disease through long experience. With other like-minded clinicians, he insisted that knowing such subjective conditions as the state of the appetite required all the tact and skill the bedside practitioner could muster. In short, appetite remained elusive and knowledge of it heavily dependent not on experimental prowess but on a trusting relationship between clinician and patient. Such an approach was a far cry from the laboratory rigor demanded by the stringent positivist canons of the age.

All this was to change at the turn of the century when, for a time at least, the "psychic factors" that scientific investigators had repeatedly invoked but just as quickly declined to pursue became primary in studies of appetite. This cerebralization of appetite had to do partly with the rising influence of psychiatry—the importance newly accorded even by a skilled laboratory physiologist like Luciani to such factors as the power of autosuggestion—and partly with the new interest in animals not as organisms to be cut open on a table but as products of evolutionary adaptation, actors in environments now under intense scrutiny within animal psychology and the nascent discipline of ethology. Yet the decisive shift toward viewing the psychic nature of appetite as a matter of keen interest rather than a vexing problem to be avoided came in physiology. This was the accomplishment of Ivan Pavlov, with whose revolution in thinking about appetite the next chapter begins.

PART FOUR

Appetite as a Scientific Object, 1900–1950

The patient asserted that she had not eaten, simply because she had *no appetite,* and for no other reasons. Loss of appetite—in sexual terms—loss of libido.

SIGMUND FREUD, "DRAFT G—MELANCHOLIA" (1895)

Nature, resourceful and unerring, has linked the seeking and finding of food, with the commencement of digestion. That this initiation of secretion should stand in closest connection with an everyday phenomenon of human life, namely appetite, might easily have been surmised. Thus appetite, so important to life and so full of mystery to science, here at length assumes a tangible existence and becomes transformed from a subjective sensation into a concrete factor within reach of physiological investigation.

I. P. PAVLOV, *THE WORK OF THE DIGESTIVE GLANDS* (2ND ED., 1910)

Often when a child expresses a great appetite for certain substances, he is told by his parents that he must not eat them; equally often he is told to eat things which are very distasteful to him. He quickly learns to distrust his own appetite; and when he comes under the influence of advertisements, he falls prey to them.

CURT P. RICHTER, "BIOLOGY OF DRIVES" (1941)

Introduction to Part Four

This study began by looking at the broad sociocultural anxieties about food and eating that emerged with the influx of luxury products into eighteenth-century Europe. In the nineteenth century, the most pressing socioeconomic issues that impinged on the medicine and science of appetite were problems of want—how to determine the minimal diet required to sustain poor inmates in public institutions or to maximize the output of ill-paid laborers. But with the spread after 1850 of ideals of value-free inquiry, medical and scientific investigations of appetite took less notice of such demands of real-world eating. This pattern continued into the twentieth century despite the ever-widening impact of industrial food—the production of standardized and branded products, distribution through national and global networks, and the increased marketing power of advertising. Changes in the economics of food were matched by important developments in food "culture"—new sites for consumption of food outside the home, modernized styles of cookery, and shifting perceptions of foods as markers of class distinctions.[1]

Although the impact of such changes on the individual experience of appetite must have been enormous, few scholars have sought to explore how the functioning of appetite has shifted in response to broad economic and sociocultural transformations.[2] Nor, given the bounds of this study, do I attempt to outline the impact of such modernizing forces on the formation and experience of appetite. Rather, I explore how and why these shifts in the economics and culture of food drew little attention from scientific and medical students of appetite. Of the investigations that I examine in part 4, only two (both by psychologists) made any mention, for example, of new techniques of advertising foodstuffs. The others unfolded by and large at a distance from the daily realities of human eating, a tendency that resulted in good part from altered conditions in the world of science itself. The research I discuss here mostly predates the coming of big science—the sweeping changes in scientific funding, organization, and institutions that occurred in the years surrounding the Second World War.[3] But even before that point, transformations both quantitative and qualitative had altered scientific practice in fundamental ways. I noted earlier the new model for work in life science supplied by physiology with its reliance on an increasingly sophisticated laboratory apparatus. After 1900 laboratories served as the privileged sites for investigating appetite in experimental physiology (chapter 10) and psychology (chapter 11). The basic assumption of work in these settings was that ingestion and digestion were in crucial ways uniform across a broad spectrum of species and that laboratory animals could supply a model for human eating. To be sure, not every investigator of appetite in the era 1900–1950 was a laboratory scientist. Researchers in the developing field of ethology observed the eating of diverse species "in the wild" or in rapidly proliferating field stations, museums, and zoos. Ethologists often vocally contested the dominance of the laboratory, but given the high value placed on precision within science generally, the experimental setting remained preeminent.[4]

Up to 1900 physicians often opposed a narrow scientization of problems of appetite, insisting that it be viewed in relation to overall states of health and well-being. But as I explore in chapter 12, after 1900 the work of physicians also changed markedly with the rapid advance of scientific medicine and the emergence of clinical research as a new hybrid style of investigation. By the turn of the century leading medical centers required a solid grounding not only in the sciences traditionally closest to medicine—anatomy and physiology—but also in cellular pathology, bacteriology, organic chemistry, and other sciences. Then, with the impetus of the First World War, state-sponsored research institutes and medico-philanthropic foundations encouraged inquiry that joined hitherto disparate styles of clinical and scientific inquiry.[5] As scientific medicine gained steadily in prestige and resources, physicians concerned with appetite tended more and more to interpret their own observations of patients in light of claims about the governing role played in ingestion by hormones, specific parts of the brain, or chemical constituents of nutrients. This meant that in rapid succession diagnoses of troubled eating based on clinical observation of mixed psychic and bodily symptoms gave way to others emphasizing strictly somatic anomalies (glandular defects, vitamin deficiencies) that could be charted and quantified with the help of innovative instruments and methods. In the process psychic and social dimensions of troubled appetite and eating were often excluded.

Another crucial development of this era was the emergence of a new kind of practitioner, the "nutritionist," whose labors focused less on correcting somatic defects believed to underlie disturbed appetite than on preventative measures aimed at ensuring "healthy eating." Drawing on the authority of the emergent field of biochemistry, nutritionists staked strong claims to a decisive expertise in respect to food and, unlike other investigators of appetite, forthrightly asserted the indispensability of their form of inquiry not only to individuals but also to public authorities. Working in new disciplinary settings such as the "home economics" departments of American land-grant universities or in governmental institutions devoted to hygiene or public health, these experts sought to mold the eating habits of collectivities—schoolchildren, immigrants, and the poor. Nutritionists also played highly visible roles in the two world wars, exhorting civilians to do their patriotic duty by exercising control over appetite and soldiers to be content with basic nutritional fare. Such developments meant that though the nature and source of appetite were still much debated, measures for controlling it went forward with the full warrant of medical science.[6]

Despite the expanding authority of scientific medicine, the nineteenth-century contest between clinic and laboratory did not come to an end. Dissatisfaction with what German critics derided as a putatively "exact medicine" encouraged the emergence of a "psychosomatic" approach to ills that, like those of appetite, intermingled troubles of body and psyche. Psychosomatic medicine, which had disparate sources in Freudianism, "independent" schools of psychoanalysis, and assorted forms of "social medicine," sought to resolve mind-body dichotomies and to encourage attention to the patient as a "whole person." Yet the drive of psychosomatic medicine to establish a unitary approach to illness quickly foundered. Psychoanalytically oriented practitioners interpreted appetite in symbolic and psychosexual terms to the neglect of bodily realities of desiring and ingesting food, whereas those leaning toward somaticism downplayed psychic factors. By 1950, then, despite increasing awareness of the complexities of eating desires, the goal of approaching appetite as a holistic phenomenon remained largely unfulfilled. Yet widespread recognition that appetite remained enigmatic did not forestall the search for instrumental interventions—chemical, psychiatric, "behavioral"—whose outcomes could be charted whether the underlying nature (mechanisms) of appetite were understood or not.

An essential feature of the changing face of science after 1900 was the inexorable rise of the American research establishment. In the late nineteenth century American science was still subordinate to European, but by the outbreak of the First World War, American researchers had begun establishing independence from European personalities and institutions. When the war ended many American researchers cut ties with Germany, a process reflecting both lingering suspicions engendered by the conflict and the growing strength of American institutions.[7] While German-American relations in science and medicine resumed during the interwar period, the triumph of National Socialism and the ensuing transformation of intellectual and cultural life in Germany opened a rift between German investigators and their counterparts in the democracies that endured until well after the Second World War. The impact of these momentous political events was felt in the sciences of appetite as in all research domains. The rise and spread of behavioral psychology, a largely American product despite the pivotal role of Pavlov, indicates the extent to which US researchers distanced themselves from European influence. A similar bifurcation opened up in ethology, where the arguments of (mostly European) "instinctivists" and (mostly American) "anti-instinctivists" often reflected larger ideological divisions over the origins and plasticity of human difference.[8]

Political and national concerns aside, one way to think about what happened to appetite after 1900 is to consider how, after long decades of appearing as an ambiguous concept in dictionaries and textbooks or as a secondary feature of investigations focused on something else (the role of gastric juice, the rapidity of digestion), appetite emerged as a definable "object" of science, a phenomenon that could be "observed and manipulated" or, as Pavlov put it, that had a "tangible existence."[9] In Pavlov's telling, appetite eluded clarification until, with the help of new methods mobilized in his laboratory, he demonstrated that the "psychic influence" of appetite was not ancillary to digestion but the essential stimulant of the whole process.[10] Here, it seemed, was a revolution in the science and, potentially, the medicine of appetite. Within a brief time, however, British physiologists showed that pancreatic secretion, also essential to the digestive process, had nothing seemingly to do with the psyche or, indeed, with nervous stimuli but was set in motion by a chemical agent they labeled a "hormone."[11] This step laid the groundwork for decades of controversy between "nervists" and "hormonalists" over the mechanisms driving ingestion and digestion. Even more powerful in undermining Pavlov's view of appetite as a psychic force was the criticism that the term "psychic" was next to meaningless, and that what was actually at issue with appetite was a reflex action linking the stomachal nerves to the cerebral cortex.[12] Acceding to this criticism, Pavlov recast his findings about appetite in the language of reflexes and, in place of eating desires, animal behavior. In so doing, he embraced key developments in the rapidly shifting discipline of psychology, in which experimentalism and behaviorism dovetailed, and appetite, only just granted "tangible existence," was displaced by objects of study—"feeding behavior," "food intake," "ingestive activity"—that entailed externally observable activity rather than desires or psychophysiological states. While not disappearing from the language of science, in these years appetite appeared more often than not in demurring quotation marks that indicated its dubious status.

There were exceptions to this pattern. Seeking a broad explanatory framework for animal behavior, an American zoologist labeled as "appetitive behavior" an initial phase of restless desire that ended in "consummatory" activities, and experimental psychologists designated as "appetites" instinctual drives to consume essential nutrients, a claim that sparked heated arguments between instinctivists and anti-instinctivists. Most important of all, appetite figured in the work of antibehaviorists schooled in German Gestalt psychology not as an outmoded notion from the past but as an essential component of "holistic" processes of searching out, choosing, and consuming food that reflected interlinked physiological, psychological, and social influences.

By 1950, then, the study of appetite was still marked by profound differences over method, terminology, and approach, indeed over whether appetite in fact constituted an object science could grasp. Frustrated by the complexities of appetite and determined to replace this seemingly indefinable object with something concrete, some researchers urged that appetite along with hunger be banished from the lexicon of science or, at the least, redefined for strictly "operational purposes." Thus, contrary to Pavlov's expectations at 1900 and despite the emergence of innovative approaches to appetite, the question of its "tangible existence" remained an open question for many researchers at midcentury.

TEN

Psyche, Nerves, and Hormones in the Physiology of Ingestion

In the later nineteenth century the main goal of physiologists who sought to clarify the urge to eat was to identify a definite bodily cause or point of origin for hunger. By 1900 three discernible camps had emerged among students of these questions. Peripheralists, holding to a line of argument that reached back to Albrecht von Haller, attributed hunger to the painful condition of the empty stomach and to corresponding stimulation of "hunger nerves"; they continued to debate the specific nervous pathways involved as well as the character (mechanical, chemical) of the stimuli involved. Contrary to the views of peripheralists, centralists such as Moritz Schiff held that all internal impressions were transmitted to the brain—by which nervous pathways was not known—creating a conscious awareness of nutritional need and serving as goads to actions such as the decision to eat. Then in the late 1890s the Lyon investigator Joanny Roux (1866–1909) vigorously promoted a general view, emphasizing the interplay of cellular needs and reflex action.[1] Roux praised Schiff for demonstrating "in magistral fashion" the error of those who put the origin of hunger in the stomach, but he also contended that Schiff had fallen into an "error just as great" in assigning hunger a seat in a cerebral center, when in fact it took its origin "in the innumerable cells of our body."[2] Roux was deeply critical of the "old psychology" that, "based almost solely on internal observation," was concerned only with the "sensation" of hunger (409). What was important was not conscious sensation but the need experienced by a cell, a need that "by reason of its specialization it is incapable of satisfying itself," causing it to "call on other cells, using the intermediary of the nervous system." This, he urged was the origin of the "*cortical nutritive reflex*" that "*gives rise . . . as an epiphenomenon to a fact of consciousness: the sensation of hunger* in the old sense of the word." In this general view, then, the feeling of hunger was "epiphenomenal"; what mattered was nutritional insufficiency at the cellular level (413, emphasis in the original).

Whatever their commitments to these competing explanations, physiologists one and all regarded experimental methods as essential in exploring the cause of hunger, but they also, not infrequently, invoked appetite as a kind of fallback concept when laboratory results were difficult to explain.[3] Such was the case already in the 1860s when Carl Ludwig, noting that severing the vagus nerves did not cause lack of interest in food, dismissed this finding as irrelevant to efforts to localize the origin of hunger given that "*psychic* reasons" could account for an animal's eating despite the absence of the hunger sensation transmitted by the vagus nerves.[4] This argument was still approvingly cited more than a half century later by peripheralists such as the American physiologist A. J. Carlson (1875–1956), who likewise urged that "the fact that an animal eats after section of the vagi tells us nothing as to presence or absence of hunger."[5] Thus, despite the desire of investigators to discern a strictly experimental approach to the problem of hunger, the implicit problem of its relation to appetite invariably returned: because there was no way to know why a vagotomized animal would still eat, the explanation could be inferred only from knowledge—introspective knowledge—of human beings, who, as Carlson put it, "may eat from appetite alone in the absence of hunger, as in the case of eating sweets or dessert at the conclusion of a dinner" (23). Once again appetite, finding no evident place in the laboratory, slipped in by the back door.

By 1900, some physiologists thought it time to give up these debates about the locus of hunger, and to attempt an explanation that blended the diverse approaches. Such was the case with Émile Bardier (1870–1949), who in 1903 contributed an article on hunger to Charles Richet's *Dictionnaire de physiologie*.[6] Surveying competing theories of hunger, he found all of them wanting in some respect and proposed, instead, a composite view that would explain hunger as a sensation in which, simultaneously, the nervous centers are excited directly by variations in the chemical-physical composition of the blood and indirectly by an excitation whose point of departure lay in the cells (16). Bardier also explored the outcome of experiments designed to specify how the hunger sensation was transmitted, reviewing decades of work on the role of the pneumogastric nerves and experiments in which branches of the sympathetic were cut. In respect to the latter he observed that "little is known of the function of this nerve" (18), and he concluded overall that "in the current state of science, it is very difficult to identify precisely the modes of transmission of hunger" (18).

One thing that was clear was that while Richet and Bardier regarded hunger as an important topic in a dictionary of physiology, the same was not true for appetite, which had no entry.[7] Bardier sought to dispose of the problem of appetite quickly by reasserting the view upheld during the Enlightenment and the early nineteenth century that it was simply the "first degree" of hunger, the "light and pleasant sensation . . . we feel at meal-time, when the barely felt need to eat is about to be satisfied" (1). He did not acknowledge that this idea of appetite as differentiated only quantitatively from hunger had also been found wanting on many grounds.[8] In any event, his summation of views of hunger aptly indicated the overall position of hunger and appetite on the eve of Pavlov's "revolution." Hunger was the focus of physiological inquiry but much contested. Appetite was at best secondary, perhaps not an independent phenomenon at all. Pavlov was to turn these valuations on their head.

Appetite and "Psychic Secretion": Pavlov's Interrupted Revolution

Although unremarked by Bardier, by the time his entry on hunger was published in 1903, developments were under way in the laboratory of the Russian physiologist Ivan Pavlov (1849–1936) that were to change the whole discussion of hunger and appetite, first by not only admitting but also privileging the "psychic" in explanations of the digestive process, and then, just as quickly, transforming the study of appetite as a state of mind into study of observable feeding behavior. Throughout the 1890s Pavlov, who had been trained both in Russia and Germany, directed work by a team of researchers that investigated the physiology of the digestive system with particular interest in the nature and cause of digestive secretions. Pavlov's "dog-technology" entailed isolation of a portion of the stomach encompassing the passageway of the vagus nerve, the "miniature" stomach, from the "main" stomach so that the organ's secretory activity could be studied without contamination by the presence of food.

7 Dogs harnessed for production of gastric juice in the Physiology Department, Imperial Institute of Experimental Medicine, St. Petersburg. Photograph, 1904. Pavlov and his associates confirmed the mid-nineteenth-century observation that animals sometimes secreted gastric juice at the mere sight of food. Wellcome Collection.

The general goal was to disprove the widely held view that "central nervous mechanisms played no role in gastric secretion."[9] Of the many results described by Pavlov and his associates, the most telling was that gastric secretion could be triggered in dogs merely by showing food to the animals, thus proving that "psychic excitation"—"the interest aroused by the sight of food, and the feeling of satisfaction and contentment derived from having it"—was at work, and that, indeed, "the greater the desire . . . the more certain and intense is the secretory effect."[10] The production of what he termed "appetite-juice" was indisputably under central nervous control, a conclusion Pavlov drew from vagotomies performed on his dogs, which showed that "all psychic influence" was absent when the nervous connections between brain and stomach were severed (112).

Similarly striking was Pavlov's observation that ingesting particular substances caused particular levels and types of secretions, a result that encouraged him to postulate the working of what he called the "mind of the glands," purposive "choices" made by the glandular bodies to secrete in appropriate amounts what was needed to deal beneficially with ingesta.[11] One memorable experiment on the salivary glands, for example, tested what occurred when first rocks and then sand were placed in a dog's mouth. With rocks, there was no salivation, since all the dog needed to do was to spit them out, but the presence of sand caused salivation to facilitate the washing down of tiny grains that remained after most of the sand was expelled. As Todes characterizes Pavlov's conclusion, only "an acute judgment rendered by the psyche" could account for such a difference.[12]

In this seminal work on digestion, Pavlov emphasized the results of the extensive experiments undertaken by his team, but he was also concerned to draw out practical applications for medicine of this new understanding of the essential role appetite played in the digestive process and in overall health. Crediting "human instinct" as a guide to "unconscious adoption of the most favorable conditions for life," Pavlov suggested that "the old and empirical requirement, that food should be eaten with attention and enjoyment" should be heeded more seriously. Noting that in "every land . . . care has always been taken to direct attention to the food [and] . . . to excite an appetite," he proposed that "every other form of eating, eating according to prescription or from persuasion, soon becomes worse than useless" (220–21). Couching his conclusions in language Todes sees as designed to "flatter" physicians by hailing their ancient clinical wisdom, Pavlov observed that knowing how to stimulate the appetite of an ailing patient had always been an important part of the physician's art but that present-day medicine wrongfully slighted the appetite. This was regrettable because "no other excitant of gastric secretion can compare, so far as quantity and quality of the juice are concerned, with the stimulus of a desire for food" (222). Although Pavlov was deeply committed to laboratory experimentation, on this matter he attributed physicians' disregard of appetite precisely to the supremacy of the experimental method: "In the pathology and clinical treatment of digestion, assistance was sought in the laboratory, but nothing was there met with relating to appetite, and consequently this factor was overlooked in medical practice" (222).

With such formulations Pavlov celebrated the importance of appetite in a way not seen in European physiology or medicine since the days of Erasmus Darwin. Sidestepping the results of decades of research on chemically determined nutritional requirements, he insisted that appetite was essential to healthy eating and that any dietary prescription that neglected appetite ignored precisely that component of ingestion and digestion that assured both enjoyment and health.[13] All the same, Pavlov admitted that appetite was a complex phenomenon whose functioning had never been satisfactorily illuminated. Acknowledging arguments that the primary phenomenon in the drive to eat was not desire but the depletion of body reserves, he considered the difference between "the conscious longing for food, that which is called 'appetite,'" and "the latent need of the organism for nourishment, the lack of nutrition, which has not yet been transformed into a concrete desire" (110). This distinction was apparent in dogs that underwent "sham feeding": although the "necessity for food exists in such cases, even before the experiment," the "juice . . . only begins to flow as soon as this need has taken the form of a distinct longing" (110). Here Pavlov touched on the difficulty that caused him, eventually, to reject the concept of appetite as a purely subjective phenomenon unknowable from the outside and thus inadmissible in analyzing experimental results with animals. So long as he equated appetite to the secretion of gastric juice, Pavlov remained on solid experimental ground; how to know or measure "distinct longing" was a different matter altogether. Pavlov's resort to an unexamined psychological language marked other crucial moments in his experimental endeavors as well. When he laid out the conditions essential to effecting "'psychic excitation of the gastric secretion,'" he first listed basic physical conditions—the dog must be healthy, time must have passed since its last meal—but then acknowledged that psychic secretion did not occur in all dogs but only in those that were "excitable" or "impressionable" and exhibited an appropriate temperament (91, 102). What the term "temperament" implied he did not consider, although its meaning and use were at the time a subject of debate among psychologists.[14]

Pavlov expended much creative energy in attempting to avoid experimental difficulties caused by the psychic factor, but he did not seek to define the psychological terms that he used or to say how an investigator could know that his experimental animals were possessed of appetite. In practice Pavlov and his team judged an animal's desire to eat by its observable motor activity (what "the dog most keenly desires—that is, judging from its efforts to obtain it," 86), but this judgment was unsystematic and, as Todes observes, did not always correspond to quantified secretory responses.[15] Ultimately, there would seem to be little difference between Pavlov's casual slipping in of psychological states and terminology to account for anomalous results and the outcome of Luigi Luciani's painstaking experimentation on the hunger artist Succi, when he turned to idiosyncrasies of character or the indefinable power of autosuggestion to explain Succi's remaining healthy despite depletion of his physiological reserves.[16]

The precise train of thinking that led Pavlov to abandon both the terminology and concept of appetite is not known, but in any event, his reversal was dramatic when he renounced the psychic interpretation of glandular secretion in favor of what one collaborator urged was "a reflex . . . entirely accomplished in the subcortical region of the brain," another "reflex at a distance," and that Pavlov himself later renamed the "conditional reflex."[17] Todes has shown that Pavlov gave up the psychic interpretation only reluctantly but eventually yielded to critics both inside and outside his circle who urged that his findings, far from demonstrating the role of the psyche in digestion, in fact lent support to the increasingly refined understanding of reflexes gained in recent decades.[18] After this turn to the language of reflex, Todes observes, "Pavlov instituted a system of fines [in his laboratory] to discourage lapses into the subjective lexicon of previous years" (244). By 1910 Pavlov had come to advocate the existence of a "food center" in the brain that was stimulated not by a state of desire but by "the chemical composition of the blood of the animal which has not eaten for several hours," blood that had acquired " 'hungry' properties." Continuing to argue against local or peripheral explanations of hunger, he urged that cutting "nerves leading from the gastro-intestinal tract" had never been seen to cause a "disappearance of the positive movement reaction of the animal to food, or, using the usual terminology, a loss of appetite."[19] Although Pavlov affirmed that in human beings the food center produced "sensations of appetite and of hunger" and that "appetite in man . . . is an indisputable fact," in respect to the all-important experimental findings with animals, he abandoned the concept of appetite for a formulation—"the positive movement reaction of the animal to food"—that aligned him with the cluster of methods known as behaviorism, which came to dominate early twentieth-century psychology and for a time crowded out other ways of thinking about the urge to eat.[20]

Pavlov's evolving views on the nature of nervous activity involved in ingestion and digestion were important in his decision to abandon study of these functions, but it seems the crucial factor was the rise of the "new humoralism," with the discovery in 1902 by the British investigators William Bayliss (1860–1924) and Ernest Starling (1866–1927) of the substance called "secretin," a "chemical messenger" that stimulated pancreatic secretion. From this point forward, a strictly nervist view of appetite and hunger, ingestion and digestion was untenable, and Pavlov, rather than moving to retool in fundamental ways, left the study of digestion behind in favor of intensified work on nervous reflexes.[21]

Overall, then, Pavlov's pronouncements on appetite, which at first seemed to constitute a dramatic shift in thinking about the significance of this psychic factor, had a mixed impact. Certainly his work and name quickly came to have wide currency. Although his book on digestion was written originally in Russian, a language rarely read in scientific circles elsewhere, it quickly appeared in German, French, and English translations.[22] And, of course, once Pavlov won the Nobel Prize in 1904 for his corpus of work on digestion this high accolade made knowledge of his contributions de rigueur for anyone claiming expertise on the subject. Todes shows that scientists in both Europe and the United States greeted Pavlov's work enthusiastically. Standard textbooks of physiology incorporated Pavlov's findings, and even those physiologists who disagreed with his specific claims regarded his methods, especially the new "dog-technology," as standard laboratory procedures.[23] As I show later, the medical response to Pavlov was more ambivalent, sometimes even hostile. In any event, the impact of Pavlov's claims about the centrality of appetite—whether conceived as psychic or reflex in character—was undercut within a few short years by findings that led to a radical reconceptualization of the causes of internal secretions and the influence of newly named "hormones," and to decades of contestation between nervous and hormonal theories of ingestion and digestion.

Nerves versus Hormones

Roughly at the same time that Pavlov postulated psychic and nervous stimulation of digestive secretion, the British physiologists Bayliss and Starling ushered in the era of the "new humoralism." Work by researchers in Pavlov's laboratory and others who hoped to specify the nervous pathways for pancreatic secretion had so far proved inconclusive, and Bayliss and Starling continued work on the problem by testing the reaction to chemical excitants at specific intestinal loci (a "loop of jejunum") where nervous connections had been severed. The result was proof that pancreatic secretion responded not to nervous but to chemical stimulation, "an entirely different order of phenomena."[24] In a 1905 address to the Royal College of Physicians, Starling drew out the implications of the discovery of secretin, emphasizing its role, and that of other such substances that would doubtless come to light, as "chemical messengers."[25] Although originally insistent that neural influence was not involved in pancreatic secretion, Bayliss and Starling were impressed by a later (1912) experimental demonstration by Pavlov's student G. V. Anrep (1891–1955) that stimulating the vagus could cause secretion of pancreatic juice. In subsequent years Bayliss and Starling seem to have diverged in assessing the relative importance of nerves and hormones.

In his 1915 textbook of physiology, Bayliss emphasized that hormones enabled "a chemical correlation of the functions of the organism to be brought about through the blood side by side with that which is the function of the nervous system."[26] And in his wartime manual on nutritional science, he paid tribute to Pavlov for his proofs that "food should be approached with *appetite,* the stimulus being conveyed through the nervous system" and echoed Pavlov's early claim that "without appetite, scarcely any gastric juice is formed."[27] Starling, however, remained an energetic advocate of the "doctrine of internal secretions," believing it, as Diana Long Hall states, "because it is too wonderful not to be true."[28] All the same, from the perspective of nervists, an era had ended with the identification of secretin, and the role that the nerves had played as the essential or exclusive agent in integrating bodily processes (what T. H. Huxley had called the "headquarters and field telegraph" of the organism) was at an end.[29]

Thanks in part to Starling's vigorous advocacy, researchers around the world set about the business of mapping the nature and role of hormones, only in a few instances seeking to integrate into a unified approach the mixed influence of nervous and hormonal action in physiological and pathological processes. The conceptualization of hormones opened up new dimensions of appetite-hunger inquiry, and indeed the next prolonged controversy in this history unfolded, from roughly 1900 to 1940, between investigators who attributed the urge to eat to the influence of hormones and those who continued to seek an understanding of appetite and hunger via the nervous system. This debate first centered on the problem of the origin of obesity. Aside from its intrinsic interest in the history of appetite, this line of investigation shows a kind of cooperation between physicians and physiologists that had been rare in the past but became increasingly important after 1900 as some physicians embraced new ideals of clinical research.[30]

Experimental work on the role of hormones in causing obesity got under way first as a response to clinical observations. Already in the mid-nineteenth century, a German physician had linked obesity to a pituitary tumor in the case of a fifty-seven-year-old woman.[31] Other such observations followed until, a half century later, the French neurologist Joseph Babinski (1857–1932) and the Austrian clinician Alfred Fröhlich (1871–1953) more or less simultaneously related cases of the syndrome now called adiposogenital dystrophy, in which obesity and sexual underdevelopment were prominent symptoms, to tumors of the pituitary gland.[32] Babinski's observation was brief and tentative, but Fröhlich, describing the case of a fifteen-year-old boy, confidently attributed his condition to a pituitary tumor discovered in autopsy. Such results spurred interest in the possibility that obesity, for which straightforward physical etiologies had been sought for some time, was primarily a glandular disorder. Partisans of the hormonal view included the Vienna pathologist Artur Biedl (1869–1933), who in 1910 published the most extensive survey to date of glandular secretions.[33] His chief interest in examining hypophyseal (pituitary) disorder was acromegaly and the associated condition of gigantism, but he also took up the problem of the glandular origin of obesity. He began his survey with "Fröhlich's type," noting the genital "hypoplasia" that had figured importantly in Fröhlich's account but presenting as "the most striking symptom" of the disease the obesity that sometimes attained "enormous dimensions" (345). Biedl readily acknowledged "the scantiness of our knowledge concerning the function of the hypophysis" and lamented that experiments undertaken with animals to determine "the scope of the hypophysal function" (348) had met with little success. Overall, however, despite the lack of any clearly defined "clinical entity" (348) or detailed knowledge of pituitary function, he concluded that the most persuasive evidence pointed to a glandular origin of the conditions under review. He ended by proposing a "pluriglandular" hypothesis, suggesting that what might well be at issue was not dysfunction of a single glandular organ but of a number of organs disturbed at once (347).

Glandular explanations of obesity received strong reinforcement from findings the American physician Harvey Cushing (1869–1939) put forward in a 1912 book on "the pituitary body."[34] Cushing argued for disturbance of the hypophysis as the origin of a range of conditions including adiposogenital dystrophy and extreme obesity originating after middle age but also, in some cases, the obverse condition of severe emaciation. He also proposed a highly significant role for the pituitary and the other "ductless glands" in "psychasthenias and neuroses."[35] In 1930 Cushing wrote to a physician in England that the pituitary "was proving to be the 'absolute monarch of so many and diverse functions in the kingdom of the body.'"[36] Obesity, it seemed, was under the sway of this glandular "monarch," although Cushing did not specify whether it resulted from disturbed eating or some kind of metabolic dysfunction. While noting the "large" appetite, "great liking for sweets," and "hunger for starches and sugars" seen in some patients, he did not emphasize these symptoms over the many others observed in these cases (31, 74, 263).

Despite the backing influential figures gave to hormonal explanations of obesity, almost from the moment that researchers began asserting the determining role of the glands others argued, instead, that damage to the base of the brain was responsible for the symptoms observed in Fröhlich's syndrome and related pathologies. In regard to the disturbed eating that featured in these conditions, this was not the first time that researchers had pointed to a specific locale in the brain as the source of the trouble. Early in the nineteenth century the phrenologist J. G. Spurzheim had claimed to know the brain site for the "desire to feed," and incidental observations made later in the century suggested that damage to specific parts of the brain resulted, at least temporarily, in disordering of the appetite. The neurologist and advocate of cerebral localization David Ferrier (1843–1928) recounted in the first edition of his celebrated work *The Functions of the Brain* (1876) that when he destroyed the occipital lobes in experimental animals the "appetite for food is abolished," although he was perplexed by one exception, an animal that after some days of refusing to eat "began to eat with great relish."[37] Then in the late 1890s the physician Stephen Paget (1855–1926) cited cases of "voracious hunger" associated with brain disease or injury, such as a boy who suffered a cerebral abscess in the "left temporo-sphenoidal region" and after trephining had a "ravenous" appetite.[38] After 1900 other researchers concentrated on the base of the brain; in 1904, the Viennese pathologist Jakob Erdheim (1874–1937) observed that tumors in this region sometimes caused obesity but in other cases caused "its opposite—lack of appetite and undernourishment."[39] Other researchers in widely dispersed locales pursued work on a little understood area of the brain "just over the pituitary" and "known to anatomists as the hypothalamus" that seemed to be "a very important center, intimately connected with the entire visceral nervous system."[40] Although this early work targeted such phenomena as dilatation of the pupil and heat regulation, the hypothalamus was soon to emerge as the searched-for center of appetite and hunger.

Thus, despite the pull of the new humoralism, by 1920 it was widely argued that lesions of the base of the brain were principally responsible for ills that hormonalists attributed to the pituitary. Key entries in the continuing debate included the 1921 report of Percival Bailey (1892–1973) and Frédéric Bremer (1892–1982) that they had succeeded in producing experimental obesity by stabbing a probe into the hypothalamus while leaving the pituitary undisturbed.[41] A year later the French neurologists Jean Camus (1872–1924) and Gustave Roussy (1874–1948) lauded these results and affirmed that "it is to lesions of the base of the brain that [we] attribute most of the symptoms and syndromes generally attributed to the hypophysis."[42] Camus and Roussy asserted that conflicting results achieved by different researchers chiefly reflected differences in method, noting that some had approached the hypophysis through the mouth, whereas Bailey and Bremer had developed a much more reliable temporal route. They drew attention to the extreme difficulty of operating on the pituitary without damaging neighboring parts in the brain and, similarly, of causing irritation to the cerebral base while leaving the pituitary intact. They then presented the results of their own experimental ablations performed on dogs, cats, and other animals, considering a range of resultant disturbances, including diabetes insipidus, genital dysfunction, and acromegaly, as well as the obesity seen in the "adiposo-genital syndrome" (626–29). Ultimately, after reviewing pertinent anatomical structures, microscopic analyses of ablated tissues, and "anatomo-clinical facts," Camus and Roussy rejected the pituitary hypothesis and proposed a renaming of the syndromes observed as "syndromes infundibulo-tubériens" to indicate the brain structures responsible (637).

Viewed from the perspective of evolving thinking about appetite and eating, it is notable how little any of this work focused specifically on ingestion as an element of the obesity that was a crucial symptom in the pathologies studied. Beginning with Fröhlich himself, these researchers paid attention to weight gain but not to the reasons for it, and they said little about eating desires or habits. This inattention to the appetite for food contrasted markedly with the interest researchers showed in the question of whether damage to the pituitary or the base of the brain disturbed the sexual appetite. Camus and Roussy, for example, noted continuing "genital activity" of dogs whose pituitary was removed but "suppression of the genital appetite" when lesions of the "base of the brain" were experimentally produced (628). When it came to obesity, however, they reported simply that such lesions caused animals to gain weight and put on fat (629). This fact not only aligns with the long-term tendency of many researchers to neglect the issue of how obesity was linked to excessive eating; it also means that the glandular-nervous controversy unfolded for some time without direct pertinence to the complex of problems surrounding appetite and hunger.

This situation changed dramatically with the intervention into the debate of American researchers who, beginning in the 1930s, explored the role of the hypothalamus in "regulating" eating and contributing, when disturbed, to pathologies of both over- and undereating. In this work a leading investigator was John R. Brobeck (1914–2009), who first worked on this subject in the laboratory of John F. Fulton (1899–1960) at the Yale Medical School; in the early 1930s Fulton had published articles suggesting that in monkeys removal of the frontal lobes produced "ravenous appetite."[43] Brobeck's first major entry into the debate over obesity was an article in which he and his collaborators asserted that Fröhlich's glandular theory was "now completely discredited and that controversy on the matter had ceased when a report appeared showing that "no obesity followed lesions restricted to the hypophysis of rats, although obesity appeared in rats with concomitant hypophyseal and hypothalamic damage."[44] According to this account, the decisive moment had arrived when Northwestern University researchers made use of a Horsley-Clarke frame (an instrument used to pinpoint locations for surgical lesions in the brain) to produce "hypothalamic obesity in rats in which, for the first time, direct hypophysial involvement was avoided."[45] Unlike previous studies, that of Brobeck and his collaborators focused especially on overeating as the cause of obesity, noting findings of an association between lesions of the hypothalamus and " 'enhanced appetite.' "[46] Using a slightly altered Horsley-Clarke instrument to produce hypothalamic lesions in female rats, Brobeck's group measured the animals' "food intake," documenting "ravenous eating behavior" and concluding that the marked increase in body weight resulted from the animals' "voracity" (833–35). Autopsy data confirmed that "many of the animals were almost incredibly obese . . . undoubtedly a result of their hyperphagia" (845). Disputing alternate theories that pinpointed not overeating but metabolic disturbance, Brobeck argued that the rats "appeared to be ravenously hungry" after the operation, "ate quickly and greedily," and evidently experienced "increased appetite" (848–49).

Brobeck continued work in this vein for some time and eventually came to be the chief American spokesman for the view that "hypothalamic mechanisms" were responsible for the regulation of eating behavior.[47] His work indicates a number of key shifts in the conceptualization of appetite, most importantly, perhaps, in his embrace of a behaviorist perspective that was intended to eliminate all subjective elements; he used terms such as appetite and satiety, he said, "in a purely behavioral sense, without prejudice as to their subjective or psychological nature."[48] Ordinarily Brobeck stated results in quantitative terms, resorting to what he seemingly regarded as an outmoded language of desire (rats that were "ravenously hungry") only when arguing generally for a relation between obesity and overeating. Such expressions of excess stood in sharp contrast to what was, for Brobeck, as for physiology generally by this point, the most important term he employed—"regulation"—to signify an automatic operation dependent on specific, well-calibrated bodily activities or devices.[49] Thus, Brobeck's hypothalamic conception, while targeting eating as one of a large number of functions under the specific control of a part of the brain, left unacknowledged the problem of the relation between activities governed by the brain but marked by greater or lesser degrees of automaticity and those propelled by sometimes violently experienced states of emotion or desire. In any event, thanks in good part to the work of Brobeck and his collaborators, the long-standing controversy dividing hormonalists and nervists seemed to many to be settled in favor of those who argued for localized "regulation of eating" in the hypothalamus and against a role for hormonal secretions. But the wheel was to turn yet again.

Appetite as "Wisdom of the Body": Walter B. Cannon

While physiologists were often sharply divided between those who favored either hormones or nerves as the source of eating proclivities and patterns, already by the 1920s some researchers sought to link the two systems into an integrated physiological whole. The Harvard physiologist Walter B. Cannon (1871–1945) was central to these efforts. In his early work on digestion, Cannon focused chiefly on "mechanical factors" but from around 1905 he began searching for the underlying causes of gastric movements and, in so doing, linked investigation of the hormonal and nervous mechanisms responsible for both ingestion and digestion.[50]

Cannon traced his interest in digestion, which he first studied in the late 1890s while a medical student at Harvard, to a teacher's suggestion that he use the "newly discovered x-rays" to observe the process of swallowing in experimental animals.[51] Working with different collaborators, Cannon began a series of experiments on the mechanics of digestion that continued for a number of years. He first studied swallowing by feeding animals food mixed with the inert powder subnitrate of bismuth and watching the movements of deglutition. Then Cannon turned to other movements of the alimentary canal, work that prompted him to look further at gastric peristalsis, and, in turn, local reflex control of this and other stomach and intestinal movements.[52] After observing the interruption of peristaltic activity in an enraged cat, Cannon became interested in the effect of strong emotion on digestive activity and this problem led him to what was to be one of the central strands of his life's work: the impact of emotions on physiological function, including the broad spectrum of activities associated with eating and digestion. Cannon first investigated this issue in relation to the functioning of the adrenal glands, specifically the production and impact of adrenaline, but in time he came to recognize the mixed action of the adrenals and the sympathetic division of the autonomic nervous system and referred the observed effects to what he began calling the "sympathico-adrenal" system.[53]

In 1912 Cannon summarized the results of his work on digestion in a Harvey Society lecture on hunger.[54] He first discussed the role of the "automatic devices" responsible for physiological regulation, which, in respect to ingestion, meant the "experiences of appetite and hunger" (130). Criticizing physiologists who confused the two, he urged that appetite was a "fundamentally different" experience from hunger, one "related to previous sensations of taste and smell of food" and therefore with "important psychic elements." In contrast, hunger was "unpleasant," often accompanied by lassitude, headache, or restlessness (131–32). In this address Cannon chose to examine only hunger (he did not say why) and asserted that, despite much theorizing, "direct proof" was needed in regard to its origin (145). He divided past theories of hunger into two groups—the "general," which he attributed to Moritz Schiff and presented as the dominant view among physiologists and psychologists, and the "local," upheld by a minority (133). Although in an early study Cannon had suggested that hunger expressed a general need of the body, by this point he had become convinced of the local view.[55] Against Schiff's assertion that hunger increases constantly as bodily need intensifies, he adduced the evidence of the hunger artist Succi, who had testified that his hunger pains disappeared after the early days of a fast (134).

After surveying other nineteenth-century arguments, Cannon turned to his own experiments on hunger, undertaken in 1911 with his student Arthur Washburn.[56] He recounted how he had urged Washburn to habituate himself to the presence of a tube extending from mouth to stomach so that they could observe the relation between Washburn's experience of hunger pangs and stomach contractions registered by means of a rubber ball placed at the end of the tube and connected to a kymograph. Washburn was seated in such a way that he did not see the recording apparatus, and the two men were delighted to observe that his expressions of hunger were precisely synchronized with the contractions recorded. Cannon noted that he was by no means the first to argue that hunger resulted from stomach contractions, yet he stated that no one had supplied "direct proof" of the kind he provided or definitively shown, as he did, that hunger was local in character and caused specifically by stomach contractions (145–49).

In light of Cannon's subsequent emphasis on "the wisdom of the body"—already in this lecture of 1912 he commented on the importance of "automatic devices" (130) that serve bodily need—it is curious that at this point Cannon acknowledged that his investigations cast no light on the cause of stomach contractions and declared that "they do not seem to be directly related to bodily need." Rather, he argued, "for present considerations . . . it is enough that they do occur, and that they are abolished when food, which satisfies bodily need, is taken into the stomach. By such indirection . . . are performed some of the most fundamental of the bodily functions" (149). Nonetheless, commenting on potential clinical applications of his findings, he did argue that understanding the link between hunger and stomach contractions clarified the nature of a disease such as bulimia ("the inordinate hunger . . . of certain neurotics," 150), which accorded with the "well-known disturbances of the tonic innervation of the alimentary canal in such individuals" (150). By an "indirection" of his own, then, Cannon suggested that the state of the bulimic's appetite was dependent on the healthful or deficient "tone" of nerves of alimentation.

After 1911, Cannon undertook no further investigations into the nature of hunger, moving instead to further work on the role of the emotions in physiological function. For some time Cannon devoted his research to what he came to call the "emergency theory" of adrenal function, the famous "fight or flight" mechanism enabled by the flow of adrenaline. In his early formulation of the emergency theory, he emphasized the specific action of the adrenal medulla, but after a time he came to postulate an intimate connection between glands and nerves of the sympathico-adrenal system.[57] This focus on nervous-glandular interrelations was connected to Cannon's developing interest in the processes constituting "homeostasis," a term he first used publicly in 1926. Stephen J. Cross and William R. Albury have linked this element of Cannon's thinking to his overall "Hippocratic" orientation, his committed belief in the *vis medicatrix naturae* (healing power of nature), and specifically to the revival in these years of the thinking of Claude Bernard.[58] Cannon's theory of homeostasis—his conviction that the internal environment of the body, Bernard's *milieu intérieur*, was regulated or held constant by automatic mechanisms—was to supply perhaps the most credible argument available to researchers who held that the living organism "knew" innately what nutritional substances it needed and naturally moved to secure them.[59]

In the later phases of his career, Cannon continued to be interested in digestion, exploring both the nutritional needs of living organisms and, especially, the impact of emotions on the digestive process. He revisited these issues when he was invited by colleagues in Detroit to deliver the Beaumont lecture for 1933. Cannon had long been an admirer of Beaumont, whose observation of Alexis St. Martin had taken place at Fort Crawford, the site of the modern Wisconsin town of Prairie du Chien, Cannon's own birthplace. In his talk Cannon explored (as Beaumont had long before) not only the mechanisms of appetite, hunger, thirst, and digestion but also larger themes such as the relation between good digestion and "bodily vigor" and between indigestion and states of "pain, worry and excitement."[60] Opening his talk, he echoed his 1912 lecture, referring to the "highly efficient devices," including appetite and hunger, by which "we are led automatically . . . to replenish our reserves" (19–20). Here again, Cannon insisted that appetite and hunger were distinct states, appetite arising "from the experience of previous pleasures" and leaving "a trace in memory . . . that invites repetition" while hunger was "a primitive, elemental sensation" that does not "invite men to eat . . . [but] drives them to do so" (21). He observed further that hunger and appetite differed "in their sensory quality, in their bodily reference, in their relation to experience and in their effects on behavior" (22).

Despite this introduction on both appetite and hunger, Cannon once again mainly treated hunger. At first glance the subsequent discussion showed few changes from Cannon's 1912 lecture. Yet Cannon's years of reflection on homeostasis in fact left their mark on this late-career exposition of hunger. One significant difference was evident in his choice on this occasion to compare the experience of hunger to that of thirst. Noting that the two had common features—"both are nagging and disagreeable experiences, associated with powerful impulses, the so-called 'drives,' which impel us to act in such ways as to obtain relief" (46)—he observed of thirst that physiologists had been misguided in this case as with hunger in denying its local character (57). Yet at the same time, in contrast to his earlier dissociation of stomach contractions from bodily need, he now stated his confirmed sense of bodily "wisdom": "The organism manages its necessary affairs automatically," acting in such a way that "when needs arise annoying or distressing sensations appear" and, in eliminating these sensations, in satisfying "real hunger and real thirst," bringing "pleasures so keen that they almost warrant the deprivation which makes them possible!" (78). It was this viewpoint that encouraged Cannon to praise organs of the body "commonly regarded as less honorable," the stomach and intestines, to which "we [here] pay . . . tribute."[61]

Cannon also demonstrated the distance he had come from 1912 when he emphasized how essential digestion was to "bodily vigor" and how thoroughly interconnected this and all other functions of the body were in establishing health. Describing the role of the nervous system in uniting the living organism, Cannon compared it to a "vast telegraph and telephone network" that "join[s] together distant regions in a large country" (82). He described the nervous system as encompassing two great divisions, one concerned with the outer world and the other with the inner world of the organism, the first voluntary and based on sounds, odors, and other sensations arriving from the external environment, the other the autonomic system that ensured the smooth functioning of the heart, digestive system, smooth muscles of the blood vessels, and abdominal organs. Surveying issues that had divided physiologists for a century and more, Cannon zeroed in on the role of the vagus nerves in transmitting the urge to eat. He argued that the results of vagotomy had long been misunderstood, especially by those who held that the operation resulted in paralysis of the esophagus and prevented swallowing. Cannon urged that these activities gradually resumed after sectioning of the vagus nerves, which, he concluded, were responsible not for motor activity but for establishing the "tone" essential to the healthy functioning of smooth muscle (88, 91–93). Tone, in turn, was a complex phenomenon governed partly by voluntary and partly by reflex activity, and perhaps most importantly, subject to habit. "It is well to remember that our bodies, including the digestive tract, become habituated to actions and processes by repeated experience" (106). Cannon thus insisted on the interconnections of voluntary and reflex activity, the workings of both the central and the autonomic nervous system, while nonetheless continuing to stress the role of automatic devices that, functioning undisturbed, ensured bodily health.

Cannon ended this discussion by looking at influences that could upset automatic and healthful activity in the body, emphasizing the influence on ingestion and digestion of emotional upsets. As always, he was less interested in what might go wrong with voluntary parts of the nervous system—the joys of taste leading to gluttony or the ill effects of people ignoring signals of the need to defecate (118–19)—than he was in troubles of the sympathetic division of the autonomic nervous system, with its high susceptibility "to a variety of conditions, among which are pain and profound emotional experience" (119). Acknowledging that digestion was at base a chemical process, Cannon nonetheless insisted that digestive troubles were not exclusively chemical in nature but dependent on "circumstances," among which the state of "appetite and hunger" were essential: "If appetite and hunger are lacking, apathy instead of interest characterizes our attitude towards the viands of the table" (153). Once again, Cannon paired appetite and hunger, but here he stressed not the elemental, primitive sensation of hunger but appetite: "a hearty appetite, gusto and pleasure in eating, are highly advantageous

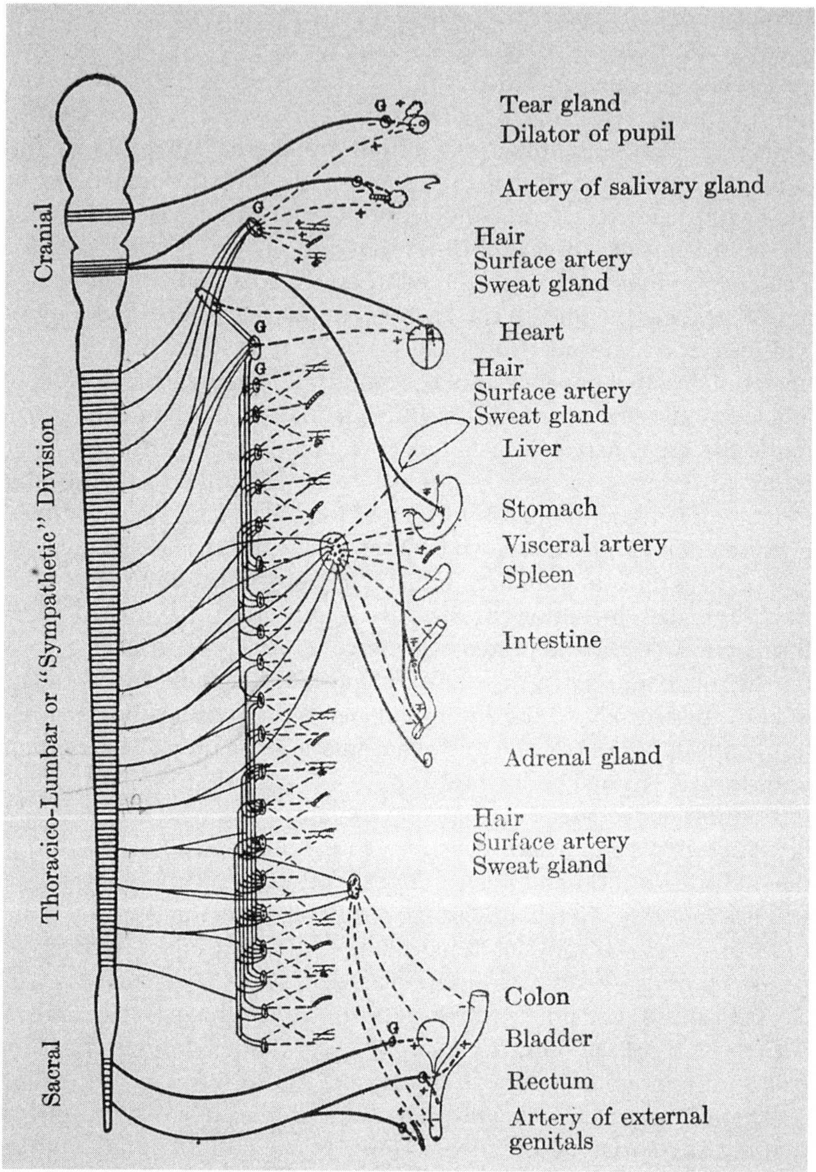

8 Walter B. Cannon, *Bodily Changes in Pain, Hunger, Fear and Rage* (New York: Appleton, 1915), "Diagram of the More Important Distributions of the Autonomic Nervous System." Cannon was less interested in voluntary eating choices than in the "automatic devices," nervous and hormonal, that regulated appetite, hunger, and digestion. US National Library of Medicine.

conditions. They are typical of abundant bodily vigor and they contribute to perpetuation of that vigor" (153–54). Thus, in the end, appetite was primary; what was enjoyed and remembered in eating established the "tonic state" that allowed the involuntary "devices" to do their work. Always concerned to show the therapeutic value of his work, Cannon took on the matter of what practical advice derived from these insights into digestive function. Strikingly, in this context, he reverted to the introspection of the self-observing physician: his own experience allowed him to testify that taking "a rational attitude toward any exciting incident" made it possible to minimize the "baneful influence" of emotions and, if all else failed, "hard physical labor" could "work off" the "excited state" that might disturb the essential function of digestion (147–48). Appetite, he seemed finally to say, might be a trusted, "automatic" regulator of ingestion, but rational choice and voluntary action must also be called on to ensure the healthful functioning of the eating and digesting human organism.

Cannon enjoyed a long friendship with Ivan Pavlov, and from his earliest work on eating and digestion, he assumed as given Pavlov's discovery that appetite was "psychic" in character and that health demanded pleasure in eating.[62] It is striking, however, that long after Pavlov himself had abjured the term "psychic" in favor of a strict language of reflex activity, Cannon continued to use the term in untroubled fashion and to insist on the intimate interconnections between the psychic and the visceral, the central and the vegetative, the pleasures of appetite and the "primitive" force of hunger. That he adopted this stance should occasion little surprise: his holism, his commitment to the power of the Hippocratic *vis medicatrix naturae* rested at base on his conviction that nature undisturbed operated beneficently. Nonetheless, the precise character of the relation between appetite and hunger, pleasure and need, was something Cannon was ultimately ill equipped to explore, a fact he himself acknowledged when he ventured, tentatively, into the ever more divided and specialized terrain of the psychic expert.[63] And while his theory of the emotions was well received by physiologists, psychologists found much to criticize in Cannon's ventures into what he himself characterized as "the twilight zone where physiology and psychology, pathology and psychiatry are likely to meet."[64] It is telling that despite his frequent suggestion that solid knowledge could rest only on "direct proof," when he spoke of appetite Cannon recited well-known proverbs (from John Ray, "New dishes beget new appetites") and adduced evidence from the Bible and Shakespeare attesting to "the extreme sensitiveness of the stomach and intestines when emotions are involved."[65] Whatever his limitations in this regard, however, Cannon did more perhaps than any other physiologist of his era both to call into question investigations of eating and digestion that concentrated on isolated structures, whether nerves or glands, and to encourage the emergence of approaches that bridged old gaps. The consequences for the study of appetite were to be far reaching, not only in the "pure science" of the drive to eat but also in the "psychosomatic" medicine that Cannon did much to encourage.[66]

Regulating Food Intake

A decade after Cannon's death in 1945, the drive to integrate "regulatory mechanisms" governing the urge to eat inspired a multidisciplinary conference sponsored by the New York Academy of Sciences devoted to the "regulation of hunger and appetite."[67] At this meeting physiologists outnumbered representatives of other fields, but physicians, psychiatrists, toxicologists, and even one expert in animal husbandry also made presentations. Despite the diversity of subject matter and theoretical preoccupation, participants largely shared the view that investigation of hunger and appetite meant study of ingestive activity rather than psychic states, however these might be defined, and that the intake of food reflected the work of specific physical mechanisms. The mechanisms considered included the much-discussed "glucostatic hypothesis" according to which ingestion was governed by levels of sugar in the blood as registered by glucose receptors located "possibly in the hypothalamic centers . . . [and] perhaps peripherally as well," along with refinements of the long-standing view that stomach contractions caused the hunger sensation that prompted ingestion while gastric distention brought satiety and caused eating to cease. One speaker saw these gastric mechanisms working hand in hand with the "conditioned reflex" of appetite ("founded on the learning or memory of" the pleasures of eating), and another focused on food "cravings and aversions," seemingly strange forms of eating behavior—cows that ate bones, chickens that consumed eggshells—that were clear responses to deficits of essential nutrients.[68]

Yet while some speakers pointed to specific mechanisms, most participants emphasized the extraordinary complexity of regulatory interactions. One physiologist noted the underlying role of "driving forces or urges" while stating that the regulation of ingestion resulted from "an integration of forces derived from special movements of the digestive tract, hormone influences, emotions, impulses from the cerebral cortex, from the hypothalamus, and perhaps from the entire body" (6). Two other speakers included a still more complex range of factors: "environmental temperature, the integrity of several hypothalamic structures, oropharyngeal stimuli, gastrointestinal distention, 'emotional' and other conditioned responses, energy requirements, and perhaps circulating metabolites, as well as genetically determined metabolic pathways."[69] Despite the overall emphasis of the meeting on physical phenomena, emotional factors also entered in, with physicians arguing, for example, that obese people ate excessively in response to depression, anxiety, or a sense of hopelessness. Finally, evidence from history and ethnography found a place: eating choices, urged one physician, often reflected seemingly irrational and unhealthful customs, ethnic preferences, or conditions created by "our complicated modern civilization."[70]

Even John Brobeck, whose work throughout the 1930s and 1940s had targeted particular locales in the hypothalamus as the primary mechanism responsible for regulating food intake, had shifted to an integrationist stance that included "an important substructure of reflex actions" involved in feeding (visual, olfactory, auditory, tactile, gustatory, and "enteroceptive"), as well as complex bodily "signals" indicating the "availability of glucose in body fluids . . . , the concentration of certain metabolites . . . , water concentration" in different "compartments within the body" as well as "sensations from the upper digestive tract associated with eating, swallowing, and the presence of food in stomach and intestine."[71] He concluded that "a variety of types of disturbances and reactions within the organism and between it and its environment" appeared undoubtedly to be involved.[72] Thus Brobeck, long the most confident defender of the role of the brain and central nervous system in governing ingestion, had by this point come to recognize—indeed, to insist on—the view that the regulation of eating was "not an isolated phenomenon" but the work of a myriad of complex regulatory processes. And although he continued to emphasize the "integrating" function of the nervous system, he urged that "the study of appetite and satiety will advance as more is learned about other regulations" (53–54), thus acknowledging the limitations of the brain focus that had inspired the better part of his own decades-long labors in the field.

By 1950 physiologists had opened new lines of approach to appetite that, in some cases, were to remain significant up to the present day. Pavlov had demonstrated experimentally the once merely curious fact that the simple sight of food could trigger gastric secretion, even if the nature of the trigger remained unclear. Other physiologists had traced to the hypothalamus or to specific glandular structures significant roles in regulating appetite. Most important, by midcentury, leading investigators had come to argue that appetite could not be explained by any single mechanism but must be linked to the work of complex and intricately interrelated processes. As these developments unfolded, appetite came into clearer focus as an object of science than ever before. Yet despite such steps toward concretizing appetite, other shifts of this era had given it a more equivocal status than before. When physiologists used the term "appetite," they often did so only once they placed it, defensively, within quotation marks, or offered statements such as Brobeck's to the effect that "appetite and satiety are used here in a purely behavioral sense, without prejudice as to their subjective or psychological nature."[73] Why investigators continued to refer to appetite at all, given their uneasiness with the term, is a question of some interest. Clearly it retained rhetorical power even when it was challenged as imprecise or indefinable. Early in the century Pavlov had confidently asserted that in man "appetite is an indisputable fact" but that "when we refer to the animal world . . . , we must limit ourselves to registering and comparing the visible phenomena."[74] A half century of terminological disputes later, this picture had changed little. Introducing the issues to be covered at the 1955 symposium, the gastroenterologist Franklin Hollander lamented that the "dual term 'hunger and appetite' . . . is the classical and most widely used" and expressed the hope that a "major consequence of th[is] conference . . . will be either a redefinition of these two words in forms suitable for operational purposes, or else that they will be discarded from our technical terminology."[75] Thus, for many researchers, appetite still had only a tenuous existence as a scientific object: interest in "ingestive behavior" and "regulation of feeding" had displaced concern for inaccessible eating desires or feelings. No doubt at the dinner table every one of these researchers would cheerily have agreed with Pavlov that human beings experience what traditionally had been termed "appetite." And Walter B. Cannon was probably not alone in knowing firsthand the troubled appetite that, as doctors had always known, often accompanied states of emotional distress. Yet finding a means to explore this uncertainly labeled phenomenon—as opposed to visible, quantifiable food intake—remained as elusive as ever, even more so perhaps, given the widely recognized involvement of a multitude of micromechanisms. Hence Hollander's demand for an approach to appetite suited to "operational purposes," a challenge that was to elicit eager responses in the coming era of big science.

ELEVEN

Appetite and the Nature-Nurture Divide: Eating Behavior in Psychology and Ethology

In the late nineteenth century physiological psychologists explored internal sensations of appetite and hunger by various means, often by interrogating their own experience. By the turn of the century, however, introspection came under challenge as the early twentieth-century "revolution" in psychology that encouraged the emergence of behaviorism got under way.[1] Behaviorism in psychology was largely an American phenomenon, a response to intellectual and social currents (philosophical pragmatism, Progressivism) specific to the United States. In other national settings it was chiefly important for the critiques it evoked, including the charge that it lacked any unifying approach or theory.[2] Yet while behaviorists came in many varieties and were divided on many issues, they shared the fundamental idea that psychology had no creditable means to investigate mind or consciousness and thus must focus on external behavior.[3] In the realm of appetite and eating this meant that attention must shift away from internal sensations or states and toward observable (and hopefully quantifiable) ingestive behavior or food intake. This step had a powerful impact not only in psychology but also in all the pertinent disciplines within what came to be known as the behavioral sciences.[4]

Yet much about this behaviorist stance was problematic. While certain phenomena fell clearly enough into the realm of the mental or psychic and were easy to avoid, the status of appetite and other inner states was unclear. Not only had these phenomena traditionally been viewed as blends of the material and the immaterial; they also had been studied chiefly as causes of behavior, a vexed matter for behaviorists, who often held that questions of causation were not part of their remit. Overall behaviorists were interested in devising experiments to determine why and how laboratory "subjects" responded, in the here and now, to particular kinds of stimuli. How the individual (animal or human) got to be what it was or came to react as it did were questions, so many researchers held, outside the scope of the behaviorist enterprise.[5]

Among other consequences of this viewpoint on causation was the studied lack of interest among behaviorists in the cluster of questions surrounding the relative power of "nature" versus "nurture" to shape eating and other patterns of behavior.[6] Nonetheless, the nature-nurture contest, which in key respects carried on the instinct-intelligence debates of the late nineteenth century, was of crucial concern to researchers in diverse fields given the post-Darwinian search for mechanisms of inheritance and broad fears about the impact of "tainted" heredity and processes of "degeneration." The rediscovery in 1900 of Mendelian laws of inheritance gave new immediacy to arguments over the relative power of genetic and environmental influences in shaping body and behavior.[7]

Although behaviorists largely ignored such questions, for a variety of reasons psychologists who investigated eating behavior intervened forcefully on the question of whether ingestive choices and patterns were innately determined or established by processes collectively known as "learning." By and large it was psychologists who argued for the innateness of eating choices who, despite behaviorist skepticism about inquiry into inner states, retained appetite as an object of study. Psychological research on appetite went forward principally within "comparative psychology," a subfield that was devoted to "animal work" as opposed to human psychology and that took as its prototypical species the white rat.[8] Much of this research belonged solidly on the nature side of the nature-nurture divide. Unlike other psychologists, comparativists often worked closely with physiologists and were strongly influenced by conceptions of inborn mechanisms responsible for regulating bodily functions. Such researchers saw appetite as the psychological expression of metabolic need and they worked to identify distinct appetites for salt and other essential nutrients. They often asserted, furthermore, that inborn appetites enabled "wise" food choices not only by animals but also by human children who were too young to have registered social and cultural influences that encouraged deleterious food habits.

The nature-nurture dispute also colored much research and debate in psychology's sister science of ethology, which consolidated and expanded after 1900. As successors to the animal psychologists of the nineteenth century, ethologists largely pursued study of animal behavior outside the laboratory. Denouncing the narrow focus of psychologists on the laboratory rat, they insisted on the necessity of intensive study, in the wild, of a broad range of animal species. Only when investigators amassed intimate knowledge of the behavior of many different animals living in complex environments, they asserted, could determinations be made about the relative influence of heredity and environment. Overall a broad division emerged in these decades between Continental and Anglo-American ethologists on questions of nature-nurture, with the former arguing for innate and fixed eating patterns and the latter for the malleable influence of learning. From the 1930s to the early postwar years, nature-nurture questions became increasingly divisive as they became entangled with the violent ideological controversies engendered by the virulent racism (extreme "hereditarianism") promoted in Germany by National Socialists.[9] Yet Germany was also home to systematic efforts to get beyond nature-nurture, some of the most promising emergent from the framework of Gestalt psychology. Although not a central theme for Gestalt theorists, appetite was of keen interest to certain Gestalt researchers, who tried to point the way beyond dichotomies of bodily and psychic, inner and exterior, innate and acquired. Notwithstanding such efforts to resolve rigid divisions, however, the issue of whether determinants of "eating behavior" were inborn or acquired remained highly contentious. If only animals had been at issue, it might have been otherwise. But all appetite researchers seem to have had, at some remove, questions about human eating in mind. The great irony of the entry of psychologists into this domain was that, to understand human appetite, they mostly studied not humans but rats. In doing so they pursued with great single-mindedness the project of substituting for variable human appetite uniform "laws" of eating behavior based on a "model" animal whose "ingestive behavior" in the laboratory could scarcely have been more remote from the ordinary circumstances of human eating.

Appetite as the Behaviorist Views It

In the United States the rise of behaviorism is chiefly associated with John Broadus Watson (1878–1958), author of the famed "behaviorist manifesto" of 1913. Watson took his doctoral degree in psychology from the University of Chicago, but he was drawn less to the functionalism taught in his home department than to styles of animal experimentation favored in neurology and physiology. Watson was an early and enthusiastic defender of animal research; his dissertation entailed study of the white rat, which was fast becoming the canonical animal of comparative psychology. In 1908 Watson was named professor of experimental and comparative psychology and director of the psychological laboratory at Johns Hopkins University. It was early in his tenure there that he published the manifesto in which he urged that psychology abandon the study of mind and consciousness and become an experimental science focused strictly on behavior.[10]

In his animal research Watson built on prior work done by "America's first comparative psychologist," Edward L. Thorndike (1874–1949), who experimented with cats, dogs, chicks, and monkeys to illuminate how animals form associations and learn forms of behavior.[11] Thorndike's starting point was his hostility toward nineteenth-century inquiries into "animal intelligence" such as those offered by the British Darwinian George Romanes.[12] Thorndike denied animals the power of reasoning, urging that activities required for their survival were learned through "trial and error" and gradually formed associations. In respect to eating and other forms of behavior "in the higher animals," Thorndike asserted that "associations and consequent habits . . . not instinct, make the animal use the best feeding grounds, sleep in the same lair, [and] avoid new dangers" (21–22). To study the formation of associations, he confined animals in cages in a state of "utter hunger" and observed their attempts to get at food placed outside the cage but within their sight by performing simple acts such as pressing a lever (27–34).

Thorndike's use of food as "reward" indicated the most significant way in which "eating behavior" was to figure in much subsequent work by behaviorist psychologists. His interest was not in the origin or nature of the animal's urge to eat but in learning processes. Thorndike acknowledged that he had "no special interest in animals," an attitude evident when he discussed hunger in his experimental subjects.[13] Defending himself against accusations that he experimented "with starving or half-starved animals," he observed that his goal in inducing "utter hunger" was procedural, establishing "the motive as nearly as possible of equal strength in each experiment with any one animal with any one act" by feeding the animal, after each successful step, only "a very small bit of food." As he explained further, "'utter hunger' means that no diminution in his appetite was noted and that at the close of the experiments for the day he would still eat a hearty meal." Thus, Thorndike thought carefully about how to manage food as reward. To gauge an animal's desire to eat, however, he saw no better way than observing its consumption of "a hearty meal," a procedure in use in physiology since the early modern era.[14]

In any event, Watson built on Thorndike's work, including the latter's use of "operant" conditioning, to debunk claims that instinct guided eating and other forms of animal behavior.[15] In his early work Watson took for granted that hereditary instincts played an important role in behavior, but he moved over time to the view that complex behavior could never be ascribed to instinct. At first Watson fought conceptions of instinct simply by asserting, as Thorndike did, that animals acted instinctively only as the first step in the trial-and-error process, but he later argued that behavior was largely reflex in character and, indeed, that Pavlov's conditioned response might afford the key to all habit formation.[16] In this connection Watson had little to say about appetite or hunger but simply assumed, in the fashion of "classic behaviorism" generally, that hunger constituted one of the primitive "needs" that animals learned to satisfy.[17] Defining the term "stimulus," Watson quickly disposed of stimuli experienced on the "inside" by offering simple mechanical explanations of the prompt to eat such as the stomach contractions that were felt "just before dinner" but ceased "as soon as food was eaten."[18] Although he did not mention Cannon and Washburn by name, Watson briefly summarized their procedure as proof that the stimulus to which eating was the response (a rhetorical maneuver that avoided both the terms "appetite" and "hunger") resulted from a simple movement of the stomachal muscles.[19]

While the use of food as reward became a mainstay in the animal experiments of behaviorists, neglect of internal states as goads to action became a hallmark of the variety of behaviorism linked to Watson's name.[20] Not long after publication of his manifesto, Watson himself abandoned animal research, instead promoting behaviorism by urging its utility in human affairs, and behaviorists generally turned their attention to practical human "applications," especially in pedagogy and education.[21] The task of illuminating ingestion in accord with behaviorist principles thus fell to those who remained committed to comparative study. Watson's own ascription of eating behavior to simple reflex movements offered little to researchers interested in the complex dynamics of animal appetite and food selection. It is unsurprising, then, that of the students he helped to train at Johns Hopkins the one who first contributed significantly to the psychology of ingestion, Curt P. Richter

(1894–1988), diverged in important ways from Watson's conception of behaviorism.

"Salt Appetite": Curt P. Richter and the Instinctive Search for Nutrients

In decades of experimental work pursued at Johns Hopkins, Richter bridged the fields of psychology, physiology, and animal behavior.[22] The son of German immigrants, Richter studied for a time in Dresden; he later earned his bachelor's degree at Harvard and, in 1921, completed his doctoral dissertation in psychology under Watson's direction. After Watson left Johns Hopkins, Richter's chief mentor was Adolf Meyer (1866–1950), the Swiss-born neurologist and psychiatrist who was among the first to embrace the albino rat as the mammal most appropriate for laboratory investigation.[23] Richter followed suit with a dissertation focused on "general and specific activities" of the rat, which included extensive discussion of hunger and eating as part of the activities rats engaged in "spontaneously," or without the intervention of external stimuli.[24] Already in this work Richter argued for the view that all animals have internal "rhythms," later to be termed "biological clocks," that regulate essential bodily activities.[25]

In collaboration with colleagues at the Psychobiological Laboratory, Richter undertook a series of studies in the 1920s intended to illuminate activities in animals "when all external stimuli have been eliminated."[26] He began by monitoring all activity of the experimental animals (chiefly rats, but also guinea pigs, rabbits, and others) and observing that activity and rest occurred in approximately two-hour intervals. He sought the explanation for this fact in the periodic action of the stomach, a phenomenon he regarded as now definitively established (307, 310). A convinced partisan of the Cannon-Washburn view that hunger was caused by stomach contractions, Richter claimed to have demonstrated that rats confined in cages and given free access to food ate on a regular schedule unless interfered with in some way by changed external circumstances (311–20).

Richter used the term "hunger" without exhibiting any self-consciousness about the extensive literature on how this term was to be defined or how hunger was to be differentiated from "appetite." Indeed, his terminology suggested that he saw these two states as essentially interchangeable. Describing a rat made restless by stomach contractions, he referred first to the "general discomfort" it felt as a result of the "hunger sensation" but then

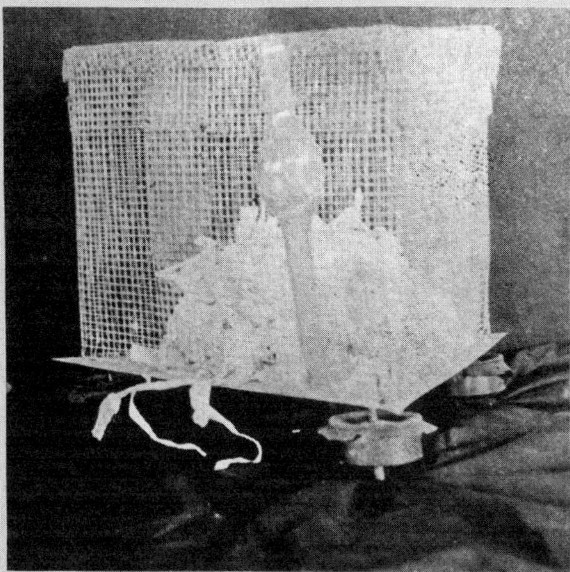

FIG. 1. ACTIVITY CAGE

9 Curt P. Richter, *A Behavioristic Study of the Activity of the Rat* (Baltimore: Williams & Wilkins, 1922). Cage designed by Richter to study "spontaneous activity" in the rat. US National Library of Medicine.

to the satisfaction of its "appetite" and its entering "a period of quiescence" (313). As Jay Schulkin has observed, Richter was a researcher characterized by a "laboratory state of mind," and he did not concern himself much with terminological issues.[27] As was to become increasingly typical of behaviorists, Richter reported his experimental results using accepted language while placing contested words such as "drive" in quotation marks.[28] In any event, the central claim of this 1927 essay was that gastric function should be regarded as a cause not just of hunger but of all bodily activity, given that his rat subjects engaged in activity of any kind only when in a hungry state. Finding it difficult to record stomach contractions in the rat using an adapted form of the Cannon-Washburn apparatus, Richter used other animals, including bullfrogs, pigeons, and even human infants, into whom, so other researchers had found, "the stomach tube could be passed quite easily."[29] Overall, his results showed a clear correlation between stomach contractions and "gross bodily activity" (316).

Although Richter stressed the effect of stomach contractions, he was already very interested in the regulative influence exerted by hormones.

He stated with great enthusiasm his finding that female rats showed a rhythmicity of their own, a four-day cycle in which running increased or declined in association with ovulation (321–22). This interest led to a series of articles published in the 1920s and 1930s, in which Richter detailed the activities of rats in which gonads, adrenal glands, thyroid and parathyroid glands, and the pituitary gland were removed in whole or in part. His work with adrenalectomized rats was to lead to one of his most notable contributions to the field that would come to be known as "behavioral endocrinology." This was his theory of an innate salt appetite that caused animals in which adrenal regulation of the salt content of the body was disturbed to engage in behavior intended to return the body to a state of equilibrium in respect to this essential mineral. Summarizing his findings in 1941, Richter paid tribute to the two physiologists—Claude Bernard and Walter B. Cannon—whose work he regarded as central to understanding processes of self-regulation. Richter described his own work as an attempt to move beyond study of "physiological regulators" to consideration of activities of the "total organism" such as eating, drinking, and nest building that animals undertook when physiological regulation was interrupted or disturbed.[30] Reporting on work done in the late 1930s, Richter argued for the existence of an innate "sodium chloride appetite," "calcium appetite," and "appetite for carbohydrates and fats."[31] In these investigations he made use of what he called the "appetite method"—depriving the animal of an essential nutrient (either by removing all or part of the pertinent glandular structures or simply by manipulating diet), and then observing behavior the animal engaged in to correct the deficiency.[32] Animals whose adrenal glands were removed, when provided a choice of food substances, eagerly sought out those containing salt, just as animals from which parathyroid tissue was removed sought sources of calcium. "How," he asked, "are the rats able to make such wise and advantageous dietary selections?" The answer lay in the principle that "after the physiological means of maintaining a constant internal environment have been removed, the organism itself makes an effort to attain this end."[33]

Wise food choices were, then, innately driven, a result of the animal's hereditary organization.[34] Richter did not entirely rule out a role for learning; a newborn rat, he observed, "sucks at everything with which its mouth comes into contact," learning to return to its mother's teats when, "on a conditioned reflex basis, an association between discomfort due to stomach contractions, feeding, and subsequent relief" was established.[35] Yet overall, Richter was little interested in processes of learning, and although he begged the question of the existence of "drives" by employing demurring quotation marks, he argued strenuously for the work of internal regulatory mechanisms and for the indispensability of instinct as a biological and psychological concept. Nor did he insist on the role of instinct only in respect to animals. Referring to the work of the Chicago pediatrician Clara Davis, who studied food selection with human infants and toddlers, Richter asserted that when the children were "given free access to a variety of natural foods" they were prompted by an internal guide to make "selections which result in normal growth and development."[36] Evidently, this internal guide sometimes went astray, and—in this instance on the basis of no experiments or research of his own—Richter asserted that poor dietary choices reflected faulty "parental guidance" and the sway of advertising: "Often when a child expresses a great appetite for certain substances, he is told by his parents that he must not eat them; equally often he is told to eat things which are very distasteful to him. He quickly learns to distrust his own appetite; and when he comes under the influence of advertisements, he falls prey to them."[37] Richter's mention of parental guidance and the impact of advertising indicates that he was well aware of the work of other investigators, far removed from his style of research, who called into question the basic assumptions and procedures of his "psychobiological" approach to appetite. To be sure, Richter's work inspired continuing experimental investigation of the number and types of "appetites for different food elements" and initiated a sustained debate over whether animals in fact enjoyed a capacity to "self-select" essential nutrients.[38] Yet at the same time critiques of his kind of investigation, elaborated in the overlapping domains of ethology and psychology, added new layers of complexity to the study of appetite and eventuated in a powerful challenge to all claims, Richter's chief among them, that eating choices were guided by innate physiological mechanisms and ultimately by instinct.

Appetite in the Wild

Early in his career John B. Watson showed serious interest in animal behavior in the wild, but he soon gave this up and established as psychology's principal objective predicting and controlling human behavior.[39] Thus his form of behaviorism did not encourage interest in behavioral variation in different species. Nor did other behaviorists in succeeding decades see value in broad-ranging comparative study. One enthusiast for the white rat asserted that "it is possible to write an essentially complete outline of the science of animal behavior without going beyond the available data on the rat," and syntheses such as B. F. Skinner's *The Behavior of Organisms* (1938) drew criticism for looking only at the behavior of white rats.[40] Nonetheless, after 1900 varieties of animal eating came to be investigated in depth by researchers whose work coalesced as the new field of "ethology." Richard W. Burkhardt Jr. has explored the intellectual and institutional development of the field, looking at work done both in Europe and in the United States by biologists, zoologists, psychologists, and specialists in subfields within these broader categories.[41] Within this complicated domain, vigorous debate unfolded over what precisely ethology or animal behavior studies should entail and what methods and concepts should be employed. In different geographical and institutional settings, the species studied and problems taken up varied widely, with topics such as courtship rituals and nesting behavior predominating. Yet eating patterns did draw interest and, as I will show, controversies that marked the field between the 1920s and 1950s were ultimately to play a major role in changing conceptions of appetite.

Within animal behavior studies the most important debate for the future of appetite was the struggle over nature versus nurture. An intervention on this theme that had broad impact on thinking about appetite was made in 1918 by the ornithologist Wallace Craig (1876–1954) when he argued for interactions between learned and innate patterns in eating behavior.[42] Unlike many of his contemporaries, Craig held that the study of animal behavior must include intensive work on closely related species, and he followed this principle himself in pursuing detailed study of different species of pigeons.[43] The essence of Craig's contribution to the nature-nurture debate was his division of behavioral patterns into two components or phases that he termed "appetitive" and "consummatory": "An appetite (or appetence, if this term may be used with purely behavioristic meaning), so far as externally observable, is a state of agitation which continues so long as a certain stimulus, which may be called the appeted stimulus, is absent. When the appeted stimulus is at length received it stimulates a consummatory reaction, after which the appetitive behavior ceases and is succeeded by a state of relative rest" (91). Craig also offered a clear-cut definition of the term "aversion," which had been widely used in medicine and psychology but not defined systematically. "An aversion," he held, "is a state of agitation which continues so long as a . . . disturbing stimulus . . . is present; but which ceases, being replaced by a state of relative rest, when that stimulus has ceased to act" on the sense organs (91). This "state of agitation" was marked by muscular tension and contractions, bodily attitudes and gestures that "are easily recognized signs or 'expressions' of appetite or of aversion" (92). Denying that actions prompted by appetite were, as some claimed, purely a result of "supposedly innate chain reflexes," Craig argued that "the reactions of the beginning or middle part of the series are not innate, or not completely innate, but must be learned by trial," whereas "the end action of the series, the consummatory action, is always innate" (92).

In this seminal article Craig made a notable advance in the task of pure description, by dividing sequences of behavior into distinct phases, but his major contribution was the effort to cut through the dichotomy between what was learned and what was instinctual. In his construction an animal was possessed of a number of basic appetites—for food, a nest, caring for young, sexual activity, and the like. He emphasized that his own investigations of appetite concerned only external behavior and said little about "the internal states [that]. . . . are probably exceedingly complex and numerous and similar to the physiological states which in the human organism are concomitants of appetites, emotions, desires."[44] Yet he did suggest generally that a mix of nervous and hormonal influences created such physiological states. Thus, he urged that animal behavior reflected both inborn physiological dispositions and actions repeated in light of experience. Craig supplied numerous examples of behavior in the dove—its learning to drink, building a nest, brooding, and copulating—that were rooted in an original appetite but then modified in accord with trial and error and the formation of "associative memory."[45]

Despite such efforts as Craig's, the struggle between instinctivists and anti-instinctivists remained a central preoccupation of animal behavior studies. Burkhardt has charted a broad division in the 1930s and 1940s between European defenders of instinct and American advocates of learning.[46] Especially forceful in arguing for the existence and broad range of instincts was the Austrian-born Konrad Lorenz (1903–1989), who, along with his close collaborator Niko Tinbergen, was responsible for the development of classical ethology. In 1935 Lorenz published an acclaimed monograph on "releasers of social behavior" that was based on observations of a range of bird species. This work presented the concept that, in respect to instinct, was his principal contribution to the study of animal behavior, that of the "releaser," the stimuli, "one or at most a very few . . . emanating from the objects in their environment" that released instinctive action.[47] In this work Lorenz attended especially to the social life of animals (his specific example was the jackdaw and its relations with five sorts of "companions"), and he insisted that instinctive behavior patterns must be thoroughly understood before any attempt was made to assess the role of learning. He also severely criticized investigators who drew their knowledge from artificial experiments on animals while neglecting to observe their behavior in natural settings.

Lorenz's later work strikingly exemplifies the ways that animal behavior studies could be used to draw conclusions about human behavior, including human food choices and the functioning of appetite. During the Second World War, Lorenz, who had enthusiastically endorsed certain features of National Socialism, was posted to Poland, where he worked as a military psychologist in Poznan. During his stay there he produced a manuscript titled "The Inborn Forms of Possible Experience," in which he discussed the putative dangers of race mixing and developed an extended analogy between the domestication of animals and the development of human civilization.[48] Lorenz equated the dangers that domestication and the attendant loss of instinct entailed for animals to the missteps of a civilization that produced overcivilized human degenerates, people immediately recognizable from such features as flabby muscles or a protruding stomach. These sentiments echoed views Lorenz had already expressed in the early 1930s, when he had argued that in healthy animals instinctive behavior patterns were "essentially identical," whereas variations in behavior resulted from illness or unnatural conditions.[49] Burkhardt reproduces a remarkable sketch Lorenz made in January 1939 to show "the degenerative effects of domestication in geese, ducks, dogs, fish, pigs, and humans," the human case illustrated by two men, one a fit, straight-backed man with a forward-thrusting chest replaced by the other, a short, stubby individual with a bulging gut. In this view human beings who remained true to their instincts ate well and appropriately, whereas those who ate in overcivilized fashion degenerated into "ugly" shapes that aroused instinctive disgust in others.[50]

Lorenz never repudiated either his instinctivism or his position that animal behavior studies could illuminate political and cultural struggles that reflected divisions over the influence of nature versus nurture. By contrast, his colleague and cofounder of ethology, Niko Tinbergen, in time modified the originally instinctivist stance of Continental ethology in favor of a view that blended instinct and learning. Primarily an ornithologist, Tinbergen also worked on the social insects, and on one occasion he drew a clear contrast between their eating behavior and that of birds. Reporting observations of a brood of horned owls, he ex-

10 N. Tinbergen, "Ueber die Ernährung einer Waldohreulenbrut (Asio otus otus (L.))." *Beiträge zur Fortpflanzungsbiologie der Vögel* 8 (1932), plate 2. An owl that adapted its hunting activity to available prey.

amined the "begging" behavior of the young, the hunting patterns of the adults, and the types of food the birds consumed. Tinbergen's interpretation of these data positioned him midway between instinctivists and anti-instinctivists: he argued that the hunting behavior of birds was inborn but that they learned to exercise it in respect to whatever prey were available, thus forming a marked contrast to the rigid, "specialized" eating instincts of insects. This study exemplified both what Tinbergen saw as appropriate method—close observation of one species in its natural habitat—and his disposition to lend greater nuance to claims about the roles of instinct and learning.[51] Nonetheless, in the sciences of appetite the most far-reaching effort to resolve rigid divisions between instinct and learning was made not by ethologists like Tinbergen but by psychologists influenced by Gestalt theory as it developed in late Imperial and Weimar Germany. Gestalt psychology differed markedly from post-Watson behaviorism in the United States; rather than targeting stimulus-response patterns, it sought holistic accounts of behavior, animal and human, that overcame dichotomies of interior and exterior, individual and social, innate and learned.

Appetite of "the Organism": The Holism of David Katz

The cultural movement in Germany from which Gestalt psychology emerged was marked by growing dissatisfaction with the materialist and positivist framework of science, especially with methodologies that privileged parts over functioning wholes. Anne Harrington has laid out the principles that informed a broad challenge to what German theorists termed "machine-thinking," emphasizing the widely resonant call to refocus science on human concerns and restore meaning to the scientific enterprise. The turning point in the move to reject "the Machine" in favor of a culture, specifically a science, committed to humanistic values came with the First World War. To critics of science, the massive loss of life and physical destruction of the war revealed the murderous potential of science and technology. The crisis of confidence brought by the war was especially acute for Germans as it entailed not only physical suffering and loss but also defeat and humiliation for the nation as a social, cultural, and political entity.[52] The ensuing era of "cultural despair" nonetheless also brought a sense to some that a return to essential principles could restore dynamism and life to a troubled people and polity. Among the cultural values invoked were scientific principles extolling the unique value of the human while recognizing that human life was inextricably tied to and dependent on nature, life at large.

The holistic movement within philosophy and science of the 1920s and 1930s included researchers in various disciplines. Its most influential philosopher was the medically and biologically trained Hans Driesch (1867–1941), who, after graduating from the University of Jena in 1891, pursued research in marine biology at the zoological station in Naples. Dissatisfied with the specialized work he had done there (on the sea urchin embryo), Driesch turned to philosophy, dedicating himself to defending the oft-decried position of vitalism and the principle of teleology.[53] In Naples, Driesch became friends with the Estonian-born zoologist Jacob von Uexküll (1864–1944), also a leading figure in the holistic movement in biology. With a series of works published between 1900 and 1940, Uexküll developed his theory of the *Umwelt* of the living organism, the multitude of perceptual cues and possible responses that established its immediate, lived environment. Uexküll drew attention away from features of a putatively objective physical reality and toward the experience of the living being, which was "the subject as well as the object of processes that influence its behavior."[54] Most important, he posited that the organism did not experience the *Umwelt* passively but, instead, with its own movements shaped "a new reality."[55]

Another influential advocate of holism was the neurologist Kurt Goldstein (1878–1965), who studied with Carl Wernicke (1848–1905), a prominent advocate of cerebral localization.[56] Goldstein largely rejected the thinking and methods of his teacher, developing a powerful critique of the concept of localization and elaborating in its place the view that the brain functions as a whole, its parts acting interdependently in ways that, though little understood, were readily demonstrable via clinical observation.[57] Goldstein argued pointedly against the division science erected between mind and body, urging the unique reality of the unified organism that strove to situate itself optimally in respect to both internal and external circumstances. In a series of studies, Goldstein stressed lived experience and subjectivity over the misleading and artificial "objective" measures favored in dominant life science. Focusing on therapeutics, he insisted that health was not an objective, measurable reality but a subjective sense of coherence and the capacity to "actualize" the strivings of the self.[58]

These were, then, some of the characteristic voices of the intellectual climate that gave rise to Gestalt psychology. The term "Gestalt"—translated literally into English as "form" or "figure"—took on specialized scientific and philosophical connotations in the late nineteenth century.[59] After 1900, it came to designate a school of psychology elaborated by the three principal figures Max Wertheimer (1880–1943), Wolfgang Köhler (1887–1967), and Kurt Koffka (1886–1941). Mitchell G. Ash observes that "Gestalt theorists asserted the primacy of perception over sensations in the constitution of consciousness, and advanced a conception of the subject as involved in, rather than separated from the world." In Gestalt thinking, he continues, "the objects we perceive are always located in what would now be called self-organizing systems—constantly changing dynamic contexts or situations, of which our phenomenal selves, too, are parts."[60] Ash examines problems of perception and behavior that preoccupied Gestalt practitioners as well as their efforts to develop innovative methodologies suited to their theoretical viewpoint. Although he focuses chiefly on the three leaders, Ash also recognizes the importance of ancillary figures who were related to the Gestalt project in fruitful ways. Among these was the psychologist David Katz (1884–1953), who, though much influenced by Gestalt theory and from time to time working in common with the principals, went his own way in selecting what he saw as compelling research topics. These included, as the title of his short work of 1932 indicated, *Hunger und Appetit*.[61]

Katz studied, as did many of the Gestalt psychologists, with the psychologist Georg Elias Müller (1850–1934) at the University of Göttingen as well as the philosopher Edmund Husserl (1859–1938).[62] In 1919 he was appointed to the faculty of the Institute for Psychology and Pedagogy at the University of Rostock, where he undertook experimental investigations in association with his wife and colleague Rosa Heine Katz.[63] In the 1920s, Katz worked on problems in the psychology of perception, pedagogical and animal psychology, and the field known as the psychology of needs (*Bedürfnispsychologie*). Katz may have been inspired to study issues in need psychology by his association with Kurt Lewin (1890–1947), a psychologist important to the Gestalt movement though standing apart from its chief theorists. In 1926 Lewin published a study of problems of motivation, including the forming of intentions as opposed to "genuine needs," phenomena that he characterized with the term "quasi-needs" (*Quasibedürfnisse*).[64] In any event, *Bedürfnispsychologie* as Katz practiced it challenged the view dominant in psychology, whether cast in traditional associationist, behavioristic, or specifically Pavlovian terms, that psychophysical activities were shaped chiefly by "learning" and accented, instead, the role of innate "drives" or "needs."[65] Why precisely Katz chose to address this set of issues by concentrating on hunger and appetite is uncertain. In an autobiographical essay written late in his career, he emphasized that in studying the "hunger drive, whose physiological aspects are sufficiently well known," he sought to clarify "the general laws of drives."[66] This retrospective explanation perhaps bespoke an effort to see coherence in a career that had moved in competing directions. In 1932, by contrast, Katz emphasized not this theoretical goal but the immediate context of the postwar years, especially lingering memories of the mass hunger Germans suffered during and after the war and of wartime debates over nutritional policy (3–4, 14–15).

Whatever his larger inspiration, Katz devoted a great deal of effort during his tenure at the Rostock institute to the cluster of issues surrounding hunger and appetite. Like other Gestalt psychologists, Katz was committed to experimentalism, and he strove to develop new methods that would clarify problems in this field that persisted, he observed, despite decades of study. Katz began by surveying results gained by virtually every method used by investigators of hunger and appetite. He drew on history, surveying eating habits in Greece and Rome and the role of food desires in the French Revolution (44–46). He used evidence from anthropology and ethology to support the contention—central to his overall argument—that human beings exhibited greater "plasticity" in eating desires and habits than any other creature (17–18, 19, 32–33,

38, 40–42). He was among the first European investigators to be well acquainted with American-style packaging and marketing of food (44–45). He knew the clinical and physiological literature on digestion, and he commented astutely on the theories of hunger advanced by Moritz Schiff, Luigi Luciani, Walter B. Cannon, and others (47–49). Yet even with this impressive range, the key findings Katz reported derived from studies he did himself or with associates. These were of two kinds: one involved research on animal eating, especially in chicks and adult chickens, and the other entailed direct questioning of children (in person and in questionnaires) about food preferences and aversions. Thus, Katz was an experimentalist who tried to interpret his own results in light of as broad a range of investigation as possible. This methodological latitudinarianism reflected his goal of viewing observed phenomena in holistic context.

For those who promoted Gestalt theory, taking a holistic viewpoint on phenomena meant, among other things, resolving dichotomies that hindered understanding of complex, lived realities. In respect to hunger and appetite, Katz worked to undermine long-established oppositions between the psychic and physical, innate and acquired, biological and cultural. He also made use of a strategy, common to much Gestalt thinking, of breaking phenomena under investigation into stages of development, focusing on changing dynamics rather than seeking settled definitions.[67] Finally, he criticized unidirectional approaches, arguing in respect to appetite that while it was true that appetite determined the choice of food, it was also true that available choices of food determined appetite (57).

Katz characterized his work on hunger and appetite as lying at the border between medicine and physiology in the "somewhat neglected field of Bedürfnispsychologie" (4), and he urged that the topic was of the highest practical significance. Complaining of the inadequacy of existing treatments of hunger and appetite, he observed that in textbooks the subjects appeared in connection either with the sensory phenomena surrounding taste or with internal sensations of the organs. Indeed, one of Katz's chief contributions was to argue against the centrality of taste phenomena in the formation of appetite. While recognizing that taste was highly significant, he denied that feelings about food, whether preferences or aversions, could be explained by charting tastes of specific foodstuffs (9–10) and insisted on the contributions of other senses, especially sight, to the process of seeking out and selecting foods (58–60). Thus, to Katz hunger and appetite had to be understood as phenomena embedded in a dynamic environment that changed with manifold influences ranging from metabolic requirements to historically determined eating customs to the emotional and mental condition of individual eaters.

A central task Katz set himself was to clarify the relation of the *Nahrungstrieb* to other drives, a problem, he said, that was of great interest not only to psychology but also to medicine, biology, and physiology (4–9). Although he initially stated that hunger was simply one drive among others, he ultimately held that hunger was the primary drive of organic beings, one that originated in an "uninterrupted storming [*Bestürmung*] of consciousness" rather than voluntary movements of the will or "quasi-needs" that lacked the force of "natural need" (13). Exploring the relation between hunger and the other "instincts," Katz rejected claims that hunger might be suspended in the interests of reproduction and contrasted its inexorability to the late appearance and early extinction of the sex drive.[68] To support this view of the primacy of hunger, Katz argued that, historically, no drive had acted as a more powerful cause of human migration. Pointing to the desperate movements of people caused by the Russian famine, he urged that the power of hunger could be seen only in situations of extremity, as it went unrecognized by "civilized people" whose needs were met by the highly efficient exchange of goods effected in modern economic systems (14–15).

How appetite fit into this picture of hunger as a powerful need—indeed, *the* most powerful of needs—was a problem Katz approached from many directions. First, he pointed to essential distinctions between the two: in appetite choice was central, in hunger less important; appetite was specific, hunger general; appetite was powerfully affected by emotion, hunger less so (6–8, 12–13, 17–18, 21). Surveying studies of animal feeding, Katz often emphasized the significance of their results not for understanding hunger or physiological need but for appetite—what animals chose to eat. Katz was especially impressed by the investigative ingenuity of entomologists, whose studies of the adaptability (or failure to adapt) of insects whose food supply was artificially altered, had yielded discoveries "of the highest significance" for a theory of appetite (18, 20). In sum, his inquiry was intended to illuminate not only the drive to eat of the hungry organism but also the role of choice in the complex behavior inspired by appetite.

Katz aimed ultimately to offer a theory of hunger and appetite that would replace what he saw as the flawed perspective of an eating psychology based on trial and error. Having recounted findings from a great range of disciplines, he proposed as the fundamental problem at issue the origins of the eating "arrangements" of animals: "How is it," he asked, "that an animal that lives alone and is not guided by any kind of nutritional tradition selects certain kinds of food and not others?" (54). The dominant theory of trial and error claimed that animals tried different foods available in their environment but kept on eating only those that assured their well-being (55). To contest this viewpoint, Katz drew directly on the insights of Gestalt psychology, offering in place of trial-and-error theory the view that animals test food within the bounds of a preexisting "field" that ensures that they are in a state of "readiness" to react to experience in a given fashion.[69] This "core phenomenon" of readiness, constituted by the play of internal stimuli, necessarily existed prior to an animal's experience of eating in the world and took the form of inborn appetites (55). Terming his own view the "avidity theory of appetite," Katz stated his debt to the language of chemistry, seeing at work in eating choices attractions and satiations ("Aviditäten und Absättigungen") that were as fully "lawlike" as those seen in chemical reactions. That such laws were operative despite the extraordinary variety of individual eating choices exhibited by both animals and humans was undeniable given that unobstructed food choices invariably led the organism to eat what it needed.[70] Unlike the one-dimensional *Probiertheorie*, avidity theory recognized "two components," charting the interplay of internal states and external environment that accounted for the dynamic changeability of hunger and appetite.[71] Internal states altered constantly depending on the chemical effects of foods consumed just as external states shifted with the animal's movements to widen or limit its food choices (55–57).

Katz regarded the results of the experiments he conducted on the eating habits of chickens as explicable only in terms of his "two-component" avidity theory. In one series of experiments he charted the eating of chickens in environments in which the same food was supplied but external factors were altered—the number and size of the piles of food, colors of surrounding objects, and social influences such as the presence of other hungry animals (26). Such factors accounted for obvious differences in eating behavior and could not be discounted, as physiologists did when they focused purely on putatively "normal" eating, which, as Katz insisted, was but one of many possible eating situations.[72] A different set of experiments sought to determine how eating behavior altered when the foods were changed rather than the external environment. Here Katz's approach resembled the feeding experiments previously undertaken by physiologists, but his goal was not to measure the physiological impact of deprivation but the changes in animals' eating behavior, especially the creation of what Katz called "special hungers" (52). Disputing all "one-sided" theories that depended on such simple mechanical actions as stomach contractions, Katz held that the types of hunger aroused by deprivation of vital food components brought on behavior that depended not on prior experience of the desired substances but, rather, on inborn needs. These experiments involved newborn chicks set loose in the farmyard, where they encountered both edible and inedible substances and engaged in "picking" activities guided by optical and other sensory signals but ultimately selected appropriate feed. In sum, the chicks ate as dictated by innate demands, a phenomenon documented repeatedly for animals who ate strangely when deprived of the ordinary foods from which they obtained essential nutrients such as cattle that ate bones to supply missing phosphorus (23, 58–60). And yet avidity theory, far from reducing eating to the simple demands of chemistry, recognized the complex interplay—in the actual search for and selection of food—of sensory stimuli, food images formed in the central nervous system, and social and affective influences, in sum the entire Gestalt of food desire and food consumption.[73]

Finally, Katz moved to show how avidity theory solved long-standing mysteries of human appetite as well as animal. In all times and places human beings selected and consumed substances that met basic human requirements for fat, carbohydrates, and proteins.[74] In this respect the inborn drive to eat was governed by chemistry. But the actual eating of human beings exhibited astonishing variety dependent on history, custom, and climate as well as "natural tendencies," appetites of a range of individuals representing a range of discernible human "types." Here Katz adduced a second variety of experiment undertaken at Rostock, inquiries into the food preferences of schoolchildren. Employing surveys and direct interrogation, Katz's research group asked boys and girls aged ten to fourteen about different categories of foods, eliciting responses he displayed in detailed charts listing simple likes and dislikes and strong food aversions. The patterns that emerged indicated the formation of distinct "bio-types" Katz designated as "leptosomic" and "pyknic," the former the slender, "spiritual" type of the vegetable eater and the latter the sturdier, more "reality-oriented" type of the meat eater, both of them formed by inborn preferences and correlated with a "body constitution" given at birth.[75] Katz also conducted interviews with the children's parents, who claimed that from their earliest years the children had selected foods they wanted and rejected others without regard to parental admonition or family tradition. Given this perspective, Katz left limited scope for intervention into the innumerable questions about nutrition confronted by psychologists and physicians. Yet in the end he urged that nothing at all could be done to redirect harmful eating habits if hunger and appetite were not recognized as inextricable blends of somatic, psychic, and cultural realities.[76]

Katz's work, like that of Richter in the United States, indicates that, despite the wide sway of behaviorists who insisted on the primacy of learning, investigators schooled in other traditions devised methods to prove the existence, even dominance, of inborn eating proclivities. Thus, the conflict of nature and nurture strongly colored thinking about the subject in the very different settings of Richter's laboratory and the outdoor and school-based inquiries pursued by Katz. For Katz and his associates the nature-nurture debate was, in any event, cut short by the rise of National Socialism, which skewed all science in the specific ideological direction demanded by Hitler and sent into exile Jewish academics like Katz and other Gestalt psychologists hostile to the regime.[77] Elsewhere in Europe the onset of the Second World War led to the dispersion or outright persecution of scientific investigators and, in the United States as well, to the redirecting of energy and resources. The war, in which combatants made use of hunger as a weapon in unprecedented ways, also provoked intensive research on starvation, much of it in unprecedented settings like the Warsaw Ghetto. What attention was paid to appetite came chiefly in wartime medicine, where physicians, nutritionists, and other researchers sought to make wartime rations palatable to soldiers and hungry civilians. The resumption of great theoretical debates had to await the war's end.[78]

The Paris Instinct Conference

When the war was over the question of nature versus nurture, which was inevitably tied to the ideological struggle between fascism and its opponents, demanded attention as postwar commentators sought to assess the impact of ideological dogma on purportedly objective claims about the determinants of human behavior. Major symposia were held to consider the relative power of instinct and environment in governing human and animal behavior. Of these the 1954 conference on instinct held in Paris brought into clear focus the scientific and ideological combat between instinctivists and their opponents.[79] Here leading representatives of Continental ethology and Anglo-American animal behavior studies met, clashing over interpretations of wide-ranging behavior patterns in species from insect to mammal.

As had always been true in animal behavior studies, eating habits drew less attention than such topics as courtship, habitat building, and migration. Yet both European and American participants argued the case for innatism by examining inborn determinants of eating behavior. The Strasbourg psychologist Gaston Viaud (1899–1961) pointed to the importance of chemical and "kinéto-tropisms" in the "detection of aliments," and his colleague the biologist Marc Klein (1905–1975) looked at the close interrelation of reproductive and nutritive instincts.[80] Cutting across the European-American divide, Curt P. Richter gave a spirited defense of the instinct concept, summarizing two decades of work on salt appetite carried out in his laboratory. Claiming that his method was "the characteristic approach of the American School of Animal Behavior," Richter urged that his work united neurological, endocrinological, and nutritional "viewpoints," thus attempting "to gather all information that contributes to understanding behavior."[81] All of the experiments he reported involved the "Norway rat" ("nutritionally speaking . . . very closely related to man," 578–79) and together demonstrated that in conditions of salt deficiency, rats instinctually acted to regulate salt intake in quantities required by metabolic need. Many of the experiments involved rats whose adrenal glands were excised, thus causing salt depletion and producing a "craving" for salt that the rats satisfied by drinking ever stronger salt solutions until they attained the "taste preference threshold" (617).

Richter did not hesitate to use such results to make claims about human appetite and dietary choices. Rats and humans, he held, enjoyed similar capacities to regulate the intake of essential dietary components with the help of both physiological and behavioral "mechanisms." Richter presented findings comparing the effectiveness of these mechanisms in domesticated and wild rats, urging that in humans, as in domesticated rats, evolution seemed to be reversing itself and fine-honed regulatory mechanisms to be deteriorating. Asserting that "we are losing our ability to taste certain substances," he held further that this was an effect of the "evolutionary process" and that such losses were "much greater among Europeans than in members of native races" (626–27). By this means the issue of appetite was squarely lodged within the larger debate on the nature and number of the instincts, their relative distribution and power in different species at different stages of evolution, and their apparent "degeneration" under the influence of domestication and, in the human case, of culture.

Other American participants were hostile to any argument built on instinct, and the Paris symposium brought into clear relief the tensions that by the mid-1950s divided instinctivists and anti-instinctivists in the United States. The latter case was laid out by the animal psychologist

T. C. Schneirla (1902–1968), curator at the American Museum of Natural History, who argued that, historically, opponents of the instinct concept had had the great merit of insisting on objectivity and experimentation and that all investigations must focus on carefully delimited behavioral patterns manifest at specific points of "ontogenetic" development.[82] The strong claims Richter put forward on the instinctual basis of feeding choices nettled the defenders of learning. The zoologist Daniel S. Lehrman (1919–1972), also of the American Museum of Natural History, drew the conferees' attention to work that undermined Richter's claims, especially a study showing that "intact rats, trained to discriminate between two concentrations of salt solution by an appropriate distribution of rewards and punishments, can learn discriminations just as fine as those shown by adrenalectomized rats."[83] In other words, there was no need to invoke instinct: learning processes could account for the rats' behavior. In the discussion that followed Richter's presentation, Lehrman lauded his exposition of "the facts of diet selection and its relationship to nutritional needs" but concluded that Richter failed to explain "the physiological and psychological mechanisms involved."[84] In contrast, the French defenders of self-regulation (and, implicitly, the physiological tradition inspired by Bernard's *milieu intérieur*), regarded Richter's findings as "données classiques," facts that illuminated not only the role of salt in mammalian physiology but also the "invincible needs" that throughout history had driven individual and collective action in defense of life (631–32). Recalling that under the French ancien régime, the salt tax was "synonymous with tyranny," Klein urged that Richter had established unassailable verities about innate physiological demands and that researchers continued to be divided only in respect to the "intimate mechanism" involved (631).

Up to 1950, thanks to the broad sway of behaviorism in its many guises, psychology and ethology both took as their chief focus behavior rather than mental or inner states. In the study of eating, "food intake" and "ingestive behavior" became normative phrases. Even researchers like Craig who retained the language of appetite deduced its presence from physical movements indicating a "state of agitation" while circumventing "internal states" he surmised to be "exceedingly complex." Similarly, Richter and the experimentalists he inspired identified appetites with metabolic needs for essential nutrients and charted their manifestation in the movements of rats in "activity cages." Sometimes strident arguments divided those who observed animals eating in laboratory cages from those who watched animals hunting and devouring prey in the wild, but both emphasized animal actions and movements, appending only as speculations mention of human appetite and eating.

Seen in this context, a psychologist like Katz, whose acknowledged focus was the functioning of specifically human appetite, was a true outlier. To Katz the puzzle of appetite could never be solved so long as study of human eating was limited by the physiologist's construal of "normal" regulative function or the behaviorist's insistence on predictable patterns of "stimulus and response." For Katz, appetite was, to adapt Marcel Mauss's famous phrase, a "total organismic fact."[85] Certainly other psychologists acknowledged the many faces of appetite. When Richter's admirer Paul Thomas Young (1892–1978) summed up experimental work on appetite of the 1920s and 1930s, he urged in a coda attention to "social and aesthetic factors in food ingestion."[86] Nonetheless, after devoting many pages to the particulars of his experiments with rats, Young said nothing about how psychologists might proceed in investigating such influences. In this respect he resembled most psychologists and ethologists of his era in pursuing sophisticated and specialized work of one kind while leaving to others, or simply to the future, the problem of how to integrate disparate research programs or even to have one approach serve as a check on the one-sidedness of another. Just as with physiologists in this period, moreover, such researchers often speculated freely on the significance of their particular approaches for the human exercise of appetite. Watson never studied human eating, but he argued that it, like all behavior, could be "conditioned" with appropriate technique. Richter investigated rats, not humans, but somehow he was sure that parents were to blame when children lost the guidance of natural appetite. Lorenz was no student of human civilization, but he nonetheless equated it to the domestication of animals in its destructive effect on the "instinct" for healthy eating. Whatever the disparities of method and theory, or the broad orientation to nature or nurture, the one constant in all this work was the project to discover truths of appetite and eating that were accessible only through refined modes of expert inquiry. Whether by-product or intentional outcome, the effect was to render unstoppable the long process by which appetite moved out of the domain of the ordinary and into the domain of science.

TWELVE

Somatic, Psychic, Psychosomatic: The Medicine of Troubled Appetite

In 1900 physicians schooled in scientific medicine anticipated that its rigorous methods would soon bring new clarity to maladies that blended ingestive or gastric disorder with trouble of mind and emotion. And indeed, after 1900 progress seemed to be under way. Perhaps most promising, nutritional science seemed by this point ready to supply clear scientific guidance about diet and, in so doing, to prevent ills linked to unhealthy eating before they took root. More precise diagnoses also seemed to be emerging as physicians using new tools replaced "nervous dyspepsia" with peptic ulcer, resolved neurasthenic symptoms into such conditions as "neurocirculatory asthenia," and reconfigured anorexia nervosa as an endocrine illness. Yet just as these steps appeared to bring new clarity, a radically different approach to such conditions gained adherents among specialists in "mental" illness, who, rejecting both the labels and methods of somaticist medicine, staked their own claims to components of prior diagnoses, reconceptualizing features of neurasthenia under such new headings as "anxiety neuroses" and thoroughly "desomaticizing" anorexia nervosa.[1]

Although this bifurcation of illness into psychic and somatic forms seemed unstoppable, already by the 1920s and 1930s efforts were under way to create a psychosomatic medicine that saw illness as a phenomenon of the "whole person." The psychosomatic movement had diverse sources and, perhaps inevitably, soon saw divisions emerge within its own ranks. Psychosomaticists disputed whether all illness entailed mixed physical and psychic components or whether certain systems or parts of the body were especially likely to involve distress of mind or feeling. Similarly, a division arose between practitioners who believed psychosomaticism had to be firmly grounded in physiological science and others who thought illness of this kind accessible principally, or only, through methods tailored to mental and emotional disturbance. In this respect Freudian psychoanalysts were the most extreme, arguing that the state of the appetite was irrelevant to illness involving disordered eating because the desire for food was actually a displaced expression of the libido.

These divisions mirrored continuing uncertainties about the nature of appetite itself with its uncertain balance of the somatic and the psychic. Yet the fact that appetite remained puzzling did not produce medical inaction in the face of appetitive ills. Rather the era saw a constant proliferation of new therapies that targeted troubled appetite—drugs and surgical procedures and, most dazzling, an array of glandular extracts made possible by the new knowledge of "hormones." But again, simultaneously with the appearance of innovative somatic remedies, new forms of psychotherapy multiplied as the influence of psychology and psychiatry expanded. By 1950, some therapies had been discredited, whereas others had gained strength, but none provided the royal road to curing troubles of appetite and eating hoped for when the century began.

Appetite and the Nutritionist: The Era of "Vitamania"

In the late nineteenth century the science of nutrition, once simply the study of the internal processes by which foods and drinks are transformed in the body, came to focus specifically on the dietary requirements of humans and animals.[2] This work was keenly pursued in Germany, especially at the University of Munich, where researchers used innovative procedures and equipment to calculate human caloric requirements. One crucial outcome of this research was the principle of the isodynamic equivalence of foods, the interchangeability of fats, proteins, and carbohydrates in meeting nutritional needs, a view that made the quantity of foods consumed the most important feature of diet. Many researchers came from abroad to study in Munich, including one who was to be preeminent in American nutritional science, W. O.

Atwater (1844–1907), who went on to train a generation of American nutritionists. Atwater's estimates of the numbers of calories in food are essentially those found on food labels today.[3]

After 1900 the quantitative view of nutrition gave way to a new, qualitative understanding of dietary constituents and bodily needs. The crucial figure in this shift was the British biochemist Frederick Gowland Hopkins (1861–1947), who argued that a diet adequate for growth, reproduction, and other vital functions did not consist purely of proteins, fats, and carbohydrates but required what he called "accessory factors."[4] Hopkins's claims first gained notice in articles published in the British newspaper the *Daily Mail* as part of a campaign to convince the British public of the superior nutritional value of "standard" over white bread.[5] This campaign unfolded throughout 1911 and, though ending inconclusively, established Hopkins's reputation and showed the growing influence of biochemistry and the new nutritional science. Within a short time Hopkins succeeded in identifying the accessory factors that he argued were essential to life and growth. Termed "vitamins" by another contributor to the field, these constituents of food and drink quickly became the subject of sustained scientific investigation as well as media and advertising efforts that went on for decades, in the United States stoking unprecedented popular enthusiasm for a dietary health product.[6]

The claims of biochemists to a revolutionary understanding of nutrition and dietary needs met at the outset with considerable skepticism from physicians, who despite a century of challenges by experimentalists continued to see themselves as the privileged arbiters of things dietetic. Resistance to Hopkins was especially strong at the University of Glasgow, where medical traditionalists rejected both the general claim that vitamins were essential components of the human diet and specific claims about the effects on health of vitamin "deficiencies." David Smith and Malcolm Nicolson have traced in detail a dispute over the etiology of rickets, a disease that Hopkins and his students attributed to a deficiency of vitamin A but that Glasgow physicians attributed to broader influences, including exercise and hygiene.[7] Other British physicians challenged Hopkins's general assertion that a healthy diet must be based on scientific findings. As Mark Weatherall notes, a professor of medicine at University College London ridiculed the pretensions of experimental scientists, declaring approvingly that the unschooled Englishman ate what his ancestors had learned to eat "centuries ago by the light of Nature and the craving of a natural appetite."[8]

Such defenses of appetite as a natural guide to eating choices did not long persist, however. By the early 1920s, physicians, while often criticizing "abuses" in therapies based on vitamins, had largely accepted that vitamins were essential to health and that vitamin therapies supplied a curative tool of extraordinary promise. From the 1920s into the 1950s, the pages of the *British Medical Journal* were filled with discussions of the mechanisms of vitamin action, the maladies in which vitamin deficiencies played significant roles, and the need to standardize vitamin units for therapeutic use. The conditions in which vitamin deficiencies appeared to figure were seemingly limitless, and although by the late 1940s calls for restraint multiplied, the importance of vitamins in dietary health remained an essential principle of nutrition.[9]

Thus, after an initial period of medical opposition, the role of vitamins in nutritional health was firmly established, as was the indispensability of nutritional science in guiding healthy eating. Promotion of such thinking often took the form of a direct attack on the guidance of appetite. The American nutritionist E. V. McCollum (1879–1967) cautioned that "the appetite is by no means so safe a guide for the proper selection of foods as has generally been supposed."[10] And Hopkins himself, surveying the position of nutritional science in 1935, lamented that "even the most civilized communities" still lacked "guidance of a more definite and reliable kind" than "instinct and appetite."[11] One practitioner who condemned "abuse" in the prescription of vitamins did argue that the ordinary British eater need not worry about the particulars of nutritional science because the average diet contained all the food constituents necessary for health, but he added the proviso that "a reasonably intelligent selection of the available foods" must be made.[12] That unintelligent or perverse eating could lead to disease both bodily and mental, a fact long recognized in traditional medicine, was confirmed and illuminated by the new nutritional science.

By the 1930s, then, physicians broadly accepted the benefits of vitamin therapy in overcoming such ancient ills as rickets and pellagra, and work went forward in regard to the therapeutic value of vitamins for many other conditions as well. But everywhere it was recognized that the greatest importance of the new nutritional science lay not in the cure of disease but in its prevention through improved diet.[13] And while professional medical organizations still urged that individuals and families take dietetic guidance from their own physicians, governmental and social authorities moved increasingly to assign the myriad tasks involved in promoting healthful eating to the care of the nutritionist. The First World War gave tremendous impetus to nutritional science and to the public authority of nutritionists. Combatant governments called on leading scientific experts to help fashion wartime policy on food production, rationing, and the appropriate diet of citizens whose duty it was to avoid waste. Hopkins collaborated on a booklet that addressed such problems, as did his close friend the physiologist W. M. Bayliss.[14] The Medical Research Council, which helped to shape nutritional policy in Britain during and after the war, funded a great deal of research on vitamins and other issues.[15] Bayliss attacked the problem of unruly (and unpatriotic) appetite head on. He accepted Pavlov's demonstration that eating "without appetite" meant the failure to produce gastric juice and thus that wartime food must be attractive while avoiding "mere ornamentation."[16] Yet he also observed, "We hear of a 'fine healthy appetite,' when the person spoken of is probably really consuming much more than he needs" (2), and concluded that, while people might grumble at seeming privations, attention to the "essential facts of the science of nutrition" would ensure that "any kind of diet . . . chosen in the British Isles will contain all that is necessary" (3–4).

Thus, from the age of vitamins onward, eating in developed Western countries became the special concern of the nutritional expert, who, from the outset, denigrated the reliability of appetite and urged that anyone concerned with "healthy eating" must yield to the dictates of the science of food. It is beyond the scope of this study to follow in detail the subsequent history of debates over nutritional guidelines and efforts to implement them on behalf of often recalcitrant beneficiaries.[17] The essential point for the history of appetite is that this private and personal form of experience became increasingly less private as nutritionists extended and defended their claims to arbitrate healthy eating.[18]

"Simple yet effective": Somatic Remedies for Disturbed Appetite

The growing prestige of nutritional science, with its promise of the seemingly miraculous curative potential of vitamins, helped to encourage the idea that disturbed appetite and eating could be cured by "simple yet effective" remedies.[19] This idea was of course not new. Throughout the nineteenth century physicians had sought to suppress or stimulate appetite by administering herbal remedies (valerian, ipecac, henbane) or others based on pepsin, the "active principle" in gastric juice.[20] One of the more curious aspects of the work on digestion in Pavlov's laboratory was the effort to replace commercial pepsins with natural gastric juice collected via his "dog technology." As Pavlov's work gained celebrity gastric juice became a therapy of choice in France, Germany, and Russia.[21]

Although arguments continued over the merits of competing versions of the therapy and the conditions for which it was best suited, there seemed to be universal agreement that remedies incorporating gastric juice improved the appetite, sometimes quickly and dramatically.[22]

Gastric juice therapy gave way in time to widespread use of glandular extracts in the treatment of troubled appetite, eating, and digestion. Therapies based on such extracts dated from the later nineteenth century, but the discovery of secretin in 1902 stimulated intensified work on hormones as treatments for a range of disorders, including both excessive and deficient appetite.[23] The Viennese clinician Wilhelm Falta (1875–1950) reported in 1925 that he used insulin with three patients who had suffered loss of appetite and grown abnormally thin; with treatment, all developed insatiable appetites and consumed great quantities of food.[24] Taking their cue from Falta, a physician-physiologist team in New Haven, Connecticut, lauded insulin as a treatment for patients who suffered failed appetite associated with depression, "anxiety states," or nervousness.[25] Describing four cases of female patients whose loss of appetite originated in emotional troubles, they recounted that with treatment the young women not only began eating again but also ate with "relish," regaining their lost capacity for pleasure in eating. Although they acknowledged that "the physiology which underlies the pleasure of eating is still obscure" (612), these practitioners confidently concluded that "insulin not only creates hunger, but also generates the pleasure which is associated with its satisfaction" (612). For these practitioners insulin and the intake of great quantities of food constituted the sole remedy for the failure of appetite despite the fact that their patients' ills originally derived from emotional distress. Other physicians who used insulin to treat anorexia nervosa administered it preparatory to therapeutic regimes centered on psychotherapy.[26] But both assumed that failed appetite was susceptible to glandular intervention.

Other glandular therapies were also employed to treat troubled eating. In 1934 the German internist Gustav von Bergmann (1878–1955) listed a score of preparations, including thyroid, adrenal, and pituitary extracts as well as insulin and preparations based on the recently discovered sex hormones. He argued for their success principally in treating *Magersucht*, the German diagnosis closest to anorexia nervosa, but also in cases of chlorosis and obesity. Bergmann saw obesity as resulting both from excessive appetite and the failure of regulatory processes controlling the storage of fat, and while he held that glandular extracts in isolation did not act effectively against obesity, he did believe they produced beneficial effects on regulatory processes.[27] Surveying different etiologies proposed for *Magersucht,* Bergmann tended to the view that the disease was "pluriglandular" in character, resulting from an overall inadequacy of hormonally controlled regulatory systems. Treating a "desperate case" of the condition in a nineteen-year-old male, Bergmann employed different forms of endocrine therapy, including transplants of testicular material from cattle. He also reported cases of twenty young girls in whom he had restored appetite and reversed severe weight loss by administering insulin, pituitary extracts, and ovarian preparations. Bergmann later moved toward a psychiatric interpretation of *Magersucht,* but he continued to see utility in endocrine therapies even if they worked primarily through "suggestion."[28]

American physicians also used thyroid extracts in treating anorexia nervosa. In 1930 a physician at the Mayo Clinic reviewed 117 cases of the illness treated in this fashion between 1917 and 1929.[29] He saw anorexia nervosa as originating in a "psychic disturbance" that caused "loss of appetite" (423), but in discussing symptoms and treatments he nonetheless emphasized somatic features—depressed basal metabolic rate, low blood pressure, and reduced gastric motility. Acknowledging that patients responded differently to thyroid treatment, he nonetheless concluded that "on the whole" they improved both mentally and physically after extended use of such preparations, which should be regarded as "a valuable adjunctive measure" in restoring "normal intake of calories."[30] Whether this happened because the appetite was restored or because of some somatic change he did not specify.

By the late 1930s and 1940s, skepticism was building about glandular etiologies and treatments. Fundamentally critical of the whole approach, of course, were those researchers who argued for the determining role of the hypothalamus and discounted endocrine influences on appetite. The hypothalamic hypothesis gained strength gradually from the 1920s onward, but the laboratory and operative successes of those who upheld it were not accompanied by anything like the therapeutic enthusiasm surrounding endocrine explanations and treatments. [31] Indeed, the literature of hypothalamic localization reveals scant concern for therapeutics beyond the negative goal of discrediting glandular treatments. The research team of Bailey and Bremer, whose 1921 article on the origin of diabetes insipidus did much to undermine claims about the wide influence exerted by the pituitary gland, argued that their work would aid clinicians but purely by clearing away "puzzling contradictions" in evidence amassed on the role of the pituitary in assorted syndromes.[32] Three decades later John Brobeck, as already noted a pivotal figure in the investigation of hypothalamic regulation, expressed skepticism about therapeutic or preventative regimens that purported to rest on his work. In 1955, Brobeck recalled the great progress that had been made in understanding the hypothalamic mechanism, but he distanced himself from the notion, widespread in both the "scientific and popular literature," that the hypothalamus is "sensitive" and could be directly affected in a variety of ways. He mentioned in particular the work of the nutritionist Norman Jolliffe (1901–1961), who claimed that this "hypothetical mechanism" constituted an "appestat" that, when defective, could be restored to effective functioning.[33] Only toward the end of his career did Brobeck, having come to see great complexity in the physiology of regulation, look to the therapeutic implications of his findings about the role of the brain in appetite and satiety. In a 1957 article, Brobeck went beyond his prior emphasis on the hypothalamus, observing that investigators had come to recognize that "hunger, appetite, and satiety" entailed complex "functions of the brain" and nervous system.[34] Summarizing recent findings on the hypothalamus—new proofs that the medial portion was involved in satiety, the lateral portions in appetite—he also pointed to work indicating the importance of activity at "the highest level of the brain, the cerebral cortex," which appeared to be responsible for "recognition of objects suitable for food, and . . . feeding responses as affected by social environment" (567). Such proofs of higher-level brain involvement in appetite and eating opened a space for therapeutic responses based on conscious, well-informed decisions such as the choice to engage in exercise ("That exercise takes away hunger seems to be known universally," 572). Brobeck's remarks thus suggested, if indirectly, that localizing appetite in the brain did not encourage somatic interventions but, rather, enduring changes in voluntary behavior.

Given the paucity of therapeutic offerings by proponents of hypothalamic regulation of appetite and food intake, the popularity of glandular extracts to treat troubles of appetite is perhaps understandable. But other therapies were also tried. Beginning with Benzedrine, amphetamines were used both to damp down and to stimulate appetite, although by midcentury these had proved either ineffective over the long term or to have dangerous side effects.[35] Surgery was another possibility, with procedures such as pyloroplasty (opening of a stricture in the pylorus) promoted as a corrective for failed appetite and weight loss.[36] Yet some practitioners were highly critical of such operations, especially as treatments for anorexia nervosa. In a 1936 lecture the Guy's Hospital consultant John A. Ryle (1889–1950) decried the "multiple operations" performed on anorexics that were "undertaken through lack of appreciation of the patient's psychological type and of the nervous basis for many or all of the symptoms."[37] Yet for Ryle, the worst abuses in treatments for anorexia nervosa resulted from what he called the "lures of endocrinology" (893). He especially decried the use of ovarian preparations, noting that although the disease was mainly seen in adolescent girls and young women, it also afflicted postmenopausal women and young men (894). Recalling the nineteenth-century description of the disease by William Withey Gull, who emphasized as its primary feature "want of appetite . . . due to a morbid mental state" (893), Ryle concluded that there was "no justification" for employing glandular preparations given that anorexia nervosa originated in "psychic trauma" or "foolish habit[s]" that led to loss of appetite and resulting starvation (899). The most effective measures, he concluded, were those physicians had always used: "explanation, reassurance, [and] distraction" (899).

Ryle's attack on endocrinology reflected the generally conservative cast of clinical medicine in interwar Britain.[38] Yet other physicians, not content with the commonsense perspective of clinicians like Ryle, also orchestrated attacks on glandular therapies. Diana Long Hall has charted the debate that unfolded in Britain, where endocrine skeptics sharply criticized the credulity of clinicians like Walter Langdon-Brown (1870–1946), who declared that continued research on the influence of hormones would prove that "we are marionettes of our glands."[39] In France, too, critics complained of the damage done by hyperbolic claims about "disorders of endocrine origin," one physician stating in 1939 that the anorexic syndrome "so well described by the older authors" must be treated as the "psycho-neuropathic" disorder it had once been understood to be.[40]

This criticism pointed to an especially contentious issue in the controversy over glandular etiologies and therapies, the relationship between anorexia nervosa and "Simmonds's disease." The latter condition was a syndrome named for the Hamburg pathologist Morris Simmonds, who in 1914 described an illness characterized by severe weight loss and thinning as well as progressive weakness, pallor, and other symptoms he attributed to destruction of the pituitary gland.[41] In succeeding years Simmonds's disease drew considerable attention. Gratified that a definite somatic cause had been identified for this puzzling condition, physicians rushed to adopt the new diagnosis and to argue that the poorly understood anorexia nervosa of the past should be resituated among conditions entailing emaciation of pituitary origin.[42] By the late 1930s, however, some clinicians had come to see rampant confusion between Simmonds's disease, at this point often known as "hypophyseal cachexia," and anorexia nervosa. This discussion reached a peak between 1937 and 1939, when some fifteen major studies conducted in the United States and Europe attempted to distinguish clearly between the two diseases. These studies were the subject of a meta-analysis published in 1942 by Roberto F. Escamilla (1905–?) and Hans Lisser (1888–?), who proposed what they saw as convincing differentiating features for each disease.[43] For anorexia nervosa these included not only such symptoms as severe weight loss and low basal metabolic rate but also their appearance in a "young unmarried female who has never been pregnant" (85). Because Simmonds's disease was resistant to treatment, they also pointed to "marked improvement or return to normal health" after therapy as a sign that the disease in question was anorexia nervosa (85). The authors further proposed that "psychopathic disturbances favor the diagnosis of anorexia nervosa," but, given that forms of psychopathology also often appeared with Simmonds's disease, they did not regard psychic disturbance as a clear benchmark (85).

Students of anorexia nervosa have presented the Escamilla-Lisser study as the end of the "pituitary era" in the history of the disease, but this analysis is misleading on several scores.[44] These authors did not in fact criticize the use of glandular extracts in the treatment of anorexia nervosa and indeed described "endocrine therapy" as one form of treatment that had clearly brought "marked improvement or return to normal health" to anorexics (85). Nor did 1942, the date of their publication, actually mark the end of either glandular etiologies or treatments of anorexia nervosa. Studies that continued to explore the relation between Simmonds's disease and anorexia nervosa appeared into the 1960s.[45] Most important, Escamilla and Lisser, who regarded diminution or disappearance of appetite as a crucial feature of both diseases, used glandular preparations in their own treatment programs (85–89). The work of these physicians suggests, then, not a wholesale rejection of endocrinology or the end of an era but a recognition that diagnosis and treatment of ills such as Simmonds's disease and anorexia nervosa were much more difficult than clinicians had previously recognized.[46] Other studies made the same point, one concluding, "It seems wise to concede that this strange clinical picture is as yet incompletely understood and that much further work remains to be done."[47]

Although historians have been inclined to ridicule or censure physicians of these years for excessive zeal in resorting to glandular extracts or dubious surgeries, it should perhaps be recalled that these practitioners were faced with illnesses that caused immense suffering of a kind that amply explains the search and hope for a somatic shortcut. This is not to discount cases of grotesque abuse but, rather, to emphasize how uncertain was the path of physicians who attempted to discern the origin and significance of failed appetite in the absence of any genuine understanding of the complex processes at work. To be sure, by the 1940s a great deal was known about appetite that had not been known when the century began. Scarcely anyone argued any longer, for example, that the urge to eat was purely a matter of motor activity in the stomach. Indeed, by this point many students of appetite regarded no single explanation of the why of eating—whether glandular, nervous, or psychological—as sufficient in itself. Thus, what is most striking about the therapeutic activity of these years is not the perceived failure of somatic responses to ills of appetite, although disappointment is certainly evident. Rather, it is the growing sense that the most urgent complexities in problems of appetite involved not intricacies of neural networks or hormonal chemistry—material substrates of the functioning of appetite now broadly accepted—but the long-deferred questions of the psyche and its relation to somatic realities. The rest of this chapter is devoted to the effort to create a psychosomatic medicine that would at long last grasp appetite—and its discontents—as integrated phenomena of mind and body.

Critiques of "Exact Medicine": Appetite beyond Pavlov

Dissatisfaction with a perceived neglect of the complexities of appetite pre-dated the struggles of the 1930s over Simmonds's disease and anorexia nervosa; in important instances it had already taken shape among physicians hostile to the work of Ivan Pavlov. Certainly some clinicians welcomed Pavlov's findings, seeing in his assertion that a good appetite was essential to healthful digestion confirmation of the long-standing clinical view that digestion was readily disturbed by psychic or nervous influences and that, conversely, psychic and nervous illnesses were often characterized by troubled ingestion or digestion or both. A reviewer in the *Lancet* praised Pavlov's study of digestion as "an admirable example of the value of physiological research as an aid to practical medicine."[48] Similarly, the Leipzig physician Oskar Fischer hailed Pavlov for establishing on a firm experimental basis the fact that digestive secretion operated under psychic influence and that the state of the appetite must figure in any attempt to diagnose digestive ills.[49] Yet medical reaction to Pavlov's work was often harsh. One Glasgow physician denounced what he called Pavlov's "principles of gluttony," which enshrined pleasure at the expense of health, and deplored "decadent scientific methods" that entailed the torture of animals while ignoring the wisdom expressed in such proverbs as "Hunger is the best sauce."[50] A more serious challenge than such re-statements of age-old fears of the pleasures of appetite came from the German physician Wilhelm Sternberg (1865–?), whose objection to Pavlov rested not on moral misgivings but on a thoroughgoing critique of the methods of the "exact medicine" Pavlov claimed to further.

Sternberg was indebted to holistically minded investigators who challenged the dominance of experimentalism and decried the subordination of bedside medicine to the purportedly exact sciences. As part of a general cultural movement to restore "wholeness" (*Ganzheit*) to the individual and society, medical holism attacked the very foundations of the experimental physiology Pavlov represented.[51] Sternberg, a now largely forgotten figure who produced scores of publications, decried the wrongheadedness of a medicine that was obsessed with X-rays and microscopes while ignoring the subjective feelings of appetite and hunger that constituted the foundation of good nutrition and good health. [52]

Sternberg's vociferations partly reflected his resentment of a medical establishment that by his own testimony undervalued his insights, but his arguments also constituted a systematic challenge to a medical science that sought to reduce the desire to eat and the function of eating to objective, measurable phenomena. Far from being a phenomenon comparable or equivalent to the simple process of chemical secretion—Pavlov's reigning dogma that "appetite is gastric juice"—appetite was caught up in intricate processes dependent on sensory stimulation, psychic disposition, and (the genuine novelty of Sternberg's argument) carefully orchestrated motor activities involved in opening and using the mouth as the primary orifice of the digestive system (34–38). Yet complex as it was in operation, appetite encouraged ingestive activities that were executed faultlessly even by that most helpless of creatures, the newborn babe. Prompted by appetite, the newborn opened its mouth: "Never in the whole history of the world has there been a case where a living being introduced food into any other orifice . . . of the body" (54). In the same way healthy adults performed instinctual actions that coordinated sensory impressions, reflexes, and muscular movements in ways that could not be learned in an individual lifetime yet were the first requirement for the maintenance of life (54).

Sternberg offered a wealth of collateral arguments to sustain his case for the instinctual and holistic character of the desire to eat and the primacy of subjective general feelings (*Gemeingefühle*). He ridiculed physiologists who attended only to the chemical constituents and calorie content of food to the detriment of a medicine that recognized appetite as the primal drive.[53] He also held that physicians hoping to solve the puzzles of appetite must heed the experience of ordinary people, who understood eating desires and needs without reflection, who knew, for example, that appetite and hunger differed fundamentally, in nature and bodily locale, and that one can feel hunger without appetite and appetite without hunger (4). The great repository of literature was also of inestimable value in comprehending appetite. It was Goethe's Klärchen, he pointed out, who said of her lover "'when he is with me I am not hungry'" (47) and Shakespeare's Henry VIII who, conveying to Wolsey the papers announcing his doom, urged him, "'Read o'er this; . . . and then to breakfast with / What appetite you have'" (15–16). Sternberg also cited the ancient philosophers, tracing his most striking claim about appetite—that it entailed not only the work of psyche and senses but of muscles as well (countering Pavlov's "appetite is juice" with "appetite is movement," 52)—to Aristotle's *De anima*. Only once it quit Aristotle, he argued, did science forget that appetite means moving and seeking and as such defines the essential difference of animal and plant (52).

Yet for Sternberg, the ultimate observer of appetite was not the simple layperson or Goethe or even Aristotle, but the physician attending patients. Those who, to understand appetite, watched animals eat or measured their digestive juices could never recognize realities obvious to the physician: that a food eagerly consumed could in an instant become distasteful; that illness exaggerated tastes, destroying both appetite and hunger; that appetite could exist in the absence of gastric secretion and even in the absence of a stomach, putatively the all-important organ in the digestive process (1–12, 18). Most important, only the physician recognized what was at stake in striving to understand the desire to eat: loss of appetite was the most deadly feature of fatal ills, and overcoming its loss was not a step in curing but, rather, the cure itself (19–20).

Other German-language physicians also gave special attention to appetite in the interwar period, often spurred by recollections of their countries' bitter wartime experience of hunger. In 1925, Arnold Durig (1872–1961), professor at the Physiological Institute of the University of Vienna, delivered a lecture to the Viennese Society of Physicians in which he urged that laboratory physiology had little to offer in respect to the nature and functioning of appetite and hunger.[54] To make his case for a broad-gauge approach to appetite, Durig discussed findings ranging from chemical investigations to ethnographic, historical, and artistic inquiries to studies of animal behavior in the laboratory and in the wild. Yet of greatest import to Durig, as to Sternberg, was clinical observation of the disturbed appetite seen in somatic ills such as tuberculosis, congenital disabilities, and nervous and psychiatric conditions, including eating pathologies as such.

At the outset Durig drew a distinction between the aims of the clinician and the physiologist, the former chiefly concerned with determining the causes and appropriate treatment for troubled appetite, the latter with the origin and nature of appetite and hunger (2). Yet a good part of his work was devoted to arguing against the primacy of any one style of inquiry or explanation. Durig reserved his harshest criticism for laboratory physiologists who reduced the complexities of appetite and hunger to simple factors—gastric secretion, stomach contractions, or an imagined "hunger hormone."[55] In respect to essential questions surrounding appetite and hunger, he argued, physiology had nothing to offer. It could not say whether hunger and appetite were merely quantitative differences of the same feeling or two distinct feelings with one directed to food in general and the other to specific desired foods. Nor did it clarify matters by classifying the two together as "general sensations" or seeking some parallel with feelings of pleasure and unpleasure (9).

Like Sternberg, Durig owed allegiance to the movement for a holistic medicine and physiology that would recognize the critical importance for health of innate dispositions. In his case, this holistic posture was rooted frankly in vitalism: the *Nahrungstrieb*, governed by purposive vital forces, was at base unerring (2–3). Certainly, he recognized that the drive to eat could be misdirected by a number of influences, especially the desire for pleasure, whether this was afforded by alcohol, excessive meat, or, as seen in pica, nonfood substances (3–4, 14). Similarly, disease could ruin the appetite, especially nervous disease since nerves conveyed or inhibited the effects of powerful emotions that created or destroyed the pleasurable images without which no appetite existed (44). Yet here again Durig decried explanations based narrowly on the work of the vagus nerves, the putative cortical or subcortical hunger center, or Pavlov's conditioned reflexes (41–42, 44–45, 49). All these were involved in the formation and functioning of appetite in interconnected ways that, ultimately, science did not know or understand. Durig urged, finally, that there was no solution to the question of the nature of the drive to eat; appetite participated in ceaseless processes of assimilation and disassimilation, but what specific forces put it into play were unknown. Reviving a language long since discarded by experimental science, Durig proclaimed that in the play of these forces lay "the great secret that nature covers with an impenetrable veil," the secret for which we have "the single, old, small, and still incomparably potent word—Life" (52).

Toward a Psychosomatic Medicine of Appetite

By the 1920s and 1930s hostility to the predominance of laboratory methods and experimentalism joined with other currents in encouraging the emergence of a psychosomatic medicine that would focus on the "whole person" and challenge the hegemony of a narrow somaticism.[56] The term "psychosomatic" as a descriptor for a general approach within medicine first gained wide currency in the 1930s. From the outset it had many competing meanings, yet one shared goal was to revalorize the subjective experience of the patient over and against the objective measures—chemical analysis, kymograph recordings of bodily events, X-ray images—that undergirded claims to certainty made by scientific medicine.

In Germany psychosomatic medicine drew heavily on the version of phenomenology developed by the philosopher Edmund Husserl, who, according to Michael Langenbach, aimed to transcend preconceived categories of experience and to gain access to "phenomena as they really are" with methods allowing for direct apprehension of subjective experience.[57] Langenbach notes that while Husserl's own efforts were self-directed—"first person" (214) in character—his insights were eagerly embraced by practitioners who sought access to the subjectivity of patients, especially those suffering from diverse forms of psychopathology. Such was the case with Karl Jaspers (1883–1969), professor of psychology at Heidelberg and author of the influential *Allgemeine Psychopathologie* (1913). In this work Jaspers attempted to show how the gap between the physician as observer and the patient as observed could be overcome by emphasizing "understanding" over "explaining," and, as one commentator has it, insisting on "a view from inside the patient based on empathy and a concerned relationship."[58] German advocates of psychosomatics were divided on important questions, especially the extent and nature of its connection to psychoanalysis, but they all sought a "unified" approach to illness that would restore respect for the individual's experience of suffering.[59]

In the United States, the sway of psychosomatic medicine, if not the label, may arguably be dated from the appearance of Walter B. Cannon's 1915 book *Bodily Changes in Pain, Hunger, Fear and Rage*. In this work Cannon explored the "remarkable alterations in the bodily economy" that occur "in conjunction with pain, hunger and the major emotions . . . [and] can reasonably be regarded as responses that are nicely adapted to the individual's welfare and preservation."[60] As noted earlier,

Cannon first became interested in the bodily impact of emotions when, working on the mechanics of digestion, he realized that stomach contractions were inhibited in an "enraged cat." A student of William James at Harvard in the 1890s, Cannon was aware of arguments within psychology over the nature of the emotions. In his first extended discussion of the physiology of emotion, he attacked his teacher's contention that bodily events, movements of nervous impulses arising from the viscera, governed emotional states, arguing instead that "visceral changes merely contribute to an emotional complex more or less indefinite."[61] In *Bodily Changes*, Cannon moved beyond the impact of emotion on digestion, taking up the effect of pain or strong emotion on adrenal secretion, levels of sugar in the blood, muscular contraction, and other bodily phenomena. He also laid out, for the first time, the results of his attempts to identify experimentally the role of the "visceral nerves" in linking emotional states to corporal changes. Cannon's "pioneer work" made a powerful impression on experimentalists, clinicians, and the public.[62] Although an experimentalist rather than a practitioner, Cannon sought to find therapeutic utility in his work and to serve as a public voice for what he saw as progressive changes in medicine, including psychosomatics.[63] His high standing in Anglo-American medicine helped shape the principal interests of psychosomaticists in the United States and England, especially the overarching concern with the physiological impact of emotion.

A different route to psychosomatic medicine was forged by the British epidemiologist and student of psychoanalysis James L. Halliday (1897–1983), who held successive positions in the Scottish health service. Like other British critics of Pasteur and bacteriology, Halliday challenged what he saw as dubious claims to certainty made by microscopists and argued for the powerful role of "emotional reaction" in the onset of illness.[64] Lamenting that the term "psychosomatic illness" was "cumbrous," he nonetheless insisted on its utility in indicating "both an aetiology and a mechanism." The causes of such illness were "external agents . . . of a special kind, . . . neither physical nor chemical nor micro-organic, but psychological"; the "mechanism" was the individual's encounter with such agents, "not directly by the outer or inner surfaces of the body . . . but indirectly, via the special senses and the integrating mechanism of the diencephalon, autonomic nervous system, and endocrine glands" (11). This "'bodily mechanism of emotion,'" he concluded, "may affect an individual profoundly, bringing about changes in chemistry, rhythm, secretion, and even structure" (11). Thus Halliday sought to link somatic phenomena to psychic states, conditions of "deprivation, frustration, upset, strain, or difficulty" (11). Writing for the *British Medical Journal*, an early venue for psychosomatic approaches, he declared that "an analysis of morbidity statistics . . . shows that in the light of modern knowledge the only possible interpretation of the maintenance of the present high sickness rates is an increase in psychoneurotic and psychosomatic illness" (11).

As these remarks indicate, the emergence of psychosomatic medicine owed a good deal to the growing influence of psychology and psychiatry within general medicine. The development of increasingly complex approaches to mental and emotional states encouraged investigators who might earlier have neglected psychic factors to turn to mental science for tools usable with patients whose psychological condition seemed essential to their disease picture. The reasons for the improved standing of psychology and psychiatry differ with national context, but some patterns hold true generally. Overall the fortunes of psychiatry had fallen low in the late nineteenth century given its identification with the abuses of asylum care and the deep divisions among its standard-bearers over both the origin and treatment of mental illness. Yet countervailing forces were at work that revived the reputation of psychiatry and, indeed, by the early twentieth century carried it to new heights. When fin-de-siècle psychiatrists embraced degeneration to explain mental illness and aberrant behavior, for example, they tapped into broad cultural anxieties about tainted heredity. Even more decisive was the role psychiatrists played during the First World War, when governments gave them responsibility for the legions of "war neurotics" whose "shakes, stutters, tics and tremors" showed up from the first days of the conflict.[65]

Whatever the fortunes of psychiatry in the larger society, within medicine the broad interest in psychosomatic ideas and practices registered the powerful impact of Sigmund Freud (1856–1939) and psychoanalysis. Certainly in the medicine of appetite disorders, Freud's thinking brought an epochal shift in conceptualization and therapeutics. This development is in many respects paradoxical, as Freud said little directly about appetite, and what he did say constituted perhaps the most radical denial of the significance of the appetite for food ever articulated.

"The Mysterious Leap from Mental to Bodily": Appetite in Psychoanalysis

Freud's first and probably most enduring statement of the relation between the sexual drive and the drive to eat appeared in his famed essay

"Three Contributions to the Theory of Sex," published in 1905.[66] In this essay Freud established the fundamentals of his libido theory, which asserted the primacy of the sex drive over all other forces shaping the psychic life and denied the significance of the appetite for food by subsuming it within the sex drive. At the outset Freud posited the existence of a "sexual instinct . . . analogous to the instinct of taking nourishment, and to hunger" (553), but this assertion quickly gave way to statements of the supremacy of the sex drive over the drive for nourishment (and other needs as well). That the sex drive was deeper seated, more instinctual than other needs, Freud stated in a number of different ways. In one formulation he emphasized the uncontrollability of sexual impulses: "The impulses of the sexual life belong to those which even normally are most poorly controlled by the higher psychic activities" (563). Similarly, he characterized the sexual instinct as the most variable in its search for gratification. Commenting on the grotesque choice of sexual objects individuals sometimes made—children, animals—Freud contrasted such phenomena to the relative stability of the object of hunger: "It throws some light on the nature of the sexual instinct, that it should permit such great variations and depreciation of its object, something which hunger, adhering more energetically to its object, would allow only in the most extreme cases" (563). Persons who engaged in sexual activities "remote . . . from the normal" were "dominated by the most unbridled of all instincts" (571).

The supremacy of the sexual instinct led Freud to conclude that because of the initial experience of pleasure that derived from breastfeeding, "gratification of the need for nourishment" transformed the mouth and lips into an "erogenous zone" (586). This primordial association of sexual pleasure with eating meant that the "psychic counter-forces" mobilized for repression of the sexual instinct could affect not only sexual expression but also fulfillment of the instinct of nourishment: "By virtue of the community of the lip-zone, the repression encroaches upon the instinct of nourishment" resulting, in "female patients," in such "disturbances in eating . . . as *hysterical globus*, choking sensations and vomiting" (584, 587). Thus, Freud argued that in all the neuroses, including those exhibited in disturbed appetite and eating, it was the sexual instinct that "supplie[d] the only constant and most important source of energy" (573). Proposing hysteria as the model for the neuroses, Freud urged that hysterical symptoms of all kinds were "substitutes . . . for a series of emotionally accentuated psychic processes, wishes, and desires," many of them entailing "*conversion* [of mental formations] into somatic phenomena" (573). Such was the "mysterious leap from mental to bodily" that deprived corporal dimensions of appetite of any real significance.[67]

An element of Freud's account of infantile sexuality that was to assume great importance in analysis and treatment of "conversion phenomena," especially as seen in female cases of anorexia nervosa, was his outline of the progress of sexual development from the "pregenital" (oral or "cannibalistic," sadistic-anal) to the "genital" (597). These phases of sexuality preceding the adult "normal" stage accompanied a process of "sexual investigation" in which the child exhibited great interest in "where children come from" (595). On this point psychoanalysis had yielded uniform results about the infantile mind: "One gets children by eating something special . . . or they are born through the bowel, like a passage" (596). The "anal stage" helped explain some of the most common neurotic symptoms, including constipation, which was rooted in the behavior of the child who sought pleasure by holding back the fecal mass, refusing what Freud called "the first 'donation'" and entering a course that led to "later eccentricity or nervousness" (589).

Concluding this discussion Freud returned to the comparison between the libido and other forms of "psychic energy," seeking once again to differentiate "sexual" from "nutritional processes" and emphasizing the role in each of "a special chemism" (611). He rejected absolutely the proposition advanced by his renegade follower C. G. Jung (1875–1961) that the libido flowed into and came to be "synonymous with psychic instinctive energy in general."[68] Rather, "libido-economics" reflected the fact that "sexual excitement is furnished not only from the so-called sexual parts alone, but from all organs of the body" (611). In short, the libido subsumed or converted all forms of bodily and psychic experience: "It is possible that nothing of any considerable significance occurs in the organism that does not also contribute its components to the excitement of the sexual instinct" (602).

It would be hard to overstate the long-term impact of Freud's views on the nature and manifestations of the sexual and nutritive "instincts." Freud's instinct theory, which went through multiple transformations, gave powerful encouragement to those who sought to conceptualize (partly in opposition to behaviorism) primordial, deep-seated "drives," "urges," or "needs." Yet in arguing for the pervasive significance of the sexual drive, Freud undercut the foundation of any theory positing innate alimentary impulses. In other essays Freud made this elision of nutritive drives in favor of the sexual wholly explicit. In a draft manuscript on melancholia, for example, he explicated "the nutritional neurosis parallel to melancholia" by arguing that "the famous *anorexia nervosa* of young girls seems . . . to be a melancholia where sexuality is undeveloped. The patient asserted that she had not eaten, simply because she had *no appetite*, and for no other reason. Loss of appetite—in sexual terms, loss of libido."[69] Similarly, in his famous report of the case of the "Wolf Man," he declared that the "neurosis in girls which occurs . . . at the time of puberty or soon afterwards, and which expresses aversion to sexuality by means of anorexia . . . will have to be brought into relation with the oral phase of sexual life."[70]

With these formulations, Freud brushed aside a century of arguments over the primacy of soma or psyche in eating desires by holding that the appetite for food was not really about food at all. This fateful move cut the drive to eat away from any physical need or desire for food and recast it as part of the individual's progress (whether successful or thwarted) through the psychosexual stages mandated by psychoanalytic thinking and, when pathological in character, as a symptom expressing deep-seated psychic conflicts. For psychosomatic physicians who sought to explain troubles of appetite with the help of Freudian psychoanalysis, the consequences were to be momentous. In their approach the appetite for food never stood as a salient force with its own dynamic or significance; instead, alimentary appetite meant sexual appetite, and disturbances of appetite and eating invariably signified unconscious sexual wishes. As Sándor Ferenczi (1873–1933) declared: "For a great number of patients who suffer from neurotic distaste for food, tendency to vomit, and other stomach disturbances, eating (the passage of a foreign body through the narrow tube of the oesophagus) signifies, unconsciously, genital incursions similar to those which are phantasied."[71] Psychoanalytic formulations of this sort had powerful implications for thinking about appetite and treatment of its disorders within psychosomatic medicine, which, despite its disparate sources and forms, almost always entailed some reference to Freud.

The Specificity of Psychosomatic Illness

Although its advocates sometimes claimed that there was a fundamental unity in psychosomatic approaches, from the beginning the movement was divided on key issues. A contentious question was whether all illness or only specific diseases or conditions were psychosomatic in character. The former view drew importantly on the work of the German clinician Georg Groddeck (1866–1934), who opened a sanitarium for "'lost cases'" in Baden-Baden in 1900.[72] Groddeck was an early admirer of Freud, but even before studying Freud's work he had begun forming the view that every illness has a symbolic dimension. Groddeck was among the first to argue that illness is a "language," a means for the unconscious to express its hidden features (that a sore throat, for example, meant a refusal to "swallow" unpalatable ideas) (87, 89). Groddeck was to become an extreme partisan of the idea that all illness develops as a defense by the "It" (a precursor of Freud's "id") against internal or external attack, directing "the activity of all organic and mental systems to protect life and avoid pain" (89).

Groddeck was an important influence on psychoanalytically oriented advocates of psychosomaticism in the United States, many of them refugees from Nazi Germany. These included the founders of the journal *Psychosomatic Medicine*, which began publishing in 1939. In the opening issue its editors asserted that "Psychosomatic Medicine is not restricted to any specific field of pathology" but was relevant to internal medicine, pediatrics, dermatology, ophthalmology, endocrinology, neurology, and psychiatry.[73] This journal was to prove catholic in approach, but its underlying assumption was that any kind of disease could be psychosomatic in character, not just the classic "functional" or "nervous" diseases, but also hay fever, tuberculosis, diabetes, and many others. By contrast, other psychosomaticists sought to identify specific diseases and conditions that were commonly emotional in origin or the result of "stress."[74] Peptic ulcer, gastritis, and other gastrointestinal ills often appeared first on the list of "certain types of chronic illness [that] defied solution" by the usual methods of medical science.[75]

Disputes over the question of whether all illness or only specific forms fell within the purview of psychosomatic medicine reflected its disparate origins and, in the American setting at least, served to divide practitioners into two broad camps. While some leading figures in the United States believed psychosomatic medicine virtually indistinguishable from psychoanalysis, others, especially researchers influenced by Cannon, asserted that it must establish a firm basis in experimental physiology and its proliferating subfields, especially neurobiology.[76] Theodore M. Brown has shown that already in the late 1930s these competing approaches came into conflict. Brown describes the efforts of the German-born and German-trained psychoanalyst Franz Alexander (1891–1964) to gain and maintain financial support from the Rockefeller Foundation for the kind of psychosomatic medicine pursued at the Chicago Institute of Psychoanalysis. He demonstrates that, while Rockefeller funding was forthcoming for a time in the mid-1930s, the head of its Division of Medical Sciences, Alan Gregg, eventually came to see Alexander's psychoanalytic approach as insufficiently scientific and urged him to promote the kind of "fruitful . . . experimental work on psychosomatic interactions" under way in hospital-based research centers.[77]

Nonetheless, whether grounded in psychoanalysis or in experimental physiology, practitioners of psychosomatic medicine agreed that, as one physician stated, "the *gastrointestinal tract is, above all other systems, the pathway through which emotions are often expressed in behavior.*"[78] Yet explanations of the connection between emotional and gastrointestinal disorders could not have been more different. From Cannon's work onward, the central importance of the "visceral nerves" was a focus of attention for those who, like him, sought to trace the link between psychologically influenced illness and functions of the sympathetic and parasympathetic nervous systems, either acting alone or in combination with glandular processes. Psychoanalysts headed in a different direction, as is evident in a much-discussed contribution by Alexander to a symposium on the psychology of gastrointestinal disorders published in 1934.[79]

Alexander opened by asserting the appropriateness of looking at the gastrointestinal tract given "the well established fact . . . that the alimentary tract is a system which the psychic apparatus uses with great predilection to relieve different emotional tensions" (501–2). Citing Freud's pathbreaking work, as well as subsequent studies by Ferenczi, Karl Abraham (1877–1925), and Ernest Jones (1879–1958), Alexander decried the use by physicians who dealt with gastrointestinal ills of the "everyday language" of "nervous exhaustion" or "unstable personality" as well as such outmoded terms as "'neurasthenia,'" without further effort to "go more deeply into the nature of the patient's psychologic situation."[80] In contrast, the Chicago group divided gastrointestinal disturbances into three classes—gastric ills, colitis, and constipation—each of which was related to distinctive "types of conflict-solution" that reflected "the three elemental tendencies of . . . *in-taking, eliminating, and retaining*" (507–8, emphasis in original). These corresponded in turn to the pregenital stages of the development of sexuality analyzed by Freud and his followers.

Of special interest in the history of appetite are Alexander's cases illustrating the "gastric type," six duodenal ulcers and three "gastric neuroses" observed in patients who were notable for repression of "intense oral-receptive tendencies, the wish to be taken care of and loved" (509). All were people who in their daily lives were efficient, productive, and disdainful of dependency; yet in all there was "in the unconscious a deep oral regression to the parasitic situation of the infant" (510) accompanied by a "sense of inferiority" (created by frustration of "infantile claims"), guilt, and fear (510–11). Alexander was particularly interested in the problem of why such psychological features manifested in peptic ulcers, a problem he referred to as "one of the great riddles of internal medicine" (501). Acknowledging the failure of efforts to define a particular "personality type" liable to peptic ulcer (one typifying "oral characters with over-compensated oral-receptiveness and oral-sadism"), he argued instead for a "typical conflict-situation which may develop in very different personalities" (513, 515). Thus, his cases ranged from that of a forty-six-year-old man who lacked ambition and "remained always mediocre in his profession" to others (the majority) who "over-compensated" for their deep craving "to be loved and to receive" by adopting an "active and responsible attitude in life," often achieving "real . . . success" (516, 511).

Assessing this contribution, Brown argues that Alexander tried (partly out of his desire for Rockefeller funding) to steer a midway course between the extreme, "symbolistic" approach of psychosomaticists like Groddeck and physiologically-grounded investigations that sought to identify specific mechanisms by which emotional tendencies produced somatic symptoms or structural damage. Alexander explained gastric ulcer in these cases by observing that, thanks to the patient's constant state of wishing to be fed—the result of "the repressed or externally thwarted receptive tendencies" (518)—the stomach was in a perpetual state of stimulation "entirely independent of the physiologic state of hunger" (519): "Those individuals who on account of the described conflict-situation have to repress . . . their over-strong receptive cravings express them in the tacit physiologic language of the stomach functions. Such a stomach behaves all the time as if it were taking or were about to take in food. . . . They want food not because of organic hunger but as a symbol of love and help" (519). Alexander hypothesized that such patients suffered from the chronic hypermotility and hypersecretion of the stomach that led to peptic ulcer (517–20). Yet having laid out this mechanism, Alexander recognized that the "psychological situation" he described was "by no means specific . . . to these cases" but in fact "only one of the many possible dynamic outcomes of the same unconscious conflict-situation" (522). This observation led him to endorse the view that illness constituted a "choice" on the part of the patient for the expression of deep-seated conflicts. "Why certain individuals choose just this physiological solution," Alexander mused, "remains an unanswered question which belongs to the dark, hitherto unsolved field of the 'choice of neurosis'" (522). The solution lay, he urged, in some combination of "unknown organic factors" coincident with "the above described psychodynamic configurations" (522). Such claims were to shape approaches to troubled eating, especially anorexia nervosa, for some time to come.

Symbol and Fantasy in Anorexia Nervosa

Alexander's name was ever present in studies published in *Psychosomatic Medicine* of anorexia nervosa, after gastric or duodenal ulcer the most discussed topic in early issues of the journal. An article by a team of three psychiatrists typifies the approach to appetite evident in this psychoanalytically grounded literature.[81] These researchers began their account by tracing the history of anorexia nervosa, recalling that William Withey Gull had seen in the disease "a repugnance for food" caused by "a perversion of the ego" but noting also that other physicians had held, to the contrary, that lost appetite was not essential to it because other symptoms persisted after the return of appetite and gain in weight (3). Their own theory assumed that, although anorexia nervosa began with complaints of failed appetite, it proved "on closer examination . . . to be a reaction of disgust towards food rather than a mere loss of appetite" (3). They also claimed that while the "somatic picture of the disease" was clear, the "psychological circumstances" giving rise to it were little understood (4). These "circumstances" included open conflict between the patient and her family (all the cases reported were female), especially the mother; a previous history of overeating and obesity (attributable at least partly to the family's "over emphasis on . . . eating"), and "peculiarities in the sphere of sex" (4). These "psychological factors," they asserted, "assume[d] a symbolic significance which has no primary relation to the problem of survival" (5). From the very beginning of life, the intake of food was invested with deep psychological meaning given the infant's dependency, and as the child developed, it took on "even more elaborate symbolizations related not only to aggressive . . . behavior, but also to patterns of sexual significance" (5). Thus, nutritive, aggressive, and sexual behavior were all conjoined from the outset.

A crucial element of the causation of anorexia nervosa described by these authors appeared when they related the "psychological mechanisms" involved in eating in children to those of "primitives" (5). A "magical belief of pregnancy through eating is a very common fantasy in children," they stated, one that "may continue . . . , either consciously or unconsciously, throughout life" (5). Such was the basis for their contention that anorexia was invariably tied to "fantasies of oral impregnation" (6). In the "neurotic who is tied with inextricable bonds to the family pattern" and to whom "the whole concept of sexuality and procreation is at this . . . primitive level," eating is "symbolically equated with sexuality, particularly with the fantasy of impregnation" (6). In such individuals, "dependent on the quantity of guilt and anxiety associated with the fantasy . . . and with aggressions within the family, there may be a complete rejection of eating" (6–7). This analysis made clear "why there is more than a mere loss of appetite and actually a feeling of disgust" (7). It also explained the obverse of disgust for food, "compulsive ritualistic gluttony," as well as the great significance of the "mother relationship" in cases where the mother overemphasized "the act of feeding" or, often enough, expressed "her own relationship to food" (7).

This elaborate theorization of anorexia nervosa was supported by two case studies, each containing extensive somatic and psychological detail. In both instances the patient originally complained of loss of appetite, gastrointestinal troubles, amenorrhea, and other somatic symptoms while nonetheless revealing on physical examination "no distress" but for weight loss (8, 11). Medical and family history followed, with attention to conflicts of the patient with family members, especially the mother. In each case the patient's sexual history was related, and all mentions of eating appeared in conjunction with psychosexual events. One patient, for example, compulsively ate large quantities of food after her first experience of kissing a boy (9). Her compulsive eating was "not an urge like hunger" but expressed her fear of pregnancy: the more she ate, the more her stomach became distended and the more she feared she might be pregnant (9). The second patient's eating patterns were also disconnected from hunger and related to a disturbed attitude toward sexuality; her "choice of . . . neurosis" illustrated especially well the interrelation of family conflicts, fear of genital sexuality, and "problems of eating and the gastrointestinal tract" (14–15). On the basis of these two cases, the authors boldly concluded that anorexia nervosa was grounded in a "specific type of fantasy," the "wish to be impregnated through the mouth which results, at times, in compulsive eating, and at other times, in guilt and consequent rejection of food" (15).

Not all psychosomaticists who were influenced by psychoanalysis pursued interpretations based on fantasies of oral impregnation. The Heidelberg psychiatrist Viktor von Weizsäcker (1886–1957) used dream analysis in creating a theory of anorexia as a pursuit of ascetic ideals by highly cultivated young women.[82] Although Weizsäcker emphasized the conflict between "demands of natural life (nutrition and reproduction)" (295) and the drive to achieve intellectual and spiritual transcendence, he said little specifically about his patients' relation to food, arguing that their "ascetic ideal" was not limited to food but extended to sexuality, clothing, sociability, and indeed any sphere of life in which deprivation could, as in the striving after ancient religious ideals, be transformed into a kind of pleasure and enjoyment (254). Yet another style of thinking about appetite in anorexia nervosa is evident in a classic British paper by Grace Nicolle, a staff physician at the Tavistock Clinic in London.[83] Though schooled in Freudian theory, Nicolle was chiefly indebted to Charcot and the French tradition of psychotherapy. In analyzing the origins of anorexia nervosa she made no mention of the theory of oral impregnation but emphasized, instead, as the psychological "motive" for the anorexic's behavior a "fanatical desire to be thin and the dread of obesity" (158), feelings that arose from the sexual fears and uncertainties of female adolescents. Although Nicolle stressed thwarted psychosexual development, she regarded the anorexic's loss of appetite (to her, a key marker of the disease) as a genuine physiological phenomenon probably traceable to a glandular disturbance that interfered "with the normal signals governing the sensation of hunger" (160).

In the 1940s and 1950s American psychosomaticists continued to uphold the Freudian view of the psychogenesis of anorexia in which troubled eating equaled troubled sexuality. One analysis of a "thirty-five-year-old, unmarried girl," for example, revealed her fantasies of "oral incorporation of the phallus [of father/analyst]" and "reactive castrative impulses toward father figures" along with an "overwhelming fear of female genitality." While the report was filled with references to ingestive and digestive symptoms, it seldom mentioned the patient's eating habits or foods, the latter appearing only as symbols for sexual organs (sausages for the penis, almonds for testicles).[84] By the mid-1950s disagreements had developed on whether anorexia nervosa developed primarily out of Oedipal or "pre-Oedipal" conflicts and on the relative importance of mother-daughter and sibling relations, but the essential features of the disease, its genesis in family conflict and "peculiarities in the sphere of sex," remained unchanged.[85]

These views from different national settings indicate that at midcentury divisions persisted in regard to the state of the appetite in anorexia nervosa. Eventually, however, it became psychiatric orthodoxy that such patients, far from suffering loss of appetite, experienced a powerful urge to eat that they struggled relentlessly to control amid the psychic conflicts that gave rise to and indeed constituted the disease. In the subtler psychogenic theories of the 1960s, American psychotherapists fixed their attention more closely on troubled family dynamics but kept to the view that failed appetite was not a significant feature of the disease.[86] Elsewhere, too, psychotherapists argued that anorexia did not mean failed appetite, discounting case studies from the 1870s–1930s that had emphasized the loss of appetite. Writing in 1992, for example, Tilmann Habermas held that patients of that earlier time had successfully engaged in "deception": they actually did experience a desire to eat but pretended they did not as part of the pattern of dissimulation characteristic of the illness.[87] In taking such a stance, clinicians opted for a view that sidestepped the psychophysical complexities of appetite in favor of an interpretation that insistently divided psyche from soma. In this fashion a disease whose principal feature was not eating became divorced from the urge or desire to eat.[88]

Is Appetite Responsible for Fat? Obesity as a Psychosomatic Disorder

While psychosomaticists denied independent influence to the state of the appetite in anorexia nervosa, those who explored the role of appetite in obesity, the "pathology" that became the dominant concern of the twentieth century, followed a more complex trajectory. In 1910 the Vienna pathologist Carl von Noorden (1858–1944) divided the condition into "endogenous" and "exogenous" varieties, a distinction that was to have amazing staying power.[89] Until the 1940s many physicians agreed that a significant proportion of cases of obesity resulted from endogenous disorder and thus belonged with "maladies of nutrition" or "metabolic diseases."[90] Another explanation held that obesity was a pathology of fat tissue, which, Gustav von Bergmann argued, had its own regulatory mechanisms whose disturbance led to exaggerated growth; in such cases a heightened drive to eat was a purely secondary adaptation to pathological changes in fat tissue.[91] Like physicians of earlier eras, many partisans of endogenous theories also argued that obesity was hereditary rather than traceable to overeating.[92] Only at midcentury did physicians begin to characterize endogenous etiologies as a myth and to assert that, with the exception of "relatively rare organic conditions," obesity resulted from "habitual overeating."[93] For obvious reasons, physicians who made a career in the "reducing" business emphasized exogenous factors, especially overeating but also the sedentary way of life associated with urban existence. William F. Christie, author of numerous works on obesity published in interwar Britain, emphasized that aberrations of appetite occurred when individuals relinquished the "very considerable degree of personal control over their adipose tissues" granted them by "Nature" and developed "fat-forming habits" that were products, especially, of conditions in the home: "Fat parents develop fat children because they overfeed and under-exercise them as they did themselves."[94]

As in the nineteenth century, physicians who identified overeating as a significant cause of obesity usually sidestepped the question of why the obese overate or how their propensity to overeat could be curbed. The French practitioner Roger Hyvert acknowledged the difficulty of reining in eating ("It takes courage"), but he was much more concerned with practical ways to help patients lose weight while avoiding potential ill effects such as loss of muscle or flaccidity of the skin. Thus, he had little to offer excessive eaters beyond recommending that they eat small amounts of food before mealtimes or consume appetite suppressants such as hyoscyamus.[95]

The failure of physicians even to pose, much less resolve, the question of why people experienced "voracious appetite" helps to explain the appeal psychosomatic medicine came to exert in this field, especially its psychoanalytic variant. Psychosomaticists were latecomers to the study of obesity. The American journal *Psychosomatic Medicine* devoted space in its early years to a variety of eating and gastrointestinal problems, but obesity was not one of them. Similarly, the first British journal in the field, *Journal of Psychosomatic Research*, which began publishing in 1956, included discussions in its opening issues of ulcerative colitis, peptic ulcer, and gastric neurosis but did not publish its first entry on obesity until 1962.[96] This paucity of attention to obesity perhaps reflected a slow transition away from endogenous etiologies. In any event, the number of psychosomaticists interested in obesity before the 1960s was very limited.

In the United States the most influential psychosomaticist to take up the problem of obesity was the German-born psychoanalyst Hilde Bruch (1904–1984), who emigrated from Germany in 1934. Bruch's first publications on obesity were focused on children. Appearing in the pediatric journal *American Journal of Diseases of Children* and *Psychosomatic Medicine*, these studies amounted to an extensive monograph on the subject. Of special interest is the final contribution, published in 1940, in which Bruch and her collaborator Grace Touraine argued that in children obesity was linked to emotional disturbances and that "increased appetite" was "an allied symptom of nervous tension and anxiety."[97] Their research involved forty obese children, eighteen girls and twenty-two boys, all from impoverished families, either immigrant (Jewish, "North" and "South European") or "American Negro" (143). Bruch and Touraine looked into multiple factors in family background—income, lodging, state of the home and furnishings, even the presence of houseplants (151–52). Noting the amount and type of foods provided in the home, they found that the provision of great quantities of food despite the families' marginal economic standing was a crucial factor in the children's obesity: "Abundance of food represented the one contribution . . . the mother could make . . . [to] a certain sense of affluence" (153). Information was also forthcoming about the parents, many of whom had experienced war, deprivation, separation from family, or loss of family members to violence or persecution. Although in this article the authors disavowed any intensive psychoanalysis, they commented at length on the psychological disposition of parents and children.[98] Their overall assessment of both mothers and fathers was severe: mothers were "self- pitying" and "refuse[d] to face realities"; fathers were ineffectual, subservient to their wives, and incapable of providing the family a sense of "firm masculine guidance."[99]

In this representation, then, obesity resulted from pathologies of family life. Although often hostile to their offspring, the mothers were obsessed with the children's physical safety, denying them independent control of their own dressing, bathing, or bowel movements. To compensate for treating their children badly, the mothers provided excessive amounts of food, often forcing food on their children: "*Common to all cases was the high emotional value with which offering and receiving of food was endowed*" (179, emphasis in the original). Overall, the mothers created "an environment of emotional confusion and insecurity," in which overeating was a principal "safety device" children employed to offset frustration and misery (180). The children's appetite was a constant theme, described by the parents with words like "marvelous" but by the investigators with terms like "voracious" (197, 199). While the parents were mostly unconcerned about or denied their offspring's obesity, they expressed worries about the children's overall health, especially their sexual development and marital future (195).

A prominent theme in this work on obesity was the lack of any evidence pointing to glandular problems in these cases and the damage done by glandular therapies. "Many parents were quite ready to accept the diagnosis of an endocrine disorder," as it conveniently "explain[ed] all the peculiarities of the children and appeared to promise a simple treatment. A few pills or injections would change the clumsy and fat patient into the child of their ambition." The notion of " 'bad glands,' " the authors complained, was "not only incorrect, but harmful in so far as it increases the . . . oversolicitous attitude of the parents, misdirects their attention and blocks the way for recognizing the underlying psychological problems" (196). Elsewhere, to support her attack on endocrine explanations, Bruch drew on criticisms lodged from the 1920s onward of underlying assumptions about the physiology of the obese. Concepts of endogenous obesity "presuppose," she urged, "that the energy requirements and oxidation processes of an obese person are different from those of a normal person," and she cited mounting evidence that "energy exchange" worked the same way in the obese as in the nonobese.[100]

Bruch's view of obesity as an emotional and psychological condition came rapidly to be shared by other psychosomaticists. In 1960 the psychiatrist Myer Mendelson, addressing a meeting of the Academy of Psychosomatic Medicine, declared that, with endocrine treatments a thing of the past, "people think of treatment of obesity almost exclusively in psychological terms."[101] This shift in thinking about the origins of obesity entailed a radical reconceptualization of the position of appetite. Up to this point both endogenous and exogenous theories of appetite had usually allowed for some blending of internal and external influences in obesity. Wilhelm Falta, for example, spoke of appetite as the core component of a "fine regulatory mechanism" that ensured a concordance between bodily need and food ingested.[102] In such thinking the derailment of appetite usually resulted from internal factors such as defective energy exchange, but the regulatory work of appetite could also falter under the influence of voluntary factors such as lack of exercise. The striking feature of Bruch's work was her move toward a full-blown psychologization of obesity in which voracious appetite had nothing to do with bodily mechanisms but was strictly a result of harmful emotional influences that developed within the "family frame." Excessive appetite was evident in the onset of obesity, but what really mattered were the emotional condition of the patient and the pathological effect of parental attitudes and behavior.

Yet tracing a peculiar psychology of the obese was, in its turn, to prove an elusive goal. As with other forms of psychosomatic illness, by the 1940s and 1950s psychotherapists and psychologists seeking a pragmatic alternative to Freudian depth psychology began searching for the specific "personality" in whom conditions such as obesity (or ulcer, or even such ruthless somatic ills as cancer) appeared.[103] But in the case of obesity, at least, this endeavor was called into question even before it got properly started. Already in 1960 Mendelson observed that "there is no necessary or inevitable connection between obesity and personality disorder. We found obese patients who were highly neurotic and obese patients who were impressively stable. Our patients were scattered over the whole diagnostic spectrum from stable to schizophrenic" (158). This judgment that no specific psychology characterized the obese was to remain dominant for decades.[104]

Thus, by 1950 the role that disturbed appetite played in obesity was still much contested. Somaticist approaches based on endocrine and metabolic dysfunction had been largely abandoned, only to be replaced by a cluster of psychological approaches eventually found wanting in their turn.[105] Obesity, it was said by just about everybody, came from overeating. But the question remained: why do people overeat? Whether appetite was indeed responsible for fat was a question that in many respects seemed no closer to solution than it had when the century began. It could scarcely have been otherwise, when appetite itself still lacked any clear definition and was ascribed to some unexplained integration of "forces derived from special movements of the digestive tract, hormone influences, emotions, impulses from the cerebral cortex, from the hypothalamus, and perhaps from the entire body."[106] Little wonder that good numbers of clinicians and scientists alike urged that the goal of comprehending appetite in its acknowledged complexity should give way to the task of redefining it "in forms suitable for operational purposes."[107] Here at last, it seemed, success might be had.

By 1900 physicians concerned with troubled appetite still confronted what one practitioner termed a "kaleidoscopic" array of perplexing conditions.[108] And while scientific medicine with its single-minded somaticism seemed poised to bring clarity on the nature of these conditions and to devise effective treatments, powerful countercurrents were at work in the new sciences of mind that posited an independent "psychogenic" dynamics of disease. Faced with the sometimes stark divide between somatic and psychic approaches to ills that evidently entailed blended bodily and mental distress, some practitioners pushed for a psychosomatic medicine that would resolve dichotomies of mind and body, subjective and objective, privileging, instead, the lived experience of the patient. And yet despite its unifying aims, psychosomatic medicine in fact restructured the somatic-psychic dichotomy, dividing into a physiological camp that emphasized objective measures of illness originating in emotional distress and a psychoanalytic camp bent on unveiling emotional and psychosexual "conflicts." Appetite, again, stood somewhere uncertainly between the two. To those who sought concrete evidence of the bodily impact of psychic trouble, patient-reported states of the appetite offered no surety comparable to quantifiable phenomena such as skin temperature or "respiratory tracings." To the psychoanalytic practitioner, however, appetite was pure epiphenomenon. Eating desires were actually sexual desires; conflicts seemingly centered on food were not about food at all.

Does all this mean that nothing was gained by a half century's work on ills of appetite and eating? Practitioners themselves made divergent claims, some asserting significant advances while others acknowledged perplexity and failure. Without doubt the greatest sense of achievement was evident among nutritionists, who not only took credit for supplying the means to eliminate ancient "deficiency diseases" but also asserted that their pronouncements replaced appetite as a reliable guide to diet even for the prosperous and seemingly well fed. This sense of high accomplishment contrasts vividly with the laments of psychoanalysts around 1950 that nothing seemed to avail against anorexia nervosa and that physicians must face the prospect of almost-certain failure, an outcome often ensured by the "mischievous manipulations" of the patients themselves.[109] Similarly, practitioners concerned with obesity acknowledged that what success they had was usually short lived, although by 1950 promises about the efficacy of new techniques, some grounded in serious research and some pure blandishments, were a constant feature of the ever-growing "reducing" industry.[110] Sober analysts in some cases saw the chief success of the era in the discrediting of hormonal treatments and thus the repudiation of what the British clinician Swale Vincent called "one of the worst forms of present-day quackery."[111]

Yet despite myriad disappointments, the medicine of appetite had made one sure gain, at least among those who came to question approaches to appetite that sought illumination from any single factor, bodily or psychic. Just as physiologists like John Brobeck, after decades of work on the local brain site of appetite, came to recognize its great complexity, many physicians increasingly came to see that ills of appetite were no simple matter to be handled by a pill or injection or, as John Ryle called for, encouragement by a solicitous doctor. Still, for every practitioner who recognized the complexities of appetite more clearly than ever, another—whether motivated by the desire to relieve suffering, the thrill of the investigative hunt, or the monetary rewards of an ever expanding industry of appetite control—disregarded the inescapable fact that appetite remained a little understood phenomenon. To control appetite meant defining it for "operational purposes," and that work gained seemingly unstoppable momentum.

EPILOGUE

Appetite after 1950

The second half of the twentieth century brought a great upsurge in medical and scientific work on appetite and eating. Long confined to medicine and the core disciplines of physiology and psychology, appetite had come to command attention from rapidly proliferating subdisciplines of these and other fields not only in biomedicine but in the behavioral and social sciences as well. The partitioning of long-standing fields into subspecialties was perhaps most notable in physiology, whose continued existence as an overarching discipline was a source of concern to leaders in the field.[1] But splintering and redistribution of effort was also evident in psychology and medicine as psychologists embraced research frameworks supplied by "biopsychology," "behavioral neuroscience," and "cognitive science," or served as practitioners of "clinical" or "behavioral" psychology, and as scores of specialized health fields fostered work on disturbed appetite, especially the ills that from the 1960s were called "eating disorders."[2] At the same time, however, interdisciplinary efforts sought to counter this fissioning process and to work toward "integrated" approaches to appetite and related subjects under such broad rubrics as "psychoneuroendocrinology."[3]

These burgeoning scientific and medical domains saw the elaboration of myriad new methods, concepts, and practices. Researchers in overlapping branches of physiology and psychology continued efforts to pinpoint more precisely the sites in the hypothalamus involved in regulating appetite, hunger, and satiety and to explore neurohormonal interactions.[4] This work intersected with currents in biochemistry and in neuroscience, both of which developed rapidly after 1950. The 1970s witnessed the identification by biochemists of peptide hormones that greatly complicated the old picture of the role played by the ductless glands in governing appetite and ingestion; for a time it seemed that the hormones ghrelin and leptin might hold the key to stimulation and suppression of the urge to eat. Another crucial development of the 1970s was the discovery of the "endogenous opioids" that came quickly to be seen as central components of "feeding behavior."[5] This work, in turn, brought heightened attention in pharmacology to drugs in the class of "opioid antagonists" that seemed to be involved in the suppression of feeding and thus to hold promise as remedies for obesity.[6] In all this work, finally, the impact was felt of the "genetic revolution" that got under way with the discovery in 1953 of the double-helix structure of DNA and that led to the rapid expansion of molecular biology.[7] The contest between instinct and learning, heredity and environment, which had been a recurring theme in the study of appetite and eating habits from the eighteenth century onward, took a dramatically new shape. The question of the extent to which genetic inheritance shapes appetite and eating behavior has been especially contentious in respect to disordered eating, but it has been central to all "integrative" approaches in basic science as well.[8]

While these developments were under way in life science, a great deal of research proceeded within psychology, whether labeled "comparative," "physiological," or "biological," on the "learning processes" credited for patterns of "food choice" and "food preference." From the 1950s to the 1970s such work continued to focus on "need states," especially the "hunger drive."[9] Concentration on hunger and need subsequently gave way to interest in the "incentive values," "rewards," and "hedonic value" of food. In layman's terms, psychology turned away from need to pleasure. From this point the enjoyments of appetite so long neglected in the physiology and psychology of eating took center stage as researchers sought to grasp how pleasure, by then linked to the operation of specific "brain mechanisms," functioned in food choice.[10] Such work, which depended largely on "neuroimaging" and pharmacological experiments, proceeded exclusively in laboratories and bore little relation to the study of animal eating in complex outdoor settings.[11] Indeed, research on the "food habits" of animals in the wild or in wildlife refuges unfolded in response to dynamics increasingly separated from inquiry into human eating. After centuries of serving as a model—negative or positive—for human eating, animals whose "natural" existence was now under threat finally came to stand as an object of interest all their own.[12]

After 1950, social scientists also became deeply involved in the study of eating habits, exploring the role of family environment and ethnic, religious, and cultural traditions. Although it would be impossible in this space to provide even minimal signposts to work in all the fields involved—sociology, economics, business, marketing, consumer behavior, food policy studies—some insight into approaches to appetite within social science may be gained from looking briefly at anthropology. From the first emergence of ethnology and anthropology in the nineteenth century, descriptions of eating habits had appeared, and once evolutionary schemes had taken hold, eating customs had served as one of the criteria serving to distinguish peoples in the hierarchy of "civilization." But for the most part the cultural significance of food and eating became major concerns only in the 1930s and 1940s, in the Anglo-American world with the development of "functional anthropology."[13] During the Second World War anthropologists and sociologists played major roles in agencies involved in managing food supplies and rationing as well as planning for postwar problems including the rehabilitation of concentration-camp inmates and prisoners of war.[14] In subsequent decades work went forward on a range of themes such as sociocultural classifications of foods, "cultural consumption patterns," "food insecurity," "eating and ritual," and many others.[15] Then, from the 1990s, in a striking shift, anthropologists began exploring how cultural forces shape not only concrete dietary choices but also the bodily and psychic experience of eating (and other) desires, creating sometimes highly particularized forms of embodiment and identities.[16] Breaking with dominant arguments about anorexia nervosa, medical anthropologists also began examining therapeutic contexts in which medical practitioners themselves function as essential components of the cultural dynamic from which this "culture-bound syndrome" emerges.[17]

After 1950 the medicine of appetite returned to its origins in at least one significant way in renewing the search for accurate classification of ills involving troubled eating. Thus, psychiatrists struggled to devise an accurate nosology of eating disorders based on clear distinguishing criteria for both classic and newer forms of troubled eating.[18] The third revised edition of the *Diagnostic and Statistical Manual of Mental Disorders* presented four diagnostic criteria for anorexia nervosa that reflected thinking current by 1987.[19] "Bulimia," radically redefined from the days when the term was synonymous with "canine hunger" and other ancient definitions, first appeared in the third (1980) edition of the *DSM* and was first designated as "bulimia nervosa" in the *DSM-III-R*. The closely related "binge-eating disorder" soon followed.[20] The first independent journals devoted to eating disorders began appearing in 1981 with publication of the *International Journal of Eating Disorders*; many more appeared in short order.[21] Perhaps it was the impetus provided by new resources and new publication venues that helps to explain the ever-widening compass of such disorders. In any event forms of eating that had never before carried the taint of the pathological—"restricted eating," "picky eating"—now came under the researcher's gaze.[22]

But the most momentous development in the medicine of troubled eating was the astonishing outpouring of work on obesity that got under way from the 1970s onward. Obesity studies have never been limited to core disciplines but instead have drawn attention in an ever-widening range of biomedical and scientific fields. The explanation that physicians, nutritionists, and scientific experts offer to explain the emergence of this "literature of oceanic proportions" is simple: in the second half of the twentieth century people in the Western world began to overeat in ever-greater numbers and with ever more dire consequences, and science has responded by seeking to understand why.[23] In the estimation of these experts, obesity has come to constitute an "epidemic," the first in history not to involve infectious disease.[24] Historians and social scientists who have studied the history of obesity are divided, however, on the theme of "crisis." Some agree with nutritional experts that an obesity crisis exists but offer longer-term chronologies and point to a multiplicity of factors beyond overeating.[25] Others see wild exaggeration of the dimensions of the obesity problem and urge that the sense of crisis is in good part generated by the experts themselves.[26]

So, where does appetite stand in all this? I showed earlier that in the first half of the twentieth century, efforts were made to dislodge appetite from the vocabulary of science as research came to focus on observable and quantifiable ingestive activity and food intake, and I noted the lament of one researcher in 1955 about the difficulty of getting rid of the old "hunger and appetite" pair left over from history.[27] In subsequent decades this situation radically altered as behaviorism came under challenge and gave way in part to "cognitive" approaches within psychology and especially to the methods and problems of the new brain science. Quantifying food intake has remained a major component of experimental and field-based study of eating behavior, but, as the titles included in a recent collection indicate, much recent work has sought to explain not only, or even primarily, "intake" or "consumption" but also the formation and mechanisms of appetite itself.[28] That this shift was already under way by the 1960s in behaviorist psychology itself is indicated by the declaration of one prominent researcher that investigations focused on reflex "feeding mechanisms" responsible for regulating food intake missed the crucial element of appetite in ingestive experience. Philip Teitelbaum argued that in fact pleasure in eating, long signified by the term appetite, was crucial in feeding choices: "Taste is a powerful reinforcement," a "psychic energizer that increases the urge to eat."[29] Interest in taste and its pleasures was also central to chemical study of tastes and smells, a field whose arrival was signified in the United States by the founding of the Monell Chemical Senses Center in Philadelphia in the 1960s.[30] Yet without doubt the change that most dramatically brought appetite back to the center of scientific study of ingestion was the discovery of the opioid neurotransmitters. From this point onward, appetite and its pleasures were no longer inaccessible psychic correlates of regulatory mechanisms but essential determinants of eating.[31]

In clinical medicine "appetite" never gave way wholly to "food intake" or "ingestive activity," although these terms did appear in medical studies as substitutes for or alongside appetite. Still, even during the behaviorist heyday physicians, nurses, and other practitioners still referred to appetite in discussing drugs used to control it or in describing symptoms of diverse diseases and conditions ranging from pancreatic fibrosis to disturbances caused by the ingestion of foreign bodies.[32] In any event, the most important challenge to the significance of appetite within medicine had come not from behaviorist dicta about the inaccessibility of subjective experience but from psychiatry, which, in interpreting and devising treatments for anorexia nervosa, had discounted eating desires as a meaningful component of the disease.

How did appetite fare in psychiatry after 1950? Like other medical and scientific disciplines psychiatry also splintered after the mid-twentieth century, as a range of research fields and therapeutic specialties emerged. By and large, however, clinical approaches to anorexia nervosa continued to regard appetite as essentially irrelevant to the disease, whose label "anorexia," the traditional term for loss of appetite, was said to be a "misnomer."[33] Once Freudian rationales for ignoring appetite faded away, new ones emerged in, for example, "family systems theory," in which anorexia was said to result from the failure of adolescent girls or young women (in this construct anorexia was normatively a female disease) to "individuate," to develop a sense of self separate from the family (normatively the mother). In such a view the state of the appetite was important neither as a cause of the disease nor as a part of treatment; sufferers had to be made to eat, of course, but their specific feelings about food (other than the desire to avoid it) had no place either in etiologies or in therapies designed to ameliorate underlying psycho-familial disturbance.[34] In recent years biopsychologists have proposed that anorexia and other eating disorders may be better understood and more effectively treated as the "brain reward systems" involved in appetite and eating are illuminated, but at the moment a great gulf separates the neuroscience and biopsychology approach to appetite from that seen in psychiatric understandings of anorexia nervosa.[35] Indeed, this chasm is one of the more striking indicators of the pronounced failure, so far, of efforts to integrate the new sciences of appetite with the medicine of eating (if not appetite) disorders.

The disappointments met by integrative efforts are also evident in the history of the first scientific journal devoted specifically to appetite, which began publication in London in 1980. On its founding, *Appetite* had three chief editors, two psychologists and one physiologist, and its editorial and advisory boards included representatives of psychology, medicine and clinical science, pharmacology, biology, zoology, biochemistry, psychiatry, anthropology, geography, food science, and "regulatory neurobiology." The editors defined a "broad mission" and "interdisciplinary approach" for the journal and emphasized its special interest in "*basic mechanisms* of appetite and ingestion," including "neural, systemic physiological, biochemical and pharmacological processes" as well as "social, cultural, and economic factors" and "*applied aspects*" (industrial, zoological, and clinical).[36] From the beginning the journal was eclectic, including reports of studies on such topics as "stress-induced eating," graphing methods ("edograms") to record chewing and swallowing movements, the effects of anorectic drugs on food intake, and many other widely varying topics. Some subjects covered would have been familiar to pre-1950 researchers, but many reflected new knowledge of hormonal and brain mechanisms and the chemistry of tastes and smells. Contributors came from a broad range of fields, including (beyond those represented on the boards) clinical nutrition, history, "chemoreception sciences," veterinary medicine, and recreational sports.

Despite such integrative features, however, when one of the journal's original editors, David Booth, assessed the contribution *Appetite* had made in its first twenty years of existence, he acknowledged that the potential shown in the early issues for "a new unity across the diversity of research into food choice and intake" had not been realized.[37] Booth charted the persistence of deep divisions that had developed from the 1940s onward—between basic and applied research, industrial and medical investigations, food science and nutrition science. Noting that food and nutrition science were often "at loggerheads," he also observed that both of these fields "largely ignored and were ignored by the cognitive-behavioural science that burgeoned postwar."[38] Booth concluded that a "theoretically coherent methodology" for the study of appetite was still lacking and noted persistent "failures to connect biological, cognitive and social approaches to appetite and food choice, and incomprehension even between those based in the same discipline."[39] Thus, paradoxically, the history of this first journal devoted to the subject indicates not the exploration of appetite as a clearly demarcated scientific object but its continuing status as a convenient catchall designation for a broad range of problems whose relation to one another no one could define. It is striking, too, that few of the research articles published in the journal have included the term "appetite" in their titles but have referred, instead, to various forms of "eating," "human feeding patterns," "food intake," "taste perception," "hypophagia" and "hyperphagia," "release of feeding," "human eating behavior," and the like. Presumably, this continuing incapacity to achieve a unifying approach to appetite explains why, when the journal did another retrospective in 2010, it simply included a numerical citation record in place of a discussion of methodology or steps toward integration.[40]

The research published in *Appetite* also indicates that for many investigators the term "appetite" still referred in practice to assorted behaviorist constructs. Appetite as "desire," "inclination," or "longing"—all designations used commonly up to the early twentieth century—drew notably less attention. Researchers in diverse fields did strive to devise reliable measuring instruments to assess the "subjective" experience of appetite.[41] And most telling for the current state of appetite science, neuroscientists have sought to circumvent the problem of subjectivity altogether by relying on neuroimaging and other objective measures, even if the significance of those measures remained open to diverse interpretation.[42] Despite such efforts, which have evoked continuing methodological debate, appetite remains a phenomenon resistant to measurement in a scientific world in which precise quantification is an unassailable value and goal. It is hard to imagine that we would be in the presence of an obesity epidemic if researchers had not offered mountains of statistics (assessments of body mass index, calculations of calories in relation to energy output) but had proposed, instead, only verbal accounts of the heightened appetite of overweight or obese eaters. Indeed, it is striking how much the old pattern, seen from the eighteenth century onward, of talking about obesity without talking about appetite and eating continues. Obesity is said, for example, to be a more or less inevitable result of massive increases in the food supply coupled with efforts by food producers to get eaters continually to "eat more" whether they really want to or not.[43] And yet it remains as clear now as it did in the days of Hippocrates that some people respond to the ready availability of food by eating a great deal while others eat moderately or little. One current formula has it that obese eaters respond to "signals" in a particular "food environment" that overwhelm their inborn regulatory devices, whereas other people respond to the same cues by eating only to the point required by bodily need.[44] Yet this is just a restatement of the underlying question of why some people eat more than others given the same opportunities, a question that remains as contested as ever.[45] In any event one conclusion that seems clear from the history of appetite is that quantifying food supplies and food intake, or investigating local eating cues, will never shed much light on why people feel about food as they do given the innumerable and ever-shifting variables—age, sex, religion, income level, culinary traditions, illness—that researchers over the centuries have found, often against their will, to be involved in the formation of eating desires.[46]

I suggested at the outset of this study that we are caught between two processes of homogenization that seek to shape our eating desires in radically opposed ways. One is the work of the modern food industries that have striven for at least two centuries to produce food in ever-greater quantities with increasingly reliable methods and, in so doing, to ensure the profitability and success of the enterprises involved. Why this process has entailed a drive toward homogenization has been the subject of a rapidly growing historical literature.[47] Among the many consequences of this homogenizing drive by the food industry has been the creation of what nutritionists call an "SFS" palate, one dependent on scientifically determined ratios of salt, fat, and sugar required to make food products that, despite other vagaries of appetite, have broad appeal to consumers.[48] The other homogenizing process has arisen as a consequence of the rise and spread of "nutritionism," the drive on the part of scientific and biomedical experts to establish norms of healthy eating that will, so they seem to envision, one day supplant all other ways of eating. I see no need to say much about what these norms are; anyone living in the developed world and tuned in to any form of communications media has heard the litany time and again.[49] This drive toward a universal style of healthy eating commenced, I have argued, in the eighteenth century, when natural philosophers began asserting for the first time that processes of ingestion and digestion were in key respects uniform from animals to humans, and that certain foods ensured optimal health and vitality while others did not. From the beginning, this process has entailed the denigration of appetite, the notion once ensconced in Western medicine but in time overthrown, that individuals desire to eat in their own fashion, in accord with their own form of health, and must be free to do so.

Contemporary nutritionists and food experts often express exasperation at historical critiques of what they see as their own laudable efforts to contribute to the preservation of health. Writing "in defense of food science," one research group recently challenged the view that biomedical and scientific research had undermined some fabled freedom to eat. These authors argued that the purportedly once-free exercise of appetite was in fact bound by limitations resulting from overall inadequacy of food supplies or the specific constraints imposed by ignorance, poverty, and recurrent food crises. They also noted the many benefits gained by scientific investigation of foodstuffs, especially the discovery of vitamins and other micronutrients that, in the developed world at least, have made deficiency diseases like pellagra a thing of the past.[50] Similarly, nutritionists have defended their field against "conspiracy theories" that see nutritional standards as a tool for undermining traditional ways of eating in the non-Western world.[51] At some level one must sympathize with these responses by doubtless well-intentioned people whose profession and work historians often cast in a troubling, even sinister, light. Yet such responses mostly miss the point of historical investigation. History is often, perhaps mainly, about unintended consequences. Not only is it notoriously difficult to get at the intentions of historical actors, but even if this were not the case, those intentions are often beside the point when we analyze historical processes that are hard to make out even in retrospect, much less at the moment they unfold. Certainly there are well-documented cases in which nefarious conspiracies by scientific experts have been deleterious to the health of individuals; the machinations of experts who served the tobacco industry are the best recent example.[52] But what end goal the physiologists, psychologists, and others I have examined here had in mind is often unknowable or irrelevant to what eventually has come to pass. The question is not, then, who conspired to bring us to the unhappy predicament of being squeezed between powerful yet conflicting forces that seek to control appetite and food choice. Rather, the question is what historical factors account for the process by which this predicament has taken shape.

I have suggested that the sciences involved in the elaboration of norms of healthy eating largely followed their own dynamics. Certainly many of the investigators I study here responded to broad political and cultural anxieties about the effects of food excess or deprivation, and a number of them participated in public initiatives to shape eating habits of individuals and collectivities. But they principally sought, I have argued, to meet standards of objectivity, certainty, and reliability that evolved within their respective domains of scientific inquiry. In the study of appetite this concern for methodological certainty had fateful consequences. As laboratory experimentation became the defining feature of physiology, researchers necessarily shifted their attention from human to animal appetite and assumed that the animal revealed the human. As quantification became a sine qua non of life science, first in experimental physiology and then elsewhere, objects under study had to be, by their nature, quantifiable. As phenomena of consciousness came to be seen as inaccessible to psychological science, its object of study had to be not consciousness but visible, measurable behavior.

For a long time in the period covered in this study, a particular kind of medicine supplied a counter to these methodological exigencies that of necessity marginalized appetite or recast it as some other entity (food intake) more manageable by accepted methods. Bedside observation, into the twentieth century often celebrated in reference to the "divine Hippocrates," offered a respected alternative to scientific methods that strove for uniform results. Clinical medicine privileged the individual and promised to grasp realities of health and sickness that scientific medicine could never comprehend. In the realm of appetite this meant, among other things, that physicians sought not to establish uniformities but to recognize how appetite varied almost infinitely given the impact of manifold "influences," however those were conceived at given historical moments. But as I have shown, this disposition to oppose clinical observation and experience to the normativities of scientific medicine came under challenge almost from the moment that this history begins. Already by 1800 a contest between a medicine based on close engagement with patients and a new medicine based on an array of innovative procedures and techniques, most importantly pathological anatomy but also statistical inquiry, microscopy, and others, was under way. After 1850 this contest became ever more acute as defenders of a traditionalist medicine focused on individuals came to recognize, and sought to forestall, the coming dominance of scientific medicine. Bedside medicine did not of course disappear. Even today practitioners still urge that a full understanding of the health and sickness of their patients requires direct experience with individuals.[53] Indeed, in recent years, thanks to factors ranging from the worrisome appeal of "alternative medicine" to direct ideological and political challenges from patients and activists, some health professionals have questioned the limitations of what is often called "high-tech medicine" and, in some cases, reasserted the value of the work of the hands-on practitioner.

Nonetheless, few would deny that mainstream medicine continues to rest chiefly on the principles and procedures of scientific medicine as it developed from the nineteenth century onward. And this means that the dominant, certainly the most lavishly funded, approaches to questions involving appetite and its problems continue to rely on the cluster of methods that by common consent are the most defensibly scientific in character and that privilege uniformity or universality over individuality and variability. It is time, then, to ask directly: what is the harm in that? Few would deny that the sciences of food and ingestion have yielded impressive breakthroughs in human knowledge, to name only the most obvious, the essential role of micronutrients in sustaining human life. Nor, I think, would many voices be heard to challenge the efficacy of the food sciences in making available vastly increased quantities of food and thus, for many people, ending the hunger that was an ever present specter in past centuries.[54] Yet recognizing these epoch-making achievements does not mean that we can or should turn a blind eye to the radically different discontents of our own world of food and appetite. Nor does it mean that in criticizing the consequences of the ideology of nutritionism, historians indulge a nostalgic but futile hope to return to some nonexistent golden age of spontaneous eating.[55] Instead, it means that our current eating predicament has emerged in the wake of advances in nutritional and food science and that, as in so many other domains of scientific achievement, progress has created its own pitfalls. How these have emerged is a problem only history can illuminate.

So, what does history suggest are the harms done by the advancing ideology of nutritionism? One that has inspired angry criticism is the stigmatization of people who find it difficult, impossible, or simply undesirable to conform to biomedical standards of healthy eating.[56] That some kind of stigma has attached to obesity throughout Western history is undeniable. Galen decried those who could not master themselves sufficiently to avoid becoming fat. George Cheyne, throughout a long life of gaining and shedding weight, castigated himself for his gluttonous ways. Physicians at every moment in the history of appetite have urged fat patients to exercise willpower. But there can be little doubt that as the obesity "crisis" has unfolded disdain for those carrying excess weight has reached a new pitch.[57] Overweight and obesity may "correlate" with ills such as diabetes and cardiovascular trouble, although there is mounting evidence that claims to this effect have been exaggerated, but being stigmatized as ugly or disgusting or "out of control" entails pain and suffering of a different kind that is no less, maybe more, acute. Nor is the pain of stigmatization something that is experienced only by individuals. Given that analyses of obesity usually point to the bad eating habits and excess weight of particular ethnic and income groups, the project of ending obesity often entails an attempt by well-educated, high-income white people to get low-income people of color to "look more like us."[58] And while it is true that multiple forces are responsible for the hostility and contempt shown to fat people, especially when these carry overtones of ethnic, income, or racial bias, it is undeniable that nutritionist discourse has played, and continues to play, a central role in shaping such attitudes.

And then there is the question of the "contagion" of disordered eating encouraged by relentless attention to the "nutritional self." Although this approach to the etiology of eating disorders has only a small place in the vast literature on the subject, there are indications that the more eating disorders are warned against, the more likely they are to spread among vulnerable groups.[59] Since at least the 1980s a great deal of attention has focused on the ill effects of "the media" in creating standards of beauty and thinness that promote anorexia nervosa in young girls and women. Social media that provide a venue for "pro-anorexia" websites are a more recent concern. But the explosion of comment and advice about achieving thinness has often been the work of nutritional experts themselves, who urge that their only concern is "health." A signal contribution of medical anthropologists in recent years has been to show, in sometimes harrowing ethnographic detail, how treatments for anorexia nervosa are often themselves implicated in the construction and maintenance of anorexia as a fixed identity. Inpatient treatment programs that include a daily regimen of weighing, calorie counting, and rewards and punishments for weight gained or lost replicate the din of advice and exhortation from nutritionists about "monitoring" the body and maintaining "control."[60]

Perhaps the most pernicious element of such advice is the suggestion that the state of one's health can be gauged by looking at one number—the all-meaningful "body mass index." Even if it is true that "correlations" can be discerned between particular BMI levels and the "incidence" of heart disease, diabetes, or other conditions, the idea that health is reducible to BMI is a dangerous one. Such claims encourage people to ignore their overall state of well-being or malaise, to neglect activities and pleasures they might otherwise pursue, and, increasingly often, to use noxious or worthless "health products" (so-called energy drinks, supplements) to "burn off the fat" and achieve the number that signifies a supposedly healthy BMI. The ill effects of such thinking may be most evident in people whose eating is seriously troubled, but they affect everyone who internalizes this kind of nutritional advice.

Last, I would urge, in company with a growing number of critics of nutritionism, that the atmosphere of crisis it has engendered has meant fantastic waste or misuse of valuable resources, financial and intellectual. The healthy-eating campaign in the United States, especially the "war on obesity," has come to resemble other failed campaigns—the war on drugs, the war on terror—that have generated panicked commentary and soaked up vast resources to little, if any, beneficial effect and often with damaging consequences.[61] Even those most committed to the promotion of healthy eating admit that the campaigns waged so far to change people's eating habits to conform with biomedical standards have dismally failed. The science journalist Bee Wilson, who holds out great hope for new techniques of "eating behavior" management to change the ways of picky eaters, anorexics, and overeaters, notes that after decades of exhortation on behalf of vegetables, people in the United States eat fewer vegetables than they did in the 1970s.[62] Similarly, a recent German study that targets the results of what the authors call a "preventative assemblage" shows that the "subjects" of interventions promoting a "healthy lifestyle" held to their traditional ways with great tenacity, refusing to embrace health "modernism."[63] Indeed, it seems that the only people changed by these campaigns are those who, for reasons of their own, decide to craft a "nutritional self" and to embrace the goals and techniques of nutritional science.[64]

Important as all these issues are, in the end, my own questioning of nutritionism stems not so much from the failures or unintended consequences of an instrumentalist approach to appetite and eating, one that establishes healthy eating as the only goal and seeks to funnel all public discourse and resources into pursuit of that goal. Rather, it has to do with the nature of health itself, which, as Georges Canguilhem once urged, constitutes only a truncated or lesser form of health if it does not entail the freedom to find one's own health in one's own way.[65] If health means something other than living without physically identifiable disease, if it means achieving to the extent possible for each individual autonomy in the exercise of self, then certainly experiencing and acting on the appetite for food is one of the most important elements of health so conceived. The central conclusion I draw from this history is that our appetite for food is formed from everything that we are: our given constitution, our first experience of being nurtured, our family environment, sexuality and social relations, cultural values, work and occupation, region, climate, "food environment," economic status, and doubtless a range of other factors appetite researchers have yet to discern. If this is in fact the case, if our eating desires are tied up with our entire nature and experience, then the likelihood that "expert" others, however accomplished they may be in mobilizing specific disciplinary or professional techniques, can guide us beneficially in the exercise of those desires seems to be minimal. This does not mean that there is no place for expertise in the realm of appetite and eating. What it does mean is that, viewed phenomenologically, no counselor or practitioner can ever be as expert in regard to the eating desires of individuals as the individuals are themselves. The fact that appetite is, as its most insightful observers have argued, a "holistic" phenomenon means that if professionals have a role in response to ills of appetite it must be one built on recognition of the inherently "superior" knowledge of the troubled eater. What specific forms such responses might take I cannot say (I am not a therapist), but that they should reckon with the steps and missteps of past efforts I think there can be little doubt.

As I said at the outset of this study, then, as a historian I have no solutions to offer to the "nutritional disarray" of our time. Historians make policy only at their peril and, if prudent, never forget that, although historical knowledge lends depth to analysis of any question, it never offers ready-made answers. This said, I nonetheless will make a last observation. From my reading of recent commentaries on the current state of appetite and its discontents, it has struck me that they often conclude with some kind of endorsement of quiet. Michael Gard acknowledges how unsatisfactory it may be to end a study by saying, "I don't know," but he says it anyway. Bee Wilson, though hopeful about results to be expected from new forms of behavioral therapy, declines to give advice but simply recounts her own experience of overcoming troubled eating. Even the nutritionists Marion Nestle and Malden Nesheim, whose work I have criticized in certain ways, acknowledge the past failures of nutritional exhortation and offer their own recommendations in the form of a "personal preference."[66] Thus, all these observers recognize, if sometimes only obliquely, that the barrage of advice offered in recent decades about managing or correcting appetite has, at best, had little beneficial effect and, at worst, has aggravated many people's anxieties about what and how to eat. Put another way, it is clear that close study of appetite and eating, conducted from different perspectives and with diverse tools, often leads to the sense that we need, more than anything, to do less advising and intervening and more listening, to voices present and past.

What does this mean, then, for those who see—or experience—the anguish caused by disordered appetite and who want something to do in response? Here I think of Jerome Groopman's comment on his medical training and his admiration for a teacher who urged students, in full contradiction of the aggressive spirit of modern science and medicine, "Don't just do something, stand there."[67] He says, in short, recognize the value in admitting uncertainty, in waiting and reflecting rather than acting for the sake of acting. In this spirit I finish with no call for a new offensive strategy against the discontents of appetite but with the suggestion that studying its history encourages us to value our own appetite and to find our own way to health and healthy eating.

Acknowledgments

This study has been a long time in the making, and I welcome the opportunity to express my gratitude to friends and colleagues who have helped me in crucial ways. I presented the first paper on themes that ultimately developed into this study of appetite in 2003 at a conference at Indiana University held in honor of William B. Cohen, whose premature death was a deep loss to all who knew and worked with him. The History Department at Indiana offered rigorous professional training in a congenial environment. Achieving this happy mix was in no small part the work of Bill Cohen.

Among the individuals who have shared knowledge and insights on problems bearing on this study, I am grateful to Sabine Arnaud, Ninette Ashley, Laura Belmonte, Joseph Byrnes, Jan Goldstein, Anita Guerrini, Joe Lunn, Michael Osborne, Sean Quinlan, Matthew Ramsey, Lesley Rimmel, Patrick Singy, Kathleen Wellman, and Anna Zeide. Colleagues who invited me to contribute to meetings or publications they organized include Peter Barker, David Cantor, Andrew Cunningham, Rina Knoeff, Carlos López Beltrán, Manon Mathias, Mary Jo Nye, Robert A. Nye, Barbara Orland, Bryant T. Ragan Jr., Emma Spary, Mary Terrall, Anne C. Vila, and Charles T. Wolfe. I thank them all for the opportunities they gave me to learn from other scholars and to hone my own arguments. I am deeply grateful to those who read and commented on parts or all of the manuscript. Robert Mayer critiqued each version and helped me immensely with problems of analysis, organization, and style. Bob Nye helped to shape this work as he has many of my earlier projects; his astute comments saved me from errors and helped me to recast my treatment of central themes. For many years Marsha Richmond has shared with me her extensive knowledge of the history of the life sciences; I am deeply grateful for her reading and evaluation of the chapters that now make up part 3. Given the aid and encouragement of these readers, I regret mistakes of fact and judgment that remain.

I have many debts to colleagues and staff in the History Department at Oklahoma State University. I thank my colleagues there for the collegial environment they created and for the dedication and integrity they exemplified as scholars and teachers. Susan Oliver and Diana Fry showed unfailing generosity, kindness, and skill in lending administrative and clerical help.

For providing financial support and assistance in locating and obtaining materials, I am indebted to the Department of History, College of Arts and Sciences, and Edmon Low Library (especially the staff in Interlibrary Services) at Oklahoma State University; the Department of the History of Science and History of Science Collections at the University of Oklahoma; the Oklahoma Foundation for the Humanities; the University of New Mexico Libraries; the College of Physicians of Philadelphia; the Rare Book and Manuscript Library at the University of Pennsylvania; the Wellcome Centre Trust for the History of Medicine; and the University of Edinburgh Library. I owe particular thanks to the staff of the History of Medicine Division of the National Library of Medicine. During my many research trips to that rich repository, the staff, especially Dr. Stephen Greenberg, offered invaluable research assistance, graciously accommodated my need to use voice-dictation equipment, and took special care in providing images from the remarkable HMD collections.

To friends and family—Julie and Larry Littlefield, Joe Lunn and Marsha Richmond, Bob and Mary Jo Nye, our daughters Eleanor and Susanna—I extend loving gratitude for the interest they kindly and patiently expressed in a project that often seemed without end. As always, my deepest debt is to Robert, whose keen wit, imagination, and intellect have enriched my life for a half century. This book is dedicated to him.

Earlier versions of parts of this book appeared in the following publications, all single-author works published under my name, Elizabeth A. Williams: "Neuroses of the Stomach: Eating, Gender, and Psychopathology in French Medicine, 1800–1870," *Isis* 98 (2007): 54–79; "Stomach and

Psyche: Eating, Digestion, and Mental Illness in the Medicine of Philippe Pinel," *Bulletin of the History of Medicine* 84 (2010): 358–86 (copyright © 2010 Johns Hopkins University Press); and "The Sciences of Appetite, 1750–1850," *Studies in History and Philosophy of Biological and Biomedical Sciences* 43, no. 2 (June 2012): 372–404.

Abbreviations

AGM	*Archives générales de médecine*
AJP	*American Journal of Physiology*
BHM	*Bulletin of the History of Medicine*
BMJ	*British Medical Journal*
DB	*Deutsche Biographie*
DESM	*Dictionnaire encyclopédique des sciences médicales*
DMW	*Deutsche medizinische Wochenschrift*
DSB	*Dictionary of Scientific Biography*
EMSJ	*Edinburgh Medical and Surgical Journal*
JAMA	*Journal of the American Medical Association*
JHB	*Journal of the History of Biology*
JHBS	*Journal of the History of the Behavioral Sciences*
JHMAS	*Journal of the History of Medicine and Allied Sciences*
MH	*Medical History*
MhJ	*Medizinhistorisches Journal*
ODNB	*Oxford Dictionary of National Biography*
PSMed	*Psychosomatic Medicine*
REGH	*Real-Encyclopädie der gesammten Heilkunde*
YJBM	*Yale Journal of Biology and Medicine*

Notes

INTRODUCTION

1. For guides to the rapidly expanding field of food history, see Kenneth F. Kiple and Kriemhild Coneè Ornelas, eds., *The Cambridge World History of Food*, 2 vols. (Cambridge: Cambridge University Press, 2000); Jeffrey M. Pilcher, ed., *The Oxford Handbook of Food History* (New York: Oxford University Press, 2012); Paul Freedman, Joyce E. Chaplin, and Ken Albala, eds., *Food in Time and Place: The American Historical Association Companion to Food History* (Berkeley: University of California Press, 2014); see also E. C. Spary and Anya Zilberstein, eds., "Food Matters: Critical Histories of Food and the Sciences," *Osiris* 35 (forthcoming).
2. On the massive changes in the goals, funding, and institutions of science that have been given this label, see Peter Galison and Bruce Hevly, eds., *Big Science: The Growth of Large-Scale Research* (Stanford, CA: Stanford University Press, 1992); James H. Capshew and Karen A. Rader, "Big Science: Price to the Present," *Osiris* 7 (1992): 2–25.
3. For a classic exposition of biomedical models of normativity structured around this antinomy, see Georges Canguilhem, *Le normal et le pathologique* (Paris: Presses Universitaires de France, 1966).
4. For broad surveys of developments in the United States, see Harvey Levenstein, *Revolution at the Table: The Transformation of the American Diet* (New York: Oxford University Press, 1988); Levenstein, *Paradox of Plenty: A Social History of Eating in Modern America* (New York: Oxford University Press, 1993); for Europe, see the essays in Jean-Louis Flandrin and Massimo Montanari, eds., trans. Albert Sonnenfeld, *Food: A Culinary History from Antiquity to the Present* (New York: Columbia

University Press, 1999), pt. 7. On some countercurrents, see Amy Trubek, "The Revolt against Homogeneity," in Freedman et al., *Food in Time and Place.*

5. J.-P. Poulain, "Eléments pour une histoire de la médicalisation de l'obésité," *Obésité* 3 (2009): 7–16. On anxieties roused in the modern United States by biomedical commentary on food, including a "national eating disorder" surrounding cholesterol, see Harvey Levenstein, *Fear of Food: A History of Why We Worry about What We Eat* (Chicago: University of Chicago Press, 2012).
6. Elizabeth A. Williams, *The Physical and the Moral: Anthropology, Physiology, and Philosophical Medicine in France, 1750–1850* (New York: Cambridge University Press, 1994).
7. Quoted in Roy Porter, *Blood and Guts: A Short History of Medicine* (New York: W. W. Norton, 2003), 42.
8. On disciplinary confrontations between physiology and psychology between 1870 and 1930, see Roger Smith, *Inhibition: History and Meaning in the Sciences of Mind and Brain* (Berkeley: University of California Press, 1992), esp. 142–62.
9. Lorraine Daston, "Introduction: The Coming into Being of Scientific Objects," in *Biographies of Scientific Objects*, ed. Lorraine Daston (Chicago: University of Chicago Press, 2000), 1–14.
10. Important exceptions are the paired works of E. C. Spary, *Eating the Enlightenment: Food and the Sciences in Paris* (Chicago: University of Chicago Press, 2012) and *Feeding France: New Sciences of Food, 1760–1815* (Cambridge: Cambridge University Press, 2014). For an analysis informed by Norbert Elias's concept of the civilizing process, see Stephen Mennell, "On the Civilizing of Appetite," in *The Body: Social Process and Cultural Theory*, ed. Mike Featherstone, Mike Hepworth, and Bryan S. Turner (London: Sage Publications, 1991), 126–56. For a brief overview of pertinent literary and medical texts, see Dietrich von Engelhardt, "Hunger und Appetit: Essen und Trinken im System der Diätetik–Kulturhistorische Perspektiven," in *Kulturthema Essen: Ansichten und Problemfelder*, ed. Alois Wierlacher, Gerhard Neumann, and Hans Jürgen Teuteberg (Berlin: Akademie Verlag, 1993), 137–49.
11. On Aristotle, see Leon R. Kass, *The Hungry Soul: Eating and the Perfecting of Our Nature* (New York: Free Press, 1994); Wendell Berry, "The Pleasures of Eating," Center for Ecoliteracy, https://www.ecoliteracy.org/article/wendell-berry-pleasures-eating.
12. Mark R. Rosenzweig, "The Mechanisms of Hunger and Thirst," in *Psychology in the Making: Histories of Selected Research Problems*, ed. Leo J. Postman (New York: Knopf, 1962), 73–143; Philip Teitelbaum, "Appetite," *Proceedings of the American Philosophical Society: Symposium on "Psychology: A Behavioral Reinterpretation"* 108, no. 6 (December 15, 1964): 464–72; F. Gallouin and J. Le Magnen, "Evolution historique des concepts de faim, satiété et appétits," *Reproduction, nutrition, développement* 27, no. 1B (1987): 109–28. Also helpful are histories of specific investigative techniques, including Lee J.

Skandalakis and John E. Skandalakis, "The History and Surgical Anatomy of the Vagus Nerve," *Surgery, Gynecology & Obstetrics* 162, 1 (1986): 75–85.

13. On the different aims and approaches in writing history of historians and working scientists, see Peter J. Bowler and John V. Pickstone, introduction to *The Modern Biological and Earth Sciences*, vol. 6 of *The Cambridge History of Science* (Cambridge: Cambridge University Press, 2009), 1–12.
14. For a discussion of eating choices as a domain of "everyday life" in which individuals contest the dominance of experts and technocrats, see Spary, *Feeding France*, 315–24.
15. J. M. D. Olmsted, *François Magendie, Pioneer in Experimental Physiology and Scientific Medicine in XIX Century France* (New York: Schuman's, 1944), 22–33; Karl E. Rothschuh, *History of Physiology*, trans. Guenter B. Risse (Huntington, NY: Robert E. Krieger, 1973), 205, 224, 232.
16. François Duchesneau, *La physiologie des Lumières: Empirisme, modèles et théories* (The Hague: Martinus Nijhoff, 1982), 361–430; Roselyne Rey, *Naissance et développement du vitalisme en France de la deuxième moitié du 18e siècle à la fin du Premier Empire* (Oxford: Voltaire Foundation, 2000); Elizabeth A. Williams, *A Cultural History of Medical Vitalism in Enlightenment Montpellier* (Aldershot, UK: Ashgate, 2003); Williams, *The Physical and the Moral*, 20–66; Peter Hanns Reill, *Vitalizing Nature in the Enlightenment* (Berkeley: University of California Press, 2005); Anne Harrington, *Reenchanted Science: Holism in German Culture from Wilhelm II to Hitler* (Princeton, NJ: Princeton University Press, 1996). See also the special issue edited by Charles T. Wolfe, "Vitalism without Metaphysics? Medical Vitalism in the Enlightenment," *Science in Context* 21, no. 4 (December 2008).
17. Harrington, *Reenchanted Science*, 160.
18. Elizabeth A. Williams, "Stomach and Psyche: Eating, Digestion, and Mental Illness in the Medicine of Philippe Pinel," *BHM* 84 (2010): 358–86.
19. For historical retrospectives by clinicians, see Hilde Bruch, *Eating Disorders: Obesity, Anorexia Nervosa, and the Person Within* (New York: Basic Books, 1973); Walter Vandereycken and Ron van Deth, *From Fasting Saints to Anorexic Girls: The History of Self-Starvation* (New York: New York University Press, 1994); Tilmann Habermas, *Heißhunger: Historische Bedingungen der Bulimia nervosa* (Frankfurt: Fischer Taschenbuch, 1990). Works by social and cultural historians, often inspired by feminism and focused chiefly on anorexia nervosa, began to appear in the 1970s; for titles and discussion, see Joan Jacobs Brumberg, *Fasting Girls: The History of Anorexia Nervosa* (New York: Vintage Books, 2000), 38–39 and 291nn59–61.
20. "Appetit," in Johann Heinrich Zedler, ed., *Grosses vollständiges Universal-Lexicon aller Wissenschafften und Künste* (Leipzig: Zedler, 1731–1754), supplement, 2:58, http://www.zedler-lexikon.de/index.html?c.
21. See, for example, the shifting lexicon surrounding hunger or appetite hormones; A. P. Goldstone, "The Hypothalamus, Hormones, and Hunger: Alterations in Human Obesity and Illness," *Progress in Brain Research* 153 (2006):

57–73; Marsha McCulloch, "Appetite Hormones," *Today's Dietitian* 17, no. 7 (July 2015): 26–30.

22. Robert Jütte, *A History of the Senses from Antiquity to Cyberspace*, trans. James Lynn (Malden, MA: Polity, 2005); Carolyn Korsmeyer, *Making Sense of Taste: Food and Philosophy* (Ithaca, NY: Cornell University Press, 1999).
23. A pivotal text was Pierre Roussel, *Système physique et moral de la femme* (Paris: Vincent, 1775).
24. Kurt Danziger, "The History of Introspection Reconsidered," *JHBS* 16 (1980): 241–62.
25. For two such accounts by physicians, see George Cheyne, *The English Malady; or, A Treatise of Nervous Diseases of All Kinds, as Spleen, Vapours, Lowness of Spirits, Hypochondriacal, and Hysterical Distempers, etc.* (London: George Strahan; Bath: James Leake, 1733); J.-P.-T. Barras, *Traité sur les gastralgies et les entéralgies ou maladies nerveuses de l'estomac et des intestins*, 2nd ed. (Paris: Béchet, 1827).
26. Mary Fissell, "The Disappearance of the Patient's Narrative and the Invention of Hospital Medicine," in *British Medicine in an Age of Reform*, ed. Roger French and Andrew Wear (London: Routledge, 1991), 92–109.
27. On "methodolatry" in psychology, see Kurt Danziger, *Constructing the Subject: Historical Origins of Psychological Research* (Cambridge: Cambridge University Press, 1990), 136–55.
28. On efforts by individual doctors to "bring the new physiology into ordinary practice," see John V. Pickstone, "Physiology and Experimental Medicine," in *Companion to the History of Modern Science*, ed. Robert C. Olby (London: Routledge, 1990), 728–42, at 738–40; on differing trends in France and Germany, see Michael Hagner, "Scientific Medicine," in *From Natural Philosophy to the Sciences: Writing the History of Nineteenth-Century Science*, ed. David Cahan (Chicago: University of Chicago Press, 2003), 49–87, at 65–66. Studies emphasizing conflict include Christopher Lawrence, "Incommunicable Knowledge: Science, Technology, and the Clinical Art in Britain 1850–1914," *Journal of Contemporary History* 20 (1985): 503–20; Lawrence, "Still Incommunicable: Clinical Holists and Medical Knowledge in Interwar Britain," in *Greater Than the Parts: Holism in Biomedicine, 1920–1950*, ed. Christopher Lawrence and George Weisz (New York: Oxford University Press, 1998), 94–111; John Harley Warner, "Ideals of Science and Their Discontents in Late Nineteenth-Century American Medicine," *Isis* 82, no. 3 (1991): 454–78.
29. John A. Mills, *Control: A History of Behavioral Psychology* (New York: New York University Press, 2000), 1–18; John M. O'Donnell, *The Origins of Behaviorism: American Psychology, 1870–1920* (New York: New York University Press, 1985), x–xi, 13–14, 200–208.
30. Or as B. F. Skinner had it, "The series of reflex acts by means of which an animal seizes, chews and swallows its food"; see Skinner, "On the Conditions of Elicitation of Certain Eating Reflexes," *Proceedings of the National Academy of Sciences* 16, no. 6 (1930): 433–38.

31. See, for example, M. I. Grossman, G. M. Cummins, and A. C. Ivy, "The Effect of Insulin on Food Intake after Vagotomy and Sympathectomy," *AJP* 149 (April 1, 1947): 100–102: "Theoretically . . . insulin hypoglycemia could act upon . . . learned food taking patterns, i.e. 'appetite', in some unknown way" (102).
32. R. J. Herrnstein, "Nature as Nurture: Behaviorism and the Instinct Doctrine," *Behaviorism* 1, no. 1 (Fall 1972): 23–52. Herrnstein traces the development of research programs that "centered not on action itself but on the impulse to action" (40).
33. Herrnstein, "Nature as Nurture," 45–49.
34. Roy Waldo Miner, ed., *The Regulation of Hunger and Appetite* (New York: New York Academy of Sciences, 1955), 3, 6, 15–16, 56, 77–78.
35. Gyorgy Scrinis, "On the Ideology of Nutritionism," *Gastronomica* 8, no. 1 (Winter 2008): 39–48. Scrinis sees a partial convergence of the two processes in the food industry's recent cooptation for marketing purposes of nutritional recommendations in regard to micronutrients and functional powers of foods (lowering cholesterol and similar claims).
36. Georges Canguilhem, "La santé, concept vulgaire et question philosophique," *Cahiers du séminaire de philosophie* 8 (1988): 119–33.

INTRODUCTION TO PART ONE

1. Cormac Ó Gráda and Jean-Michel Chevet, "Famine and Market in Ancien Régime France," *Journal of Economic History* 62, no. 3 (September 2002): 706–33; Cormac Ó Gráda, *Famine: A Short History* (Princeton, NJ: Princeton University Press, 2009).
2. For broad changes in the early modern era, see Jean-Louis Flandrin, "Dietary Choices and Culinary Technique, 1500–1800," in Flandrin and Montanari, *Food*, 403–17. Works covering eighteenth-century developments include Spary, *Eating the Enlightenment*, and Ursula Heinzelmann, *Beyond Bratwurst: A History of Food in Germany* (London: Reaktion Books, 2014), 131–61.
3. On luxury consumption and the debates it generated, see Maxine Berg and Helen Clifford, eds., *Consumers and Luxury: Consumer Culture in Europe, 1650–1850* (Manchester: Manchester University Press, 1999); Maxine Berg and Elizabeth Eger, eds., *Luxury in the Eighteenth Century: Debates, Desires and Delectable Goods* (Basingstoke, UK: Palgrave Macmillan, 2003); Daniel Roche, *A History of Everyday Things: The Birth of Consumption in France, 1600–1800*, trans. Brian Pearce (New York: Cambridge University Press, 2000), esp. 54–78.
4. Maxine Berg and Elizabeth Eger, "The Rise and Fall of the Luxury Debates," in Berg and Eger, *Luxury in the Eighteenth Century*, 7–27; Roche, *History of Everyday Things*, 72–78.
5. Spary, *Eating the Enlightenment*, 25–26, 195, 226–27, 234–37; see also Anne C. Vila, *Enlightenment and Pathology: Sensibility in the Literature and Medicine of*

Eighteenth-Century France (Baltimore: Johns Hopkins University Press, 1998), 192–94, 196, 212–16.

6. On religious disputes, see Ken Albala, *Food in Early Modern Europe* (Westport, CT: Greenwood Press, 2003), 199–207; Gerald J. Gruman, *A History of Ideas about the Prolongation of Life* (Philadelphia: American Philosophical Society, 1966), 68–74.
7. Anita Guerrini, *Obesity and Depression in the Enlightenment: The Life and Times of George Cheyne* (Norman: University of Oklahoma Press, 2000); Rina Knoeff, *Herman Boerhaave (1668–1738): Calvinist Chemist and Physician* (Amsterdam: Royal Netherlands Academy of Arts and Sciences, 2002), 199–204.
8. On this new genre, see Laurence Brockliss and Colin Jones, *The Medical World of Early Modern France* (Oxford: Clarendon Press, 1997), 463–64, 726, 752–53; Jean-Pierre Peter, "Aux sources de la médicalisation, le regard et le mot: Le travail des topographies médicales," in *Populations et cultures: Études réunies en l'honneur de François Lebrun* (Rennes: Amis de François Lebrun 1989), 103–11; Jan Brügelmann, "Observations on the Process of Medicalization in Germany, 1770–1830, Based on Medical Topographies," *Historical Reflections/Réflexions historiques* 9 (1982): 131–49.
9. J. H. Baron and A. Sonnenberg, "Alimentary Diseases in the Poor and Middle Class in London 1773–1815, and in [the] New York Poor 1797–1818," *Alimentary Pharmacology and Therapeutics* 16 (2002): 1709–14; Williams, "Stomach and Psyche."
10. Michel Foucault, *The History of Sexuality*, vol. 1, *An Introduction*, trans. Robert Hurley (New York: Vintage, 1990), 140–41, 143–44. For recent views of biopower, see Paul Rabinow and Nikolas Rose, "Biopower Today," *Biosocieties* 1 (2006): 195–217; Roger Cooter and Claudia Stein, *Writing History in the Age of Biomedicine* (New Haven, CT: Yale University Press, 2013), 183–204.
11. Ken Albala, *Eating Right in the Renaissance* (Berkeley: University of California Press, 2002), 7, 31–32, 34–35.
12. Rebecca L. Spang, *The Invention of the Restaurant: Paris and Modern Gastronomic Culture* (Cambridge, MA: Harvard University Press, 2000), esp. 35–39.
13. R. Passmore, "William Cullen and Dietetics," in *William Cullen and the Eighteenth-Century Medical World*, ed. A. Doig et al. (Edinburgh: Edinburgh University Press, 1993), 167–85, at 170, 175, 184; Albala, *Eating Right*, 6, 84, 104–6.

CHAPTER ONE

1. For this phrase, see Owsei Temkin, *Hippocrates in a World of Pagans and Christians* (Baltimore: Johns Hopkins University Press, 1991), 71.
2. Jacques Jouanna, *Hippocrates*, trans. M. B. DeBevoise (Baltimore: Johns Hopkins University Press, 1999), 156–58, 161–66, 233–37; Vivian Nutton, *Ancient Medicine*, 2nd ed. (London: Routledge, 2013), 96–97, 102; L. J. Rather, "The 'Six Things Non-Natural': A Note on the Origin and Fate of a Doctrine

and a Phrase," *Clio medica* 3 (1968): 335–47; William Coleman, "Health and Hygiene in the *Encyclopédie*: A Medical Doctrine for the Bourgeoisie," *JHMAS* 29 (1974): 399–421.

3. All references are to *Hippocrates*, Loeb Classical Library, vols. 1–10, various translators (Cambridge, MA: Harvard University Press, 1923–2012). Parenthetical citations first provide the title of and passage from the Hippocratic treatise and then the volume and page number in the Loeb series.
4. On the role of the senses in bedside observation, see Jouanna, *Hippocrates*, esp. 106–7, 200–201, 249, 291–303, 306–7.
5. On the physician's use of reason and on "hidden diseases," see Jouanna, *Hippocrates*, 302, 307–22.
6. Antoinette Stettler, "Zeichen lesen und Zeichen deuten: Zur Geschichte der Medizinischen Semiotik," *Gesnerus* 44 (1987): 33–54; Wolfgang U. Eckart, "'Und setzet eure Worte nicht auf Schrauben': Medizinische Semiotik vom Ende des 18. bis zum Beginn des 20. Jahrhunderts—Gegenstand und Forschung," *Berichte zur Wissenschaftsgeschichte* 19 (1996): 1–18.
7. On the philosophical background, see W. H. S. Jones, introduction to *Nutriment*, Loeb I:337–41, at 337.
8. On dietary habits, see *Regimen in Acute Diseases* XVIII, Loeb II:85–87; Jouanna, *Hippocrates*, 328–31.
9. The classic exposition is in Henry E. Sigerist, *A History of Medicine*, vol. 2, *Early Greek, Hindu, and Persian Medicine* (New York: Oxford University Press, 1961), 223–28, 298–316. Jouanna argues that Hippocratic physicians anticipated the values of Western humanism by making no distinction among patients, yet he recognizes "the wall of silence" that separated, for example, female patients and male physicians. See Jouanna, *Hippocrates*, 115–16, 123.
10. Moderation had both moral and climatic meanings; see Jouanna, *Hippocrates*, 214–15. On countertraditions that sanctioned revelry and excess "removed from the decencies of daily life," see Andrew Dalby, *Siren Feasts: A History of Food and Gastronomy in Greece* (London: Routledge, 1996), 16–20, at 19.
11. Michel Foucault, *The History of Sexuality*, vol. 2, *The Use of Pleasure*, trans. Robert Hurley (New York: Vintage Books, 1990), 71–77.
12. The physician's expertise came into play in reference to particular foodstuffs. *Regimen II*, for example, describes varieties of wheat and their presumed bodily effects (drying, laxative, nourishing, and so on); *Regimen II* XLII, Loeb IV:311–15.
13. On slaves as a special category in ancient dietetics, see Fritz Steckerl, ed., *The Fragments of Praxagoras of Cos and His School* (Leiden: E. J. Brill, 1958), 29.
14. Foucault, *Use of Pleasure*, 50–51.
15. *Ancient Medicine* V, Loeb I:21; Jouanna, *Hippocrates*, 232–35.
16. On Hippocratic ethnography, see Jouanna, *Hippocrates*, 211–37, and on larger attitudes toward the food customs of "barbarians," see Dalby, *Siren Feasts*, 21–22.

17. An English translation is available in Rebecca Flemming and Ann Ellis Hanson, "Hippocrates' *Peri Partheniôn* ('Diseases of Young Girls'): Text and Translation," *Early Science and Medicine* 3, no. 3 (1998): 241–52. Helen King has offered a valuable reinterpretation of the "disease of virgins" later known as chlorosis (in English "green sickness"), and I follow her reading closely; see Helen King, "Green Sickness: Hippocrates, Galen and the Origins of the 'Disease of Virgins,'" *International Journal of the Classical Tradition* 2, no. 3 (1996): 372–87; another version of this discussion appears in Helen King, *Hippocrates' Woman: Reading the Female Body in Ancient Greece* (London: Routledge, 1998), 188–204.
18. Flemming and Hanson, "Hippocrates' *Peri Parthenion*," 250.
19. King, "Green Sickness," 386.
20. Cited in King, "Green Sickness," 384, 374.
21. King argues that early modern physicians (as well as the seventeenth-century midwife Jane Sharp) read the Hippocratic text in light of the Galenic model of the female body, specifically the idea that "meat is the food which is most completely converted to blood" and should be avoided by pubertal girls in danger of illness from excess and obstruction of blood ("Green Sickness," 384–85).
22. *Regimen in Health* VI, Loeb IV:53. An alternative view of the connection between food intake and reproduction was that menstruation carried off dietary excess and thus contributed to women's superior health overall; King, *Hippocrates' Woman*, 50–51. On early modern discussions of the salutary role of menstruation, see Michael Stolberg, "A Woman's Hell? Medical Perceptions of Menopause in Preindustrial Europe," *BHM* 73, no. 3 (Fall 1999): 404–28.
23. King shows that the Hippocratic texts discuss not a disease called hysteria but only hysteric symptoms associated with the "roaming womb," including loss of appetite; King, *Hippocrates' Woman*, 216. The uncontrolled appetite of obese women also formed a bar to their fulfilling their most important role, conceiving and bearing a child; see *Aphorisms*, V:XLVI, Loeb IV:171. Into the twentieth century physicians asserted that fat women were often sterile; see Salomon Davidovici, *Obésité et gestation* (Paris: Jouve, 1933).
24. Anne Carson, "Putting Her in Her Place: Woman, Dirt, and Desire," in *Before Sexuality: The Construction of Erotic Experience in the Ancient Greek World*, ed. David M. Halperin, John J. Winkler, and Froma I. Zeitlin (Princeton, NJ: Princeton University Press, 1990), 135–69.
25. Steckerl, *Fragments of Praxagoras*, 14.
26. Although Aristotelian physics was overthrown by the late seventeenth century, physiologists and physicians drew on Aristotle into the nineteenth century; see L. W. B. Brockliss, "Aristotle, Descartes and the New Science: Natural Philosophy at the University of Paris, 1600–1740," *Annals of Science* 38 (1981): 33–69; Brockliss, "The Moment of No Return: The University

of Paris and the Death of Aristotelianism," *Science & Education* 15 (2006): 259–78; William Coleman, *Biology in the Nineteenth Century: Problems of Form, Function, and Transformation* (Cambridge: Cambridge University Press, 1971), 18, 69, 96–97; Edwin Clarke and L. S. Jacyna, *Nineteenth-Century Origins of Neuroscientific Concepts* (Berkeley: University of California Press, 1987), 19, 20, 102, 321.

27. "On the Soul," in *The Complete Works of Aristotle, The Revised Oxford Translation*, 2 vols., ed. Jonathan Barnes (Bollingen Series 71, pt. 2) (Princeton, NJ: Princeton University Press, 1984), 1:641–92; Fernando Vidal, *The Sciences of the Soul: The Early Modern Origins of Psychology*, trans. Saskia Brown (Chicago: University of Chicago Press, 2011), 24 and 24n8.
28. Vidal, *Sciences of the Soul*, 23–25, 29, 30–33, 85n65.
29. Jütte, *History of the Senses*, 41–42.
30. Kass, *Hungry Soul*, 47, 48.
31. Kass, *Hungry Soul*, 155; *Nicomachean Ethics*, in *Complete Works of Aristotle*, 2:1729–1867, at 1118a23–b5 (2:1765).
32. Michael Boylan, "The Digestive and 'Circulatory' Systems in Aristotle's Biology," *JHB* 15, no. 1 (Spring 1982): 89–118, at 117; on what nineteenth-century physiologists indebted to Aristotle would call the "appetite" of tissues, see Charles A. Culotta, "Tissue Oxidation and Theoretical Physiology: Bernard, Ludwig, and Pflüger," *BHM* 44 (1970): 109–40.
33. On these themes in vitalist physiology, see chapter 3, "The Physiology of Appetite: Mechanists and Vitalists."
34. On the modern contest between appetite as trained or inherent, see part 4.
35. Philip J. Van der Eijk argues that Aristotle's thinking was sometimes that of a metaphysician and sometimes that of a physician; see Van der Eijk, "Aristotle's Psycho-physiological Account of the Soul-Body Relationship," in *Psyche and Soma: Physicians and Metaphysicians on the Mind-Body Problem from Antiquity to Enlightenment*, ed. John P. Wright and Paul Potter (Oxford: Clarendon Press, 2000), 57–77.
36. "If ten pounds are too much for a particular person to eat and two too little, it does not follow that the trainer will order six pounds; for this also is perhaps too much for the person who is to take it, or too little"; Aristotle, *Nicomachean Ethics*, 1106a29–1106b6 (2:1747).
37. Galen, *On the Usefulness of the Parts of the Body*, 2 vols., trans., intro., and commentary by Margaret Tallmadge May (Ithaca, NY: Cornell University Press, 1968), 1:208–9.
38. The vagus nerves formed part of Galen's sixth pair of cranial nerves, a classification that held into the seventeenth century. The modern system of twelve pairs, including the vagus as the tenth, was formulated by Samuel Soemmering in 1778; see Skandalakis and Skandalakis, "History and Surgical Anatomy of the Vagus Nerve"; on Galen's model of the nervous system, see Clarke and Jacyna, *Nineteenth-Century Origins of Neuroscientific Concepts*, 29–30.

39. Owsei Temkin, *Galenism: Rise and Decline of a Medical Philosophy* (Ithaca, NY: Cornell University Press, 1973), 55, 121, 156n55, at 55; Nutton, *Ancient Medicine*, 238–40.
40. *A Translation of Galen's Hygiene (De sanitate tuenda)*, trans. Robert Montraville Green, intro. Henry E. Sigerist (Springfield, IL: Charles C. Thomas, 1951), 7 (emphasis added).
41. One Enlightenment physician wrote that "digestion has a general and essential influence on the whole animal economy"; G.-F. Venel, "Digestion," in *Encyclopédie*, 4:999–1003, at 999.
42. Temkin, *Galenism*, 25, 73, 91–92, 168; Nutton, *Ancient Medicine*, 240–41.
43. Galen wavered on plants, asserting in *De usu partium* that "plants, of course, cannot feel want" (1:208–9) but in *Hygiene* that like other living beings plants enjoyed "innate appetites" (7). On appetite and feeling in plants, see François Delaporte, *Nature's Second Kingdom: Explorations of Vegetality in the Eighteenth Century* (Cambridge, MA: MIT Press, 1982), 20–21, 150, 157–60.
44. Cited in Robert J. Richards, *Darwin and the Emergence of Evolutionary Theories of Mind and Behavior* (Chicago: University of Chicago Press, 1987), 20–21 and 21n3.
45. Galen, *On the Powers of Foods*, in *Galen on Food and Diet*, ed. Mark Grant (London: Routledge, 2000), 118, 127, 129, 134, 144–45, 201n5.
46. Galen, *Hygiene*, 18–19. Temkin observes that "Galen allows a remarkable latitude for health, between ideal health on the one hand and illness on the other, and he thus finds a place for the various temperaments of man" (*Galenism*, 102).
47. Galen, *Hygiene*, 18–19, 53, 75, 195–96. In this work Galen included among the non-naturals "sleep and waking, rest and motion, . . . hunger, thirst, eating, drinking, satiety or deficiency of food or drink; and in addition baths, anger, worry, grief, and all such things which in an instant change the disposition with each decline" (19). On the non-naturals in Galen's treatise *Ars medica*, see Temkin, *Galenism*, 102 and 102n19.
48. Galen, *Hygiene*, 53, 92, 172; *On the Powers of Foods*, in Grant, *Galen on Food and Diet*, 85, 98, 134, 143; for Galen's overall view of women ("in every activity and mode of learning inferior to man"), see Nutton, *Ancient Medicine*, 241.
49. Nutton, *Ancient Medicine*, 1–17, at 12.

CHAPTER TWO

1. On Hippocrates in early modern Europe, see Wesley D. Smith, *The Hippocratic Tradition* (Ithaca, NY: Cornell University Press, 1979), 13–28; Brockliss and Jones, *Medical World of Early Modern France*, 93–94, 441, 464–65.
2. *Epidemics I and II*, Loeb I:139–287; *Epidemics 2, 4, 5, 6, 7*, Loeb VII; for case reports ("observations") styled after "la médecine hippocratique," see Théophile Bordeu, "Recherches sur les maladies chroniques," in *Oeuvres com-*

plètes de Bordeu, ed. A.-B. Richerand (Paris: Caille, 1818), 797–929, at 797, 849–69.

3. For a concise overview, see Engelhardt, "Hunger und Appetit."
4. Of many sources, see Allen Debus, *The French Paracelsians: The Chemical Challenge to Medical and Scientific Tradition in Early Modern France* (Cambridge: Cambridge University Press, 1991); Theodore M. Brown, *The Mechanical Philosophy and the "Animal Oeconomy"* (New York: Arno Press, 1981).
5. Cited in Walter Pagel, "Van Helmont's Ideas on Gastric Digestion and the Gastric Acid," *BHM* 30, no. 6 (1956): 524–36, at 526.
6. Michael Ettmüller, *Ettmullerus abridg'd: or, A compleat system of the theory and practice of physic* (London: Printed for E. Harris, 1699), 57; for a fuller discussion, see Ken Albala, "Weight Loss in the Age of Reason," in *Cultures of the Abdomen: Diet, Digestion, and Fat in the Modern World*, ed. Christopher E. Forth and Ana Carden-Coyne (New York: Palgrave Macmillan, 2005), 169–83.
7. René Descartes, *Treatise of Man*, trans. and commentary by Thomas Steele Hall (Cambridge, MA: Harvard University Press, 1972), 68–69.
8. According to Hall, Descartes "radically altered" the list of internal senses accepted by medieval authors, establishing instead two internal senses. The first comprised "'hunger, thirst, and all other natural appetites," whereas the second encompassed "'joy, sadness, love, anger, and all the other passions." The two types depended on different nerves, the first those serving the stomach and the latter those leading to the heart, diaphragm, and other "internal parts"; Hall, commentary on *Treatise of Man*, 68n113. See also François Azouvi, "Quelques jalons dans la préhistoire des sensations internes," *Revue de synthèse*, 3rd ser., 113–14 (January–June 1984): 113–33.
9. Theodore M. Brown, "Medicine in the Shadow of the *Principia*," *Journal of the History of Ideas* 48, no. 4 (1987): 629–48.
10. Archibald Pitcairn, *The Philosophical and Mathematical Elements of Physick* (London: n.p., 1718), 78, 236; on Pitcairn's hydraulicist system, see Brown, *Mechanical Philosophy*, 192–237; on Pitcairn's relation to Newton, see Anita Guerrini, "Archibald Pitcairne and Newtonian Medicine," *MH* 31, no. 1 (1987): 70–83.
11. Brown, "Medicine," 632–33.
12. Christian Schlöder, Maximilian Schochow, and Florian Steger, "Alter und Altern in der ärztlichen Praxis Friedrich Hoffmanns (1660–1742)," *MhJ* 49, 3 (2014): 237–59, at 241, 255. George Cheyne used his knowledge of "micromechanisms" to help justify his recommendation of a "severe 'lowering' regime," but he also proposed to patients the maxim "Let your appetite be your rule"; see Steven Shapin, "Trusting George Cheyne: Scientific Expertise, Common Sense, and Moral Authority in Early Eighteenth-Century Dietetic Medicine," *BHM* 77, no. 2 (2003): 263–97.
13. Gary Hatfield argues against claims that psychology emerged as a science only with the adoption of experimentalism in the late nineteenth century, observing that a "science of the mind or of mental phenomena," frequently

called "psychologia," was widely taught in the early modern period; see Hatfield, "Remaking the Science of Mind: Psychology as Natural Science," in *Inventing Human Science: Eighteenth-Century Domains*, ed. Christopher Fox, Roy Porter, and Robert Wokler (Berkeley: University of California, 1995), 184–231, at 184. On the psychology of Christian Wolff, who investigated the appetitive and other faculties without reference to physiology or neurology, see Vidal, *Sciences of the Soul*, 93.

14. David Hartley, *Observations on Man, His Frame, His Duty, and His Expectations*, 3 vols. (London, 1749; rpt., J. Johnson, 1791), 1:159, 2:225. On Hartley's psychology, see Richard C. Allen, *David Hartley on Human Nature* (Albany, NY: State University of New York Press, 1999); Roger Smith, "The Background of Physiological Psychology in Natural Philosophy," *History of Science* 11 (1973): 75–123, at 81.
15. Philippe Hecquet, *De la digestion et des maladies de l'estomac, suivant le système de la trituration & du broyement* (Paris: François Fournier, 1712), cited in Spary, *Eating the Enlightenment*, 24.
16. Laurence Brockliss argues that such physiological constructs partook of pious ideals of austerity, simplicity, and comprehensibility; L. W. B. Brockliss, "The Medico-Religious Universe of an Early Eighteenth Century Parisian Doctor: The Case of Philippe Hecquet," in *The Medical Revolution of the Seventeenth Century*, ed. R. K. French and Andrew Wear (Cambridge: Cambridge University Press, 1989), 191–221.
17. Cited in Knoeff, *Herman Boerhaave*, 206.
18. Johanna Geyer-Kordesch, "Georg Ernst Stahl's Radical Pietist Medicine and Its Influence on the German Enlightenment," in *The Medical Enlightenment of the Eighteenth Century*, ed. Andrew Cunningham and R. K. French (New York: Cambridge University Press, 1990), 67–87, at 68; C. L. Dumas, *Principes de physiologie, ou introduction à la science expérimentale, philosophique et médicale de l'homme vivant*, 4 vols., 2nd ed. (Paris: Déterville, 1806), 1:161–64.
19. For Sauvages's view of ills of appetite, see "Maladies of Appetite in Eighteenth-Century Nosologies." Although Sauvages admired Stahl and other animist theorists, his medical theory was an unclassifiable mix of animism, Newtonianism, and vitalism; see Williams, *Cultural History of Medical Vitalism*, 80–111.
20. Edwin Clarke and C. D. O'Malley, *The Human Brain and Spinal Cord: A Historical Study Illustrated by Writings from Antiquity to the Twentieth Century* (Berkeley: University of California Press, 1968), 158–62, 165–69.
21. Medical physiology is discussed in chapter 3.
22. On the shift from traditional semiotics to nosology, see Jacalyn Duffin, "Jodocus Lommius's Little Golden Book and the History of Diagnostic Semeiology," *JHMAS* 61, no. 3 (2006): 249–87.
23. François Boissier de Sauvages, *Nosologie méthodique, dans laquelle les maladies sont rangées par classes, suivant le systême de Sydenham, & l'ordre des Botanistes*, 3 vols., trans. Nicolas (Paris: Hérissant, 1771); William Cullen, *Nosol-*

ogy, or, A Systematic Arrangement of Diseases, by classes, orders, genera, and species; with the Distinguishing Characters of Each, and outlines of the systems of Sauvages, Linnaeus, Vogel, Sagar, and Macbride, trans. from Latin of William Cullen (Edinburgh: Printed by C. Stewart for William Creech, 1800); for background, see William F. Bynum, "Nosology," in *Companion Encyclopedia of the History of Medicine*, 2 vols., ed. W. F. Bynum and Roy Porter (London: Routledge, 1993), 1:335–56.

24. These "species" of disease appeared in Sauvages's class 8, "Folies," order 2, "Bizarreries." His terminology varied; the terms translated here follow the "Précis de la huitième classe—Des folies," *Nosologie méthodique*, 2:587–88.
25. Sauvages illustrated the species marked by *envies bizarres* with examples of strange eating among country women and "Hottentots"; Sauvages, *Nosologie méthodique*, 2:674.
26. Cullen, *Nosology*, 163; Cullen's shifting descriptions of troubles of appetite may be searched in the consultation letters digitized at "Cullen Project: The Consultation Letters of Dr. William Cullen (1710–1790) at the Royal College of Physicians of Edinburgh," available at http://www.cullenproject.ac.uk/search/.
27. New codes for behavior at the table, assumed to be an expression of appetite, were also of prime importance; see Albala, *Food in Early Modern Europe*, 199–207; Norbert Elias, *The Civilizing Process: The History of Manners and State Formation and Civilization*, trans. Edmund Jephcott (Oxford: Blackwell, 1994), esp. 68–105; Mennell, "On the Civilizing of Appetite."
28. On the shifting terminology of overeating and overweight, see George A. Bray, "Obesity: Historical Development of Scientific and Cultural Ideas," *International Journal of Obesity* 14 (1990): 909–26. French physicians did vocally challenge religious tradition on the healthfulness of the Lenten fast; see Spary, *Eating the Enlightenment*, 22–26. On "diseases of civilization," which physicians traced to the excesses and complexities of modern existence, see Carl Haffter, "Die Entstehung des Begriffs der Zivilisationskrankheiten," *Gesnerus* 36, nos. 3–4 (1979): 228–37; Charles E. Rosenberg, "Pathologies of Progress: The Idea of Civilization as Risk," *BHM* 72, no. 4 (1998): 714–30.
29. Obesity was said to be especially common in Britain and Holland. For titles in English, French, and German, see Bray, "Obesity"; Michael Stolberg, "'Abhorreas pinguedinem': Fat and Obesity in Early Modern Medicine (c. 1500–1750)," *Studies in History and Philosophy of Biological and Biomedical Sciences* 43, no. 2 (June 2012): 370–78. On corpulence in relation to "codes of social and moral integrity," see Lucia Dacome, "Useless and Pernicious Matter: Corpulence in Eighteenth-Century England," in *Cultures of the Abdomen: Diet, Digestion, and Fat in the Modern World*, ed. Christopher E. Forth and Ana Carden-Coyne (New York: Palgrave Macmillan, 2005), 185–204.
30. Jaucourt, "Faim canine (Med.)," in *Encyclopédie*, 6:376–77, at 377a. Jaucourt traced the term "bulimia" to the ancient physician Erasistratus of Alexandria.

The distance between Jaucourt's thinking and modern ideas of bulimia is evident in his claim that it afflicted especially "those who voyage in cold countries" and thus suffer the ill effects of "refrigerating corpuscles" (6:377).

31. "Faim, Appétit (Gram. Syn.)," in *Encyclopédie*, 6:373.
32. Bray, "Obesity," 913–16. On images of the body that fostered these ideas of the cause of obesity, see Dacome, "Useless and Pernicious Matter."
33. A general rule was to leave the table while still having an appetite; specific recommendations for weight loss varied widely. Some limited only quantity, some variety, while others prescribed specific foods—white meats, coarse bread, greens and other vegetables, or the much-debated "milk diet." Such programs reflected the continuing influence of ancient dietetics but also local and national dietary traditions; see Bray, "Obesity," 916; Stolberg, "'Abhorreas pinguedinem'," 375; Guerrini, *Obesity and Depression*, 119–21, 126–27, 135–36.
34. Flemyng urged overweight patients to eat soap to clean away unnecessary fat from the body; Bray, "Obesity," 916; Dacome, "Useless and Pernicious Matter," 188.
35. Reflecting his long-term study of the bodily actions of salts, oils, waters, earths, and metals, Cullen attributed "excessive corpulence" to "the increase of oil in the cellular texture of the body"; Cullen, *Nosology*, 139; William F. Bynum, "William Cullen (1710–1790), Chemist and Physician," *Oxford Dictionary of National Biography*, https://doi.org/10.1093/ref:odnb/6874.
36. "Dyspepsia. Loss of appetite, nausea, vomiting, flatulence, eructation, rumination, heartburn, pain in the stomach; at least some or more of these at once occurring, generally with costiveness, and without any other disease either of the stomach or other parts"; Cullen, *Nosology*, 107. See also Ian Miller, *A Modern History of the Stomach* (London: Routledge, 2016), 12–13.
37. Anne C. Vila, *Suffering Scholars: Pathologies of the Intellectual in Enlightenment France* (Philadelphia: University of Pennsylvania Press, 2018), 20–45.
38. On the nature and role of vital forces in eighteenth-century physiology, see chapter 3, "The Physiology of Appetite: Mechanists and Vitalists."
39. Vandereycken and van Deth, *From Fasting Saints to Anorexic Girls*, 47–73; Tilmann Habermas, "Historical Continuities and Discontinuities between Religious and Medical Interpretations of Extreme Fasting: The Background to Giovanni Brugnoli's Description of Two Cases of Anorexia Nervosa in 1875," *History of Psychiatry* 3 (1992): 431–55.
40. Berg and Eger, introduction to *Luxury in the Eighteenth Century*, 1; Spary, *Eating the Enlightenment*, 198, 209–10.
41. On antifeminism, see Lieselotte Steinbrugge, *The Moral Sex: Woman's Nature in the French Enlightenment* (New York: Oxford University Press, 1995).
42. Jean Astruc, *Traité des maladies des femmes*, 6 vols. (Paris: Cavelier, 1761–1765); Astruc explained diverse female ills as the result of "sympathies that the womb has with almost all parts of the body" (4:76).

43. Roussel, *Système physique et moral de la femme*, 2; E.-P. Chauvot de Beauchêne, *De l'influence des affections de l'âme dans les maladies nerveuses des femmes* (Paris: Méquignon, 1781), 12–13.
44. Jean Starobinski, "Sur la chlorose," *Romantisme* 31 (1981): 113–30; on eating in times of famine, see Piero Camporesi, *Bread of Dreams: Food and Fantasy in Early Modern Europe*, trans. David Gentilcore (Chicago: University of Chicago Press, 1989), esp. 62, 101, 104, 107, 122.
45. The French physician Pierre Hunauld attributed one variety of the vapors, *vapeurs à la mode*, to indolence, luxury, and "new ways of living," urging that this ill could be avoided if people ate and slept in accord with "true needs" (194–98); see Hunauld, *Dissertation sur les vapeurs et les pertes de sang* (Paris: Jean-Noel Leloup, 1756), xvi–xviii, 76, 194–98. For a fuller discussion, see Vila, *Enlightenment and Pathology*, esp. 228–40.
46. Guerrini, *Obesity and Depression*, 123; Elizabeth A. Williams, "Hysteria and the Court Physician in Enlightenment France," *Eighteenth-Century Studies* 35, no. 2 (2002): 247–55.
47. On the trend toward seeing hypochondria as "being nervous in a manly way," see Vila, *Suffering Scholars*, 153; on Samuel Taylor Coleridge, whose chronic ailments of the gut resembled modern irritable bowel syndrome, see George Rousseau, "Coleridge's Dreaming Gut: Digestion, Genius, Hypochondria," in *Cultures of the Abdomen: Diet, Digestion, and Fat in the Modern World*, ed. Christopher E. Forth and Ana Carden-Coyne (New York: Palgrave Macmillan, 2005), 105–26, at 105.
48. On the "feeble and bizarre" appetite of vaporous women, see Beauchêne, *De l'influence des affections de l'âme*, 71, 73.
49. Philippe Pinel, *Nosographie philosophique ou méthode de l'analyse appliquée à la médecine*, 2 vols. (Paris: Crapelet, 1798), 2:30; Pinel, *Traité médico-philosophique sur l'aliénation mentale*, 2nd ed. (Paris: J. A. Brosson, 1809; rpt., New York: Arno Press, 1976), 485–87.
50. Antoine Le Camus, *La médecine de l'esprit*, 2 vols., new ed. (Paris: Ganeau, 1769), 1:243–45, 1:262–63.
51. One tragic, oft-repeated story was that of the young English physician William Stark, who died after testing extreme dietetic regimens on himself; see William Stark, *The Works of the late William Stark, M.D.: Consisting of Clinical and Anatomical Observations, with Experiments, Dietetical and Statical, Revised and Published from his Original MSS by James Carmichael Smyth, M.D., F.R.S.* (London: J. Johnson, 1788).
52. On uses made of Hippocrates across theoretical divisions in early modern medicine, see the essays in David Cantor, ed., *Reinventing Hippocrates* (Aldershot, UK: Ashgate, 2002).
53. On the origins of the term "physiology" and its relations with medicine in the later eighteenth century, see Georges Canguilhem, "La constitution de la physiologie comme science," in *Études d'histoire et de philosophie des*

sciences (Paris: J. Vrin, 1979), 226–73. On the role of experimentalism, see Hubert Steinke, *Irritating Experiments: Haller's Concept and the European Controversy on Irritability and Sensibility, 1750–90* (Amsterdam: Rodopi, 2005); John E. Lesch, *Science and Medicine in France: The Emergence of Experimental Physiology, 1790–1855* (Cambridge, MA: Harvard University Press, 1984).

CHAPTER THREE

1. Keith Thomas, *Man and the Natural World: A History of the Modern Sensibility* (New York: Pantheon, 1983), esp. 92–120.
2. On institutional settings for biological study outside medicine, see Jacques Roger, *Les sciences de la vie dans la pensée française au XVIIIe siècle*, new ed. (Paris: Albin Michel, 1993), 34–37, 170–84. On the role of gardens, the academies, and other institutions, see N. Jardine, J. A. Secord, and E. C. Spary, eds., *Cultures of Natural History* (Cambridge: Cambridge University Press, 1996). For an emphasis on natural history that developed "outside any institutional context," see E. C. Spary, "The 'Nature' of Enlightenment," in *The Sciences in Enlightened Europe*, ed. William Clark, Jan Golinski, and Simon Schaffer (Chicago: University of Chicago Press, 1999), 272–304, at 273.
3. René-Antoine Ferchault de Réaumur, "Sur la digestion des oiseaux," in *Mémoires de l'Académie Royale des Sciences* (1752): 266–307, 461–95; on Réaumur's career, see Jean Torlais, *Un esprit encyclopédique en dehors de 'L'Encyclopédie': Réaumur*, 2nd ed. (Paris: Blanchard, 1961). For more detail on his experiments on digestion, see Spary, *Eating the Enlightenment*, 46–48.
4. [René-Antoine Ferchault de] Réaumur, *Art de faire éclorre, et d'élever en toute saison des oiseaux domestiques de toutes especes, soit par le moyen de la chaleur du fumier, soit par le moyen de celle du feu ordinaire*, 2nd ed., 2 vols. (Paris: L'imprimerie royale, 1751).
5. Georges-Louis Leclerc de Buffon, *Histoire naturelle, générale et particulière: Avec la description du Cabinet du Roy*, 15 vols. (Paris: Imprimerie Royale, 1749–1767), 4:3–110, available at http://www.buffon.cnrs.fr; Michèle Duchet, *Anthropologie et histoire au siècle des Lumières: Buffon, Voltaire, Rousseau, Helvétius, Diderot* (Paris: Francois Maspéro, 1971), 229–80; Jacques Roger, *Buffon: A Life in Natural History*, trans. Sarah Lucille Bonnefoi, ed. L. Pearce Williams (Ithaca, NY: Cornell University Press, 1997), esp. 151–83, 240–56.
6. On Buffon's admiration for Aristotle, especially his use of the comparative method, see Roger, *Buffon*, 87.
7. On Buffon's relation to Locke and Descartes, see Roger, *Buffon*, 82–84, 153–55, 158–59, 247–49.
8. Buffon, *Histoire naturelle*, 4:21–41; Buffon acknowledged that animals might consume poisons if these were disguised as appetizing foods (4:49).
9. Buffon, *Histoire naturelle*, 4:24. On Buffon's secular idea of the human soul, conceived as "one very simple, very general, very constant form . . . [of]

thought," see Buffon, "Histoire naturelle de l'homme," *Histoire naturelle,* 2:434–35; Roger, *Buffon,* 153–63, 250–56.

10. Buffon, "Variétés dans l'espèce humaine," *Histoire naturelle,* 3:371–530.
11. John S. Spink, "'Sentiment,' 'sensible,' 'sensibilité': Les mots, les idées, d'après les 'moralistes' français et britanniques du début du dix-huitième siècle," *Zagadnienia Rodzajów Literackich,* 20 (1977): 33–47; Vila, *Enlightenment and Pathology,* 30–42.
12. See the discussion below of vitalist objections based on the uniquely human physical-moral relation.
13. "Faim, appétit (Gram. Syn.)," in *Encyclopédie,* 6:373; Jaucourt, "Faim, f.f. (Physiol.)," in *Encyclopédie,* 6:373–76; Aumont, "Faim (séméiotique)," in *Encyclopédie,* 6:376.
14. Rey, *Naissance et développement du vitalisme*; Williams, *Cultural History of Medical Vitalism*; Williams, *The Physical and the Moral,* 20–66. Comparative studies include François Duchesneau, "Vitalism in Late Eighteenth-Century Physiology: The Cases of Barthez, Blumenbach and John Hunter," in *William Hunter and the Eighteenth-Century Medical World,* ed. W. F. Bynum and Roy Porter (Cambridge: Cambridge University Press, 1985), 259–95; Reill, *Vitalizing Nature in the Enlightenment.*
15. J.-C.-M.-G. Grimaud, *Cours complet de physiologie,* 2nd rev. ed., 2 vols., ed. Étienne Lanthois (Paris: Egron, 1824), 1:14–15, 48–49, 94, 2:228–29; Grimaud, *Mémoire sur la nutrition* (Montpellier: J. Martel, 1787); Grimaud, *Second mémoire sur la nutrition: Supplément au Mémoire françois sur la nutrition* (Montpellier: Jean Martel, 1789); Charles-Louis Dumas, *Principes de physiologie, ou Introduction à la science expérimentale, philosophique et médicale de l'homme vivant,* 4 vols. (Paris: Crapelet, Déterville, 1800–1803), 1:61–63, 152, 330–31.
16. Steinke, *Irritating Experiments,* 93–124.
17. "Les nerfs ne sont pas les seules parties du corps animal auxquelles la faculté de sentir soit attachée. Une réflexion bien simple qui se présente d'abord contre cet opinion, c'est que la Sensibilité est différente dans chaque partie du corps, et que l'animal doit à chacun de ses organes, des sensations d'un ordre particulier"; Grimaud, *Cours complet de physiologie,* 2:170–71.
18. Théophile Bordeu, *Recherches anatomiques sur les glandes, et sur leur action* in *Oeuvres complètes de Bordeu,* 2 vols., ed. A.-B. Richerand (1751; Paris: Caille, 1818), 1:45–208, at 184. In Bordeu's construction appetite stimulated the parotid glands just as internal "convulsions" did the secretion of milk or semen (1:127–28). For more detail on ideas of interior sense that developed in reaction against mechanist psychophysiology, see Azouvi, "Quelques jalons dans la préhistoire des sensations internes."
19. For a later survey of mechanist theories of the urge to eat, see P. Rullier, "Faim," in *Dictionnaire de médecine,* 8:487–96.
20. Albertus [Albrecht von] Haller, *First Lines of Physiology,* 2 vols., intro. Lester S. King (1786; New York: Johnson Reprint, 1966), 2:81–82. Haller offered a

fuller discussion in his immensely influential *Elementa physiologiae corporis humani* (1757–1766); all citations are to the German translation: *Anfangsgründe der Phisiologie des menschlichen Körpers*, 8 vols., ed. Johann Samuel Haller (Berlin: Christian Friedrich Voss, 1759–1776).

21. Haller, *First Lines*, 2:62. Haller explicitly rejected the "error" made by the ancients in attributing the creation of appetite to the flow of a "sour juice or black bile" from the spleen into the stomach; Haller, *Anfangsgründe der Phisiologie*, 6:234.
22. Anton Julius Carlson, *The Control of Hunger in Health and Disease* (Chicago: University of Chicago Press, 1916), 17–18.
23. Lazzaro Spallanzani, *Expériences sur la digestion de l'homme et de différentes espèces d'animaux*, trans. and intro. Jean Senebier (Geneva: Barthelemi Chirol, 1783); see also Felice Grondona, "Des expériments de Lazzaro Spallanzani sur la digestion aux applications thérapeutiques du suc gastrique," *Clio medica* 8, no. 4 (1973): 285–94.
24. Spallanzani, *Expériences*, 246–51. On similar experiments performed by the Swiss naturalist Henri-Albert Gosse, see the introduction to Spallanzani's treatise by Jean Senebier, "Considérations sur la méthode suivie par Monsieur l'Abbé Spallanzani dans ses expériences sur la digestion," i–cxlix, at cxxii–cxlii.
25. To explore this matter, Spallanzani again experimented on himself. After he swallowed "five tubes of wood . . . filled with . . . corrupted flesh," the evacuated residue showed "not the least trace of putrefaction." He concluded that in both humans and animals the gastric juices "not only prevent putrefaction but restore rotted flesh [that is] soaked in them"; Spallanzani, *Expériences*, 313.
26. Senebier, "Considérations," cxx.
27. For the view of Spallanzani taken by the American physiologist William Beaumont, see chapter 5, "Digestive Experiments: A Place for Appetite?"; A.-F. Jenin de Montègre stated that experimental results could not be "rigorously applied to men in general as the facility of digesting certain aliments depends . . . on conditions particular to each person"; Montègre, *Expériences sur la digestion dans l'homme, présentées à la première class de l'Institut de France, le 8 septembre 1812. Suivies du Rapport des Commissaires nommés par l'Institut* (Paris: Le Normant, 1814), 5–6. The commentators on his 1814 memoir observed, moreover, that the chemical nature of the gastric juice remained unclear and the "cause of digestion" unknown; *Rapport de Mrs. Berthollet, Cuvier, and Thenard*, in *Expériences*, 53–54.
28. On the early nineteenth-century struggle between chemists and physiologists for authority over problems of animal digestion and nutrition, see Frederic L. Holmes, *Claude Bernard and Animal Chemistry: The Emergence of a Scientist* (Cambridge, MA: Harvard University Press, 1974), esp. 141–59.
29. On twentieth-century theories of hunger, see chapter 10.

30. Laporte, "Appétit," in *Encyclopédie méthodique, Médecine*, 3:198. Debate over whether appetite was only a phase of hunger, differing from it in quantity rather than quality, was also to continue into the twentieth century (see chapter 10).
31. Haller, *First Lines*, 1:264. Later investigators were more optimistic; on subsequent efforts to identify the mechanisms of distinct tastes, see Jütte, *History of the Senses*, 232–33; Korsmeyer, *Making Sense of Taste*, 68–102.
32. Bordeu, *Recherches anatomiques sur les glandes*, 1:163; Williams, *The Physical and the Moral*, 33–36.
33. Grimaud, *Mémoire sur la nutrition*, 98–100, 128–29, 185–90, quotation at 98; Grimaud, *Second mémoire sur la nutrition*, 1 and 1n5.
34. Haller, *Anfangsgründe der Phisiologie*, 6:234 (this section is mispaginated, returning to page 225 after page 240).
35. Haller, *First Lines*, 2:32–55; on the religious context of Haller's thinking, see Erich Hintzsche, "Der Hunger in physiologischen Lehrbüchern von Haller bis Valentin," *Gesnerus* 20 (1963): 33–46, at 33–34. Hintzsche notes Haller's belief that hunger was part of God's plan to draw both humans and animals out of their natural indolence (34).
36. On "contrary" foods, see Grimaud, *Mémoire*, 1:105; compare Hippocrates, *Aphorisms*, II:XXXVIII, Loeb IV:117–19. On the wisdom of the people, see Grimaud, *Mémoire*, 1:118.
37. Senebier, "Considérations," lxxxvii–xci, quotation at xc.
38. On the obscurity of digestion, see Grimaud, *Cours complet de physiologie*, 1:94; John Hunter, "Some Observations on Digestion," in *Works of John Hunter, F.R.S*, ed. James F. Palmer, 4 vols. (London: Longman, 1835–37), 4:81–83; Montègre, *Expériences*, 14.
39. On Redi, see J. F. Blumenbach, *The Institutions of Physiology*, trans. John Elliotson, 3rd ed. (London: Longman, Hurst, Rees, Orme, and Brown, 1820), 212; on Haller, see Hintzsche, "Der Hunger in physiologischen Lehrbüchern," 35–36.
40. Senebier, "Considérations," lxvii–lxviii.
41. Dumas, *Principes de physiologie* (1800–1803), 4:59–66.

INTRODUCTION TO PART TWO

1. Colin Jones and Rebecca Spang, "*Sans-culottes, sans café, sans tabac*: Shifting Realms of Necessity and Luxury in Eighteenth-Century France," in *Consumers and Luxury: Consumer Culture in Europe, 1650–1850*, ed. Maxine Berg and Helen Clifford (Manchester: Manchester University Press, 1999), 37–62.
2. Olmsted, *François Magendie*, 51–52, 147.
3. Richard W. Burkhardt Jr.,"Ethology, Natural History, the Life Sciences, and the Problem of Place," *JHB* 32, no. 3 (Winter 1999): 489–508; Burkhardt, "Darwin on Animal Behavior and Evolution," in *The Darwinian Heritage*, ed.

David Kohn and Malcolm J. Kottler (Princeton, NJ: Princeton University Press, 1985), 327–65; Paul L. Farber, "Discussion Paper: The Transformation of Natural History in the Nineteenth Century," *JHB* 15, no. 1 (Spring 1982): 145–52.

4. Richard W. Burkhardt Jr., "The Zoological Philosophy of J. B. Lamarck," in J. B. Lamarck, *Zoological Philosophy: An Exposition with Regard to the Natural History of Animals*, trans. Hugh Elliot (Chicago: University of Chicago Press, 1984), xxxi; Burkhardt, "Darwin on Animal Behavior," 332.
5. See chapter 8, "Feeding in Insects and Other Animals."
6. On nineteenth-century psychiatry, see Jan Goldstein, *Console and Classify: The French Psychiatric Profession in the Nineteenth Century* (New York: Cambridge University Press, 1987); Eric J. Engstrom and Volker Roelcke, eds. *Psychiatrie im 19. Jahrhundert: Forschungen zur Geschichte von psychiatrischen Institutionen, Debatten und Praktiken im deutschen Sprachraum* (Basel: Schwabe, 2003); W. F. Bynum, Roy Porter, and Michael Shepherd, eds., *The Anatomy of Madness: Essays in the History of Psychiatry*, 3 vols. (London: Tavistock, 1985).

CHAPTER FOUR

1. Rey, *Naissance et développement du vitalisme*, 360.
2. On radicalism in English scientific societies of the late eighteenth century, see Robert E. Schofield, *The Lunar Society of Birmingham: A Social History of Provincial Science and Industry in Eighteenth-Century England* (Oxford: Clarendon Press, 1963). First associated with the Lunar Society, Darwin later helped found the Derby Philosophical Society, which perpetuated his ideas, especially on progressive evolutionary development. See Paul Elliott, "Erasmus Darwin, Herbert Spencer, and the Origins of the Evolutionary Worldview in British Provincial Scientific Culture, 1770–1850," *Isis* 94, no. 1 (March 2003): 1–29.
3. Desmond King-Hele, *Erasmus Darwin* (New York: Charles Scribner's, 1963), 39–42, 109.
4. Maureen McNeil, *Under the Banner of Science: Erasmus Darwin and His Age* (Manchester: Manchester University Press, 1987), 123.
5. On Galen's physiology and view of appetite, see chapter 1, "'Innate Appetites': Galenic Physiology and Dietetics." My discussion of Bichat's thinking is based on the following works, to be cited in the text as follows: Xavier Bichat, *Anatomie générale, appliquée à la physiologie et à la médecine*, 4 vols. (Paris: Brosson, Gabon, 1801) (hereafter *AG*); *Recherches physiologiques sur la vie et la mort* (Paris: Brosson, Gabon, 1800); references here are to Xavier Bichat, *Physiological Researches upon Life and Death*, trans. Tobias Watkins (Philadelphia: Smith and Maxwell, 1809) (hereafter *Researches*). Bichat specifically rejected the "old" (Galenic) framework of animal, vital, and natural functions, saying it had no foundation (*AG*, 1:c).

6. Bichat's scheme drew on Aristotle and the Montpellier physiologist J.-C- M.-G. Grimaud. On Grimaud, see chapter 3, "The Physiology of Appetite: Mechanists and Vitalists." Although Bichat was highly critical of the latter (*AG*, 1:c–ci), Grimaud's influence on his thinking is clear; see Rey, *Naissance*, 355–57; Elizabeth A. Williams, "Of Two Lives One? Jean-Charles-Marguerite-Guillaume Grimaud and the Question of Holism in Vitalist Medicine," *Science in Context* 21, no. 4 (2008): 593–613.
7. These five properties were animal sensibility, animal contractility, organic sensibility, sensible organic contractility (or "irritability"), and "insensible organic contractility"; see *AG*, 1:xxxv–xlv, lii–lxxii; *Researches*, 69–93.
8. Clarke and Jacyna, *Nineteenth-Century Origins of Neuroscientific Concepts*, 320–27.
9. Bichat acknowledged that, given his radically changed approach, he did not adequately address the specific physiological functions, including digestion, traditionally covered in such a work (*AG*, 1:lxxix–lxxx, xcix).
10. Greater detail on the position within the two lives of the pharynx, esophagus, stomach, and intestines appeared in a work whose initial volumes were written by Bichat but that was completed by M.-F.-R. Buisson (vols. 3 and 4) and P.-J. Roux (vol. 5); see *Traité d'anatomie descriptive*, 5 vols. (Paris: Brosson, Gabon, 1801–1803), 3:379–449.
11. *Researches*, 41. Compare the view of Bichat's contemporary Franz-Josef Gall, who asserted, "Plus on avance en âge, plus on attache ordinairement de prix à la bonne nourriture, et plus elle devient nécessaire"; F.-J. Gall and G. Spurzheim, *Anatomie et physiologie du système nerveux en général, et du cerveau en particulier*, 5 vols. (Paris: F. Schoell, 1810–1819), 1:156.
12. See chapter 5 on the work of François Magendie and other French physiologists strongly influenced by Bichat.
13. Roy Porter, "Erasmus Darwin: Doctor of Evolution?" in *History, Humanity, and Evolution: Essays for John C. Greene*, ed. James R. Moore (Cambridge: Cambridge University Press, 1989), 39–69, at 41, 44.
14. Erasmus Darwin, *Zoonomia, or, The Laws of Organic Life*, 2 vols., 2nd ed., corrected (London: J. Johnson, 1794, 1796) (all references are to this edition); Darwin, *Phytologia; or, The Philosophy of Agriculture and Gardening* (London: J. Johnson, 1800).
15. Darwin, *Zoonomia*, 1:32–36. On the origins of vitalism in England, see Theodore M. Brown, "From Mechanism to Vitalism in Eighteenth-Century English Physiology," *JHB* 7, no. 2 (Autumn 1974): 179–216.
16. Porter, "Erasmus Darwin," 48–49. With this view Darwin diverged markedly from physiologists who linked habit not to volition but to continuous movement in the body (such as the circulation of the blood) in the absence of a moving force. See R. K. French, "Sauvages, Whytt and the Motion of the Heart: Aspects of Eighteenth-Century Animism," *Clio medica* 7 (1972): 35–54, esp. 42, 49–50.

17. Darwin, *Zoonomia*, 1:51. Although not inclined to establish rigid lines of division among humans, animals, and vegetables, Darwin did see "voluntarity" as a particularly human characteristic: in undertaking voluntary action humans were "always employed about the *means*," while brutes "are almost perpetually employed or about their present pleasures, or their present pains" (1:59). Yet he also saw some "voluntary powers" in plants, as in the circular movement of vine tendrils (1:103–4).
18. Edward S. Reed, *From Soul to Mind: The Emergence of Psychology from Erasmus Darwin to William James* (New Haven, CT: Yale University Press, 1997), 40.
19. On Hartley, see chapter 2, "Chemistry and Mechanics of Appetite."
20. Darwin, *Zoonomia*, 1:12. On attention as strenuous mental effort requiring "strict discipline," see Daston, "Attention and the Values of Nature in the Enlightenment," in Daston and Vidal, *Moral Authority of Nature*, 100–126, at 114.
21. "All those internal motions of animal bodies, which contribute to digest their aliment, produce their secretions, repair their injuries, or increase their growth, are performed without our attention or consciousness. They exist as well in our sleep, as in our waking hours, as well in the foetus during the time of gestation, as in the infant after nativity, and proceed with equal regularity in the vegetable as in the animal system"; Darwin, *Zoonomia*, 1:136–37.
22. Schofield, *Lunar Society*, 397–401; McNeil, *Under the Banner*, 105–15.
23. Clarke and Jacyna, *Nineteenth-Century Origins*, 320–21; on Bichat's impact in Germany, especially via the conduit of "his follower [J. C.] Reil," see Erwin H. Ackerknecht, "The History of the Discovery of the Vegetative (Autonomic) Nervous System," *MH* 18 (1974): 1–8, at 3.
24. Norton Garfinkle, "Science and Religion in England, 1790–1800: The Critical Response to the Work of Erasmus Darwin," *Journal of the History of Ideas* 16, no. 3 (June 1955): 376–88. Porter suggests that Darwin's thinking was rejected at least in part because of his "physiological vitalism," which attributed "'appetencies'" to matter; Porter, "Erasmus Darwin," 59.
25. On the return of "traditional metaphysics," see Reed, *From Soul to Mind*, 34–37, quotation at xi; on the backlash in psychology, see also Jan Goldstein, *The Post-Revolutionary Self: Politics and Psyche in France, 1750–1850* (Cambridge, MA: Harvard University Press, 2005). For Darwin's importance to Herbert Spencer, see Elliott, "Erasmus Darwin," and for his impact on German psychology, see Hatfield, "Remaking the Science of Mind," 209.
26. An older view that Bichat was hostile to experiment has been largely discarded; see Rey, *Naissance*, 323. For contrasting views of the significance of experiment to Bichat, see William R. Albury, "Experiment and Explanation in the Physiology of Bichat and Magendie," in *Studies in History of Biology*, ed. William R. Coleman and Camille Limoges (Baltimore: Johns Hopkins University Press, 1977), 1:47–131; Lesch, *Science and Medicine in France*, 77–79, 122–23.

27. On these themes in the work of Magendie, see chapter 5, "Need and Desire in the Physiology of François Magendie."
28. Darwin, *Zoonomia*, 1:111.
29. One of the better studied realms of Darwin's impact is his influence not on scientists but on the English romantic poets, especially Percy Bysshe Shelley, "a deep student of science and medicine" who was obsessed with the influence of diet; see Reed, *From Soul to Mind*, 46–47; King-Hele, *Erasmus Darwin*, 133. Darwin was also admired in Germany by those who sought to establish a unified science of living beings. On the respect showed to Darwin by the German physiologist Johannes Müller, see chapter 5, "Appetite Local or Organismic? *Müller's Elements of Physiology.*"
30. Elizabeth A. Williams, "Food and Feeling: 'Digestive Force' and the Nature of Morbidity in Vitalist Medicine," in *Vital Matters: Eighteenth-Century Views of Conception, Life, and Death*, ed. Helen Deutsch and Mary Terrall (Toronto: University of Toronto Press, 2012), 203–21.

CHAPTER FIVE

1. "Rapport fait à l'Académie des Sciences au nom de la Commission dite *de la gélatine*," *Comptes-rendus hebdomadaires des séances de l'Académie des Sciences*, 2nd semester, 13 (1841): 237–95, at 265; on the Gelatine Commission, see Frank Stahnisch, "Den Hunger standardisieren: François Magendies Fütterungsversuche zur Gelatinekost 1831–1841," *MhJ* 39 (2004): 103–34; on German efforts, see Mark R. Finlay, "Early Marketing of the Theory of Nutrition: The Science and Culture of Liebig's Extract of Meat," in *The Science and Culture of Nutrition*, ed. Harmke Kamminga and Andrew Cunningham (Amsterdam: Rodopi, 1995), 48–74.
2. See the discussion of biopower in the introduction to part 1.
3. Hans Jürgen Teuteberg and Günter Wiegelmann, *Der Wandel der Nahrungsgewohnheiten unter dem Einfluß der Industrialisierung* (Göttingen: Vandenhoeck & Ruprecht, 1972); Jack Goody, "Industrial Food: Towards the Development of a World Cuisine," in *Food and Culture: A Reader*, ed. Carole Counihan and Penny Van Esterik (New York: Routledge, 1997), 338–56; Spary, *Feeding France*, esp. 201–2, 223–24, 321–22.
4. On Lavoisier's own research into digestive processes, which led to his claim that healthy animals "constantly return each day, at the end of digestion, to the same weight they had on the previous day," see Spary, *Feeding France*, 122–24.
5. Timothy Lenoir, *The Strategy of Life: Teleology and Mechanics in Nineteenth-Century German Biology* (Chicago: University of Chicago Press, 1982), 1–3, 112; Laura Otis, *Müller's Lab* (Oxford: Oxford University Press, 2007), 23.
6. Bichat, *Anatomie générale*, 1:xli–xlii.
7. Müller, *Müller's Elements of Physiology*, 1:45; on this pivotal text, see "Appetite Local or Organismic? *Müller's Elements of Physiology.*"

8. Clarke and Jacyna, *Nineteenth-Century Origins*, 5–6.
9. Michael Gross, "The Lessened Locus of Feelings: A Transformation in French Physiology in the Early Nineteenth Century," *JHB* 12, no. 2 (1979): 231–71.
10. On the origin of this term and its synonym "coenaesthesis," see Jean Starobinski, "Le concept de cénesthésie et les idées neuropsychologiques de Moritz Schiff," *Gesnerus* 34 (1977): 2–20. On the place of the *sens intime* in anti-Lockean philosophy, see Georges Gusdorf, *Les sciences humaines et la pensée occidentale*, vol. 7, *Naissance de la conscience romantique au siècle des Lumières* (Paris: Payot, 1976), 300–316.
11. On Magendie's training, see Olmsted, *François Magendie*, 14–18. François Magendie, "Quelques idées générales sur les phénomènes particuliers aux corps vivans" (1809); an English translation is appended to Albury, "Experiment and Explanation," 107–15, quotations at 107, 115; see also Lesch, *Science and Medicine in France*, 92–97.
12. In 1813 Magendie published papers on vomiting, swallowing, the effect of emetics, and the work of the esophagus; Olmsted, *François Magendie*, 17–18, 50–55.
13. François Magendie, *Mémoire sur le vomissement, lu à la première classe de l'Institut de France* (Paris: Crochard, 1813), 1–48.
14. Magendie, *Vomissement*, 2–9. Hubert Steinke sees Magendie's account of Haller's vomiting experiments as "incomplete" and argues that his method was actually close to Magendie's own; Steinke, *Irritating Experiments*, 79.
15. "Rapport fait à l'Institut, d'un Mémoire de M. Magendie sur le Vomissement, extrait du procès-verbal de la séance du lundi 1er mars 1813," 25–48 (printed with *Vomissement*).
16. François Magendie, "Mémoire sur les propriétés nutritives des substances qui ne contiennent pas d'azote," *Annales de chimie et de physique* 3 (1816): 66–77.
17. Olmsted, *François Magendie*, 67–69; Holmes, *Claude Bernard*, 7.
18. François Magendie, *Précis élémentaire de physiologie*, 2 vols. (Paris: Méquignon-Marvis, 1816–17). Subsequent editions appeared in 1825, 1834, and 1836; all citations are to the first edition, unless otherwise stated.
19. On these claims, see chapter 3, "The Physiology of Appetite: Mechanists and Vitalists."
20. P.-J.-G. Cabanis, "Rapports du physique et du moral," in *Oeuvres philosophiques de Cabanis*, 2 vols., ed. Claude Lehec and Jean Cazeneuve (Paris: Presses Universitaires de France, 1956), 1:174; Martin S. Staum, *Cabanis: Enlightenment and Medical Philosophy in the French Revolution* (Princeton, NJ: Princeton University Press, 1980).
21. For Magendie's classification of types of internal sensations, see *Précis*, 1:137–38.
22. Gross, "Lessened Locus of Feeling."
23. In this case Magendie sought to elude the difficulty of pinpointing internal states by defining nausea to mean "not only the desire to vomit but also the sensible phenomena that signify this need" (*Vomissement*, 14n1). Simi-

larly, he often equated sensation to pain, accepting the reality of sensation in a vivisected animal only if it was apparent in external actions such as convulsions. See Olmsted, *François Magendie*, 147.

24. Magendie, *Précis*, 1:151–52n1; see also Magendie, *Leçons sur les fonctions et les maladies du système nerveux*, 1:16.
25. He said the same of the passions, referring to "internal sensations that [having] no seat . . . cannot be explained" (*Précis*, 1:185).
26. Nikolaus Mani, "Die wissenschaftliche Ernährungslehre im 19. Jahrhundert," in *Ernährung und Ernährungslehre im 19. Jahrhundert*, ed. Edith Heischkel-Artelt (Göttingen: Vandenhoeck and Ruprecht, 1976), 22–75; Holmes, *Claude Bernard*, 160–78.
27. Holmes, *Claude Bernard*, 32, 38–39, 41–42, 47, 56–59, 62–71, 80–95.
28. J. B. Boussingault, *Rural Economy, in Its Relations with Chemistry, Physics, and Meteorology, or, Chemistry applied to Agriculture*, intro. and trans. George Law (New York: Orange Judd, 1865), 433, 448, 449. Interestingly, in his personal experience Boussingault acknowledged psychological influences on appetite when, describing to a correspondent the preparation of slops for experimental pigs, he observed that he had lost his appetite "as a result of the deep disgust which the odor of the kitchen impressed on me"; cited in Holmes, *Claude Bernard*, 100.
29. Gallouin and Le Magnen, "Evolution historique des concepts de faim, satiété et appétits," 115.
30. Tiedemann and Gmelin, angered that the academy failed to recognize glaring differences between their own work and that of the Frenchmen, refused this recognition; see Nikolaus Mani, "Das Werk von Friedrich Tiedemann und Leopold Gmelin: 'Die Verdauung nach Versuchen' und seine Bedeutung für die Entwicklung der Ernährungslehre in der ersten Hälfte des 19. Jahrhunderts," *Gesnerus* 13 (1956): 190–214, at 199.
31. Friedrich Tiedemann and Leopold Gmelin, *Die Verdauung, nach Versuchen*, 2 vols. (Heidelberg: Gross, 1826–1827); except where indicated, references are to the French translation made by A.-J.-L. Jourdan, *Recherches expérimentales, physiologiques et chimiques sur la digestion, considérée dans les quatre classes d'animaux vertébrés*, 2 vols. (Paris: J.-B. Baillière, 1826–1827).
32. François Leuret and J.-L. Lassaigne, *Recherches physiologiques et chimiques pour servir à l'histoire de la digestion* (Paris: Huzard, 1825).
33. Mani, "Das Werk," 199–200.
34. A typical comment on appetite appeared in their account of a dog that was opened for removal of pancreatic juice and ate a great deal the next day, its appetite not "sensibly diminished"; Tiedemann and Gmelin, *Recherches expérimentales*, 1:28.
35. Leuret and Lassaigne, *Recherches*, 211. Researchers performed vagotomies to learn not only about persistence of appetite but also the secretion of gastric juice or speed of digestion; studies since 1800 that indicated at least some measure of control by the vagus nerves included J.-J.-C. Legallois, *Expériences*

sur le principe de la vie (Paris: D'Hautel, 1812); Gilbert Breschet, Henri Milne-Edwards, and Pierre Vavasseur, *Recherches expérimentales sur les fonctions du système nerveux: 1er mémoire, De l'influence du système nerveux sur la digestion stomacale* (Paris: Migneret, 1823).

36. J.-L. Brachet, *Recherches expérimentales sur les fonctions du système nerveux ganglionaire, et sur leur application à la pathologie* (Paris: Gabon, 1830); John Reid, "An Experimental Investigation into the Functions of the Eighth Pair of Nerves, or the Glosso-Pharyngeal, Pneumogastric, and Spinal Accessory," *EMSJ* 49 (1838): 109–76 and 51 (1839): 269–330.
37. Friedrich Tiedemann, *Physiologie des Menschen*, vol. 3, *Nahrungs-Bedürfniss, Nahrungs-Trieb und Nahrungs-Mittel des Menschen* (Darmstadt: Carl Wilhelm Leske, 1836); subsequent references are to the third volume. The second of the three planned volumes of this work was never published; see Vladislav Kruta, "Friedrich Tiedemann," *DSB* 13:402–4.
38. On one early formulation of the concept of drive, see Tobias Cheung, "Hermann Samuel Reimarus' Theorie der 'Lebensarten' und 'Triebe,'" *Sudhoffs Archiv* 90, no. 2 (2006): 143–60. Haller used the term *Trieb* for both the urge to eat and to drink, but he did not lend it any special significance; see Haller, *Anfangsgründe*, 6:247, 252. To my knowledge, Tiedemann was the first to use the term *Nahrungstrieb* in a book title.
39. William Beaumont, *Experiments and Observations on the Gastric Juice and the Physiology of Digestion*, with a biographical essay by William Osler (1833; rpt., New York: Dover, 1959) (all references are to this edition).
40. William Osler, "William Beaumont: A Pioneer American Physiologist," in Beaumont, *Experiments and Observations*, vii–xl; see also Walter B. Cannon, *Digestion and Health* (New York: W. W. Norton, 1936), 11–15.
41. Beaumont, *Experiments and Observations*, 126, 163; Beaumont stated at the outset that his own investigations confirmed the "doctrines (with some modifications) taught by Spallanzani" (preface, 5–6).
42. Ronald L. Numbers and William J. Orr Jr., "William Beaumont's Reception at Home and Abroad," *Isis* 72, no. 4 (December 1981): 590–612, quotation at 610.
43. Nicolas Blondlot, *Traité analytique de la digestion considérée particulièrement dans l'homme et dans les animaux vertébrés* (Paris: Fortin, Masson, 1843). On the "quasi-simultaneous" use of the technique by the Russian V. A. Bassov, see Canguilhem, "La constitution de la physiologie comme science," 236.
44. Friedrich H. Bidder and Carl Schmidt, *Die Verdauungssäfte und der Stoffwechsel: Eine physiologisch-chemische Untersuchung* (Mitau: Beyer, 1852); Franklin C. Bing, "Friedrich Bidder (1810–1894) and Carl Schmidt (1822–1894)—A Biographical Sketch," *Journal of Nutrition* 103, no. 5 (May 1973): 637–48; on Pavlov, see chapter 10.
45. For a brief account of Müller's career, see Otis, *Müller's Lab*, 3–4; Johannes Müller, *Müller's Elements of Physiology*, 4 vols., trans. William Baly, intro. Nicholas J. Wade (Bristol: Thoemmes Press, 2003); Wade, introduction to

Müller, *Müller's Elements of Physiology,* 1:v–xxii; this text is a reprint of the 1840–1843 English edition, which was based on Müller's text of 1833–1837. Müller added material to it until 1840; see Johannes Steudel, "Johannes Peter Müller," *DSB* 9:567–74.

46. Cited in Lenoir, *Strategy of Life,* 103, 104–5.
47. Müller, *Müller's Elements,* 2:532. Müller referred the interested reader to the more extensive treatment of these matters in Tiedemann's textbook of human physiology (2:532n).
48. "Instinct has taught him to add oil or butter to farinaceous substances. . . . [and] to fatten animals, with the view of procuring the oleaginous in conjunction with the albuminous principle"; quoted in Müller, *Müller's Elements,* 2:529.
49. Müller, *Müller's Elements,* 2:530. At this point Baly interpolated the results obtained by Reid in Scotland, who had argued that in some animals at least eating resumed after temporary anorexia; on Brachet and Reid, see "Digestive Experiments: A Place for Appetite?"
50. Müller, *Müller's Elements,* 1:204. For an account of the "discovery" of the vegetative nervous system that ignores Bichat and sees Müller as initiating the process, see William R. Woodward, "Johannes Müller, Hermann Lotze, Jakob Henle und die Konstruktion des vegetativen Nervensystems," in *Johannes Müller und die Philosophie,* ed. Michael Hagner and Bettina Wahrig-Schmidt (Berlin: Akademie Verlag, 1992), 155–71.
51. Müller, *Müller's Elements,* 2:545, and, for disagreements with Bichat, 2:816*–817* (on brain or viscera as the seat of the passions) and 3:896 (on muscle contractility not requiring the influence of the nerves). (The pages at the end of volume two are mispaginated; the asterisks are in the text and indicate repetition of page numbers 787–848).
52. Müller, *Müller's Elements,* 3:1069; Otis, *Müller's Lab,* 9.
53. Müller, *Müller's Elements,* 2:712. On this point, Müller noted the superiority of clinical over experimental evidence when he stated that patients' reports "prove, much more clearly than experiments can do, that these nerves have sensitive endowments" (2:712).
54. Müller, *Müller's Elements,* 4:1369, and on Gall, 2:835*–836*.

CHAPTER SIX

1. Farber, "Discussion Paper." On the centrality of physiology in the "epic" story of nineteenth-century medicine and life science, see Hagner, "Scientific Medicine," 51, 63–68.
2. Review of Leuret and Lassaigne, *Recherches physiologiques* and Tiedemann and Gmelin *Die Verdauung nach Versuchen, EMSJ* 27 (1827): 417–25 and 28 (1827): 364–79; the reviewer preferred the work of Leuret and Lassaigne because it required no knowledge of "organic chemistry" (419).
3. Numbers and Orr, "William Beaumont's Reception," 594–95; Holmes, *Claude Bernard,* 269–70.

4. Medical traditionalists in France also questioned the use of statistical analysis and, in some instances, the procedures of pathological anatomy; see Williams, *The Physical and the Moral*, 131–34, 156–59. On the social and professional foundations of medical traditionalism in Britain, see Lawrence, "Incommunicable Knowledge."
5. Cited in Hagner, "Scientific Medicine," 65. Hagner observes that such sentiments were strongest in Germany, where "the Romantic ideal of pure science" persisted (65).
6. See, for example, the comments of Blondlot on the "disgusting" disorder of "merycisme," the practice of regurgitating and chewing food again; Blondlot, *Traité*, 61.
7. Thomas Neville Bonner, *Becoming a Physician: Medical Education in Britain, France, Germany, and the United States, 1750–1945* (Baltimore: Johns Hopkins University Press, 1995), 142–57.
8. J.-L. Leroux, "Dissertation sur l'appétit et la faim" (thesis, École de Médecine de Paris, 1804); 6–7; J.-A. Lamotte, "Propositions sur la faim" (thesis, École de Médecine de Paris, 1804), 5–8; F.-A. Bayle, "Considérations physiologiques et pathologiques sur la faim" (thesis, Faculté de Médecine de Paris, 1816), 7–8. Bayle held that it was hunger rather than appetite that sometimes led eaters astray; as an "imperious need" or "blind penchant," hunger could drive individuals to consume "the most harmful" substances (9).
9. Bayle, "Considérations," 14; Lamotte, "Propositions," 8; Leroux, "Dissertation," 22.
10. Leroux, "Dissertation," 8, 12, 17; Lamotte, "Propositions," 10.
11. Bayle, "Considérations," 18.
12. Leroux, "Dissertation," 8, 16.
13. Bayle, "Considérations," 8.
14. Leroux, "Dissertation," 22–23.
15. Leroux, "Dissertation," 9, 15, 16, 19, 30; Lamotte, "Propositions," 11; Bayle, "Considérations," 11, 15, 19, 20, 24. Leroux and Bayle appended lists of Hippocratic aphorisms, Lamotte excerpts from *Prognosis*, bk. 1.
16. On the use of literary devices in medical narrative, see Sabine Arnaud, *L'invention de l'hystérie au temps des Lumières (1670–1820)* (Paris: École des Hautes Études en Sciences Sociales, 2014), esp. 64–88.
17. From 1795 to 1808 the journal was titled *Journal der practischen Arzneykunde und Wundarzneykunst*.
18. C. W. Hufeland, "Höchst merkwürdiges Beyspiel einer noch jezt fortdauernden, beynahe fünfzehn monatlichen Enthaltung von Speisen und Getränken bey einem sechzehenjährigen Mädgen," *Journal der practischen Arzneykunde und Wundarzneykunst* 8, part 2 (1799): 191–95 at 194; the term "female deceivers" (Betrügerinnen) appears in Tiedemann, *Nahrungs-Bedürfniss*, 41.
19. E. Müller, "Geschichte einer langen Enthaltsamkeit von allen Nahrungsmitteln," *Journal der practischen Arzneykunde und Wundarzneykunst* 24, pt. 2 (1806):

154–59; E. Schmalz, "Zwei Geschichten von Frauen, welche angeblich sehr lange ohne Nahrung zu sich zu nehmen ihr Leben fortgesetzt haben," *Journal der practischen Heilkunde* 69, supplement (1829): 216–35.

20. Benjamin Granger, "Some Account of the *Fasting Woman* at Titbury, Who Has at Present Lived above Two Years without Food," *EMSJ* 5 (1809): 319–26; Granger, "On Unusual Cases of Anorexy," *EMSJ* 9 (1813): 157–59; Granger, "Account of the Second Watch of the Reputed Fasting-Woman," *EMSJ* 9 (1813): 323–26. For French and German cases reported between 1803 and 1844, see Vandereycken and van Deth, *From Fasting Saints to Anorexic Girls,* 255, 265.
21. French works ordinarily stressed the perverse eating of chlorotic girls and pregnant women, but they also noted men who "eagerly seized occasions for eating raw meat, blood, rats, insects, and even excrement"; Mouton, "Appétit," in *Dictionnaire des sciences médicales* 2:258–60, at 259.
22. Pinel, *Nosographie philosophique*; on the Paris Clinical School, see Erwin H. Ackerknecht, *Medicine at the Paris Hospital, 1794–1848* (Baltimore: Johns Hopkins University Press, 1967); Caroline Hannaway and Ann La Berge, eds., *Constructing Paris Medicine* (Amsterdam: Rodopi, 1998).
23. Barras, *Traité sur les gastralgies et les entéralgies*; see also Jules Grange, "Essai sur les névroses de l'estomac" (thesis, Faculté de Médecine de Paris, 1845). A review of the fourth edition of Barras's work called it "the most complete and practical work we have" on gastralgia, whose causes, "like that of all the neuroses, have to this point escaped the researches of men of the art"; *Gazette médicale de Paris*, 2nd ser., 13, no. 16 (April 19, 1845): 253–54.
24. Jean-François Braunstein, *Broussais et le matérialisme: Médecine et philosophie au XIXe siècle* (Paris: Méridiens Klincksieck, 1986).
25. Barras, *Traité,* 41; see the case studies at 30–123 ("gastro-enteralgic" hypochondria is at 72–73).
26. On the influence of sex hormones on eating patterns in experimental animals, see chapter 11, "'Salt Appetite': Curt P. Richter and the Instinctive Search for Nutrients."
27. Barras, *Traité,* 245–47. A founding text on the peculiarities of women's appetite and eating, including a capacity for "prolonged abstinence" and "bizarre tastes" in food, was J.-L. Moreau de la Sarthe, *Histoire naturelle de la femme, suivie d'un traité d'hygiène appliquée à son régime physique et moral aux différentes époques de la vie,* 3 vols. (Paris: Duprat, 1803), 1:136–58.
28. A.-F. Chomel, *Des dyspepsies* (Paris: Victor Masson, 1857), 2–4. Chomel recounted cases of patients who suffered troubles of appetite every day or even at every meal (76–77, 94, 102, 189).
29. Student authors often suggested personal circumstances that prompted their interest in dyspepsia; see John Harris, "An Essay on Dyspepsia" (thesis, University of Pennsylvania, 1816), 2; Andrew J. Thomas, "An Essay on Dyspepsia" (thesis, University of Pennsylvania, 1853), 22. See also Gert Brieger,

"Dyspepsia: The American Disease? Needs and Opportunities for Research," in *Healing and History: Essays for George Rosen*, ed. Charles E. Rosenberg (Kent: Dawson Science History Publications, 1979), 179–90.

30. Cyrus B. Wood, "On Indigestion or Dyspepsia" (thesis, University of Pennsylvania, 1818); John T. Boyd, in "An Inaugural Dissertation on Dyspepsia" (thesis, University of Pennsylvania, 1810), began with dyspepsia but later referred simply to "nervous disease." William S. Alexander, "An Inaugural Dissertation on Dyspepsia" (thesis, University of Pennsylvania, 1811), shifted at one point to "gastrodinia." On distinguishing dyspepsia from "hypochondriasis," see Harris, "Essay," 9. After recalling labels used by different authorities, Stokes P. Ivey commented that none of them gives a "correct idea of the disease"; see Ivey, "An Essay on Dyspepsia" (thesis, University of Pennsylvania, 1837), 21.
31. For Cullen's view of dyspepsia, see chapter 2, "Maladies of Appetite in Eighteenth-Century Nosologies." References to Cullen appear in Boyd, "Inaugural Dissertation," 7; Alexander, "Essay on Dyspepsia," n.p.; Harris, "Essay on Dyspepsia," 4, 5, 9; Ivey, "Essay on Dyspepsia," 21. On Cullen's standing among Philadelphia students, see Whitfield J. Bell Jr., "Philadelphia Medical Students in Europe, 1750–1800," in *The Colonial Physician and Other Essays* (New York: Science History Publications, 1975), 41–69, at 44, 51, 53–55.
32. Symptoms included disturbed appetite (deficient, voracious, or "vitiated"); nausea or vomiting; stomach distention or pain; eructations (sometimes acid); heartburn; fetid or deficient evacuations; dry, white, or furry tongue; epigastric heat or pain; costiveness or diarrhea; cardialgia and heart palpitations; pyrosis; gastrodynia; colic in the bowels; muscle weakness; vertigo; headache; and troubled sight. For psychic or emotional connections, see Jacob Jeanes, "Essay on Dyspepsia" (thesis, University of Pennsylvania, 1823), n.p.; Boyd, "Inaugural Dissertation," 2, 14; Harris, "Essay on Dyspepsia," 5–6; Ivey, "Essay on Dyspepsia," 1, 13; Thomas, "Essay on Dyspepsia," 6, 22.
33. Harris, "Essay," 5–6; Ivey, "Essay," 14.
34. Alexander, "Essay," n.p.; Harris, "Essay," 10–11.
35. Alexander, "Essay," n.p.; Harris, "Essay," 7.
36. Ivey, "Essay," 15–16.
37. Jeanes, "Essay," n.p.
38. Baron and Sonnenberg, "Alimentary Diseases in the Poor and Middle Class"; *EMSJ* reported the numbers of cases of these and other diseases from the Carey Street Dispensary (also called the Public Dispensary) in London from 1805 to 1816 and from the New Town Dispensary in Edinburgh (founded September 1, 1815) from 1816 to 1819 (and later sporadically). The first report from New Town included 52 cases of dyspepsia (September 1, 1815, to March 1, 1816); "Report of Diseases Treated at the New Town Dispensary, Edinburgh," *EMSJ* 12 (1816): 245–49. The following summer 102 cases were reported (June 1 to September 1, 1816); *EMSJ* 13 (1817): 117–18.

39. Review of J. A. Paris, *A Treatise on Diet* (London, 1826), *EMSJ* 27 (1827): 163–73, at 171.
40. Review of John Reid, *Essays on Insanity, Hypochondriasis, and Other Nervous Affections* (London, 1816); *EMSJ* 12 (1816): 467–73, at 469–72; review of Thomas Trotter, *A View of the Nervous Temperament* (Newcastle, 1807); *EMSJ* 3 (1807): 473–82, at 473; see also John Armstrong, "Brief Hints Relative to the Improvement of the Pathology and Treatment of Those Chronic Diseases Usually Termed Nervous," *EMSJ* 11 (1815): 416–18, at 417.
41. Review of John Howship, *Practical Remarks upon Indigestion* (London, 1825), *EMSJ* 25 (1826): 393–403.
42. Goldstein, *Console and Classify*, 80–89.
43. Dora B. Weiner, *Comprendre et soigner: Philippe Pinel (1745–1826), La médecine de l'esprit* (Paris: Fayard, 1999), 17, 51, 58–59, 148, 151, 243–46.
44. Philippe Pinel, "Mémoire sur la manie périodique, ou intermittente," *Mémoires de la Société médicale d'émulation* 1, year 5 (1797): 94–119, at 100; Pinel, *Nosographie philosophique*, 2:29–30; Williams, "Stomach and Psyche."
45. Philippe Pinel, *Traité médico-philosophique sur l'aliénation mentale, ou la manie* (Paris: Caille and Ravier, 1800), xxii–xxiv. The second edition of this work, titled *Traité médico-philosophique sur l'aliénation mentale*, was much altered; on the differences between the two, see Weiner, *Comprendre et soigner*, 298–311.
46. Pinel condemned the "severe diet" some physicians recommended in cases of mental alienation and insisted that adequate food was basic to the success of "well-ordered hospices"; Pinel, "Mémoire sur la manie périodique," 109–10n1. On dietetic changes Pinel encouraged at Bicêtre and Salpêtrière, see Dora B. Weiner, *The Citizen-Patient in Revolutionary and Imperial Paris* (Baltimore: Johns Hopkins University Press, 1993), 263, 270, 272.
47. Pinel, "Mémoire sur la manie périodique," 109–10; Pinel, *Nosographie philosophique*, 2:22–23, 30, 49; Pinel, *Traité médico-philosophique sur l'aliénation mentale* (1809), v, 49, 296–97n1, 297.
48. Pinel introduced his new classification based on physiological function in the third edition of *Nosographie philosophique* (1807) and retained it in the editions of 1810, 1813, and 1818; see José M. López Piñero, *Historical Origins of the Concept of Neurosis*, trans. D. Berrios (Cambridge: Cambridge University Press, 1983), 44–49, whose discussion is based on the 1818 edition.
49. Goldstein, *Console and Classify*, 124–47, 174–76; René Semelaigne, *Les pionniers de la psychiatrie française avant et après Pinel* (Paris: J.-B. Baillière, 1930), 124–40; Jacques Postel and Claude Quétel, eds., *Nouvelle histoire de la psychiatrie*, rev. ed. (Paris: Dunod, 1994), 140–43, 162–71.
50. J.-E.-D. Esquirol, "Folie," in *Des maladies mentales, considérées sous les rapports médical, hygiénique et médico-légal*, 2 vols. (Paris: J.-B. Baillière, 1838), 1:1–158, at 75–76.
51. J.-E.-D. Esquirol, *Des passions considerées comme causes, symptômes et moyens curatifs de l'aliénation mentale* (1805; rpt., Paris: Librairie des deux-mondes, 1980), 65–68; Esquirol, *Des maladies mentales*, 1:336–97, at 397.

52. Esquirol, *Maladies mentales*, 1:359, 548, 558, 567. Esquirol's attention to diet was emulated by alienists he trained, including J.-P. Falret and J.-E.-F. Mitivié, who in the 1840s pushed for dietary reforms at the Salpêtrière. See Williams, "Neuroses of the Stomach," 62n21.
53. Esquirol, "Terminaisons critiques de la folie," in *Maladies mentales*, 1:336–97 (my numbers include only cases Esquirol treated himself and described individually).
54. Discussing food refusal in thirty-seven women and twenty-two men seen at Charenton, for example, Émile Desportes attributed the sexual imbalance to women's sedentary habits, religious preoccupations, and inherent tendencies to melancholy and *inappétence*; Emile Desportes, "Du refus de manger chez les aliénés" (thesis, Faculté de Médecine de Paris, 1864), 9–10.
55. On Georget, see Semelaigne, *Les pionniers*, 188–96; Étienne-Jean-Georget, *De la folie: Considérations sur cette maladie* (1820; rpt., New York: Arno Press, 1976).
56. Georget, *De la folie*, 135–36. Georget argued elsewhere that Bichat had gone too far in denying the brain any control over digestion (an overstatement of Bichat's view) but acknowledged that little was known about the brain's influence over "organs not directly subordinate" to it; see Georget, *De la physiologie du système nerveux, et spécialement du cerveau: Recherches sur les maladies nerveuses en général, et en particulier sur le siège, la nature et le traitement de l'hystérie, de l'hypochondrie, de l'épilepsie et de l'asthme convulsif*, 2 vols. (Paris: J.-B. Baillière, 1821), 2:93–95n1.
57. A.-L.-J. Bayle, *Traité des maladies du cerveau et de ses membranes* (1826; rpt., New York: Arno Press, 1976), 386–89, 388, 394.
58. François Leuret, "Note sur une nouvelle sonde destinée à l'alimentation des aliénés," *Gazette médicale de Paris*, 2nd ser., 13, no. 34 (August 23, 1845): 540–41 and no. 35 (August 30, 1845): 555–56. See also Antoine-Émile Blanche, "Cathétérisme oesaphagien chez les aliénés" (thesis, Faculté de Médecine de Paris, 1848); Martial Pellevoisin, "De l'alimentation forcée chez les alienés" (thesis, Faculté de Médecine, Strasbourg, 1862).
59. Korsmeyer, *Making Sense of Taste*.
60. Review of H. Bardsley, *Observations on Mental Derangement*, in *EMSJ* 22 (1824): 279–86, at 286. On Lawrence, see Adrian Desmond, *The Politics of Evolution: Morphology, Medicine and Reform in Radical London* (Chicago: University of Chicago Press, 1989), 117–21, 255–58. On moral causation, see Alexander Morison, *Outlines of Mental Diseases* (Edinburgh: MacLachlan and Stewart, 1824), 66–71, excerpted in Vieda Skultans, *Madness and Morals: Ideas on Insanity in the Nineteenth Century* (London: Routledge and Kegan Paul, 1975), 32–33.
61. Elizabeth A. Williams, "Gags, Funnels and Tubes: Forced Feeding of the Insane and of Suffragettes," *Endeavour* 32, no. 4 (2008): 134–40.

INTRODUCTION TO PART THREE

1. Goody, "Industrial Food"; Gabriella M. Petrick, "Industrial Food," in *The Oxford Handbook of Food History*, ed. Jeffrey J. Pilcher (New York: Oxford University Press, 2012), 258–78; Derek J. Oddy and Alain Drouard, eds., *The Food Industries of Europe in the Nineteenth and Twentieth Centuries* (Burlington, VT: Ashgate, 2013).
2. In 1908 one London commentator expressed shock "that hunger was so rife 'in this city of incalculable wealth and riches'": cited in James Vernon, *Hunger: A Modern History* (Cambridge, MA: Harvard University Press, 2007), 22.
3. Karen Offen, *European Feminisms, 1700–1950: A Political History* (Stanford, CA: Stanford University Press, 2000), esp. 131–32, 141–42.
4. On broad trends in the discipline, see Richard L. Kremer, "Physiology," in *The Cambridge History of Science*, vol. 6, *Modern Biological and Earth Sciences*, ed. Peter J. Bowler and John V. Pickstone (Cambridge: Cambridge University Press, 2009), 342–66.
5. John V. Pickstone, *Ways of Knowing: A New History of Science, Technology and Medicine* (Chicago: University of Chicago Press, 2000), esp. 50–53.
6. On this debate, see Richards, *Darwin*; Robert Boakes, *From Darwin to Behaviourism: Psychology and the Minds of Animals* (Cambridge: Cambridge University Press, 1984).
7. Hagner, "Scientific Medicine."

CHAPTER SEVEN

1. On German physiologists' "intentionally work[ing] at a distance from the clinic," see Hagner, "Scientific Medicine," 65.
2. On Tiedemann, see chapter 5, "Digestive Experiments: A Place for Appetite?"
3. On Moleschott's training and career, see Frederick Gregory, *Scientific Materialism in Nineteenth Century Germany* (Dordrecht: D. Reidel, 1977), 80–99; Harmke Kamminga, "Nutrition for the People, or the Fate of Jacob Moleschott's Contest for a Humanist Science," in *The Science and Culture of Nutrition, 1840–1940*, ed. Harmke Kamminga and Andrew Cunningham (Amsterdam: Rodopi, 1995), 15–47; A. M. Geist-Hofman, "Jacob Moleschott," in *DSB* 9:456–57.
4. Jacob Moleschott, *Die Physiologie der Nahrungsmittel: Ein Handbuch der Diätetik* (Darmstadt: Carl Wilhelm Leske, 1850). A revised edition (discussed below) appeared in 1859; except where indicated, references here are to the 1850 edition.
5. Justus Liebig, *Die organische Chemie in ihrer Anwendung auf Physiologie und Pathologie* (Braunschweig: Vieweg, 1842).
6. Gregory, *Scientific Materialism*, esp. 86–87, 89–90.

7. On nineteenth-century conceptions of metabolism, see Hannah Landecker, "Food as Exposure: Nutritional Epigenetics and the New Metabolism," *Bio-Societies* 6, no. 2 (2011): 167–94, at 170–73.
8. Moleschott, *Physiologie der Nahrungsmittel*, 78–80. On Tiedemann's attention to torturing hunger, see chapter 5, "Digestive Experiments: A Place for Appetite?" Moleschott also cited the German Ur-source, Haller's *Elementa physiologiae* (80n). On Haller, see chapter 3.
9. Moleschott's authorities on the formation of food habits ranged from Herodotus to modern experts on the acclimatization of Europeans to foods available in the colonies; Moleschott, *Physiologie der Nahrungsmittel*, 564–71.
10. Mani, "Die wissenschaftliche Ernährungslehre," 39–41.
11. Jane O'Hara-May, "Measuring Man's Needs," *JHB* 4, no. 2 (September 1971): 249–73, esp. 254–55, 261–63.
12. O'Hara-May, "Measuring," 250–51.
13. On the benefits of breast milk and dangers of sweets and excessive meat, see Moleschott, *Physiologie der Nahrungsmittel*, 554, 556–57.
14. Jacob Moleschott, *Physiologie der Nahrungsmittel: Ein Handbuch der Diätetik*, 2nd ed. (Giessen: Ferber, 1859), 538.
15. For the comparable section in the 1859 edition, see Moleschott, *Physiologie der Nahrungsmittel*, 1859 ed., 178–84.
16. Kamminga, "Nutrition for the People," 23–29.
17. Kamminga argues that Moleschott has been forgotten by the history of science because his "humanist" science of nutrition had political values at its heart as opposed to the purportedly value free and objective nutritional standards that won out; Kamminga, "Nutrition for the People," 34–40.
18. Rothschuh, *History of Physiology*, 205. Timothy Lenoir argues that Ludwig's scientific practice, while striving to eliminate external concerns from the explicit framing of scientific problems, nonetheless served the interests of "modern industrializing states"; see Lenoir, "Science for the Clinic: Science Policy and the Formation of Carl Ludwig's Institute in Leipzig," in *The Investigative Enterprise: Experimental Physiology in Nineteenth-Century Medicine*, ed. William Coleman and Frederic L. Holmes (Berkeley: University of California Press, 1988), 139–78, at 173.
19. Rothschuh, *History of Physiology*, 206; Merriley Borell, "Instrumentation and the Rise of Modern Physiology" *Science and Technology Studies* 5, no. 2 (1987): 53–62, at 53–54.
20. Rothschuh, *History of Physiology*, 204–12; George Rosen, "Carl Ludwig and His American Students," *BHM* 4 (1936): 609–50.
21. Citations are to Carl Ludwig, *Lehrbuch der Physiologie des Menschen*, 2nd rev. ed., 2 vols. (Leipzig: C. F. Winter, 1858–1861).
22. Coleman, *Biology in the Nineteenth Century*, 129. Moleschott referred to the blood as the *Muttersaft*; Moleschott, *Physiologie der Nahrungsmittel* (1859 ed.), 2.
23. C. Ludwig, "Neue Versuche über die Beihilfe der Nerven zur Speichelabsonderung"; first published in 1851, this essay was reproduced in *Abhandlungen*

über den Speichel, ed. Max von Frey (Leipzig: Wilhelm Engelmann, 1890), 3–25.

24. On this theme, see Anson Rabinbach, *The Human Motor: Energy, Fatigue, and the Origins of Modernity* (Berkeley: University of California, 1992), 125–26.
25. T. L.W. Bischoff and Carl von Voit, *Die Gesetze der Ernährung des Fleischfressers durch neue Untersuchungen* (Leipzig: C. F. Winter, 1860); cited in Ludwig, *Lehrbuch*, 2:674–76. On Voit's preeminent position in the study of metabolism, see Frederic L. Holmes, "The Formation of the Munich School of Metabolism," in *The Investigative Enterprise: Experimental Physiology in Nineteenth-Century Medicine*, ed. William B. Coleman and Frederic L. Holmes (Berkeley: University of California Press, 1988), 179–210, at 194–202, 206.
26. Ackerknecht, "History," 4; F. H. Bidder and A. W. Volkmann, *Die Selbstständigkeit des sympathischen Nervensystems durch anatomische Untersuchungen nachgewiesen* (Leipzig: Breitkopf and Härtel, 1842). Although Bidder and Volkmann acknowledged Bichat's priority, they contrasted the speculative character of his work (and that of his German disciple Reil) to their own precise anatomical demonstrations of distinctions between the two systems (2–3).
27. Rosenzweig, "Mechanisms of Hunger and Thirst," 92.
28. Ludwig's hostility to vitalism also governed his thinking about tissue respiration. Despite experimental results indicating that tissues respire on their own rather than depending on oxygen provided by blood, Ludwig continued to support the "blood-site" theory of tissue oxidation. See Culotta, "Tissue Oxidation and Theoretical Physiology." Culotta contrasts Ludwig's thinking to the vitalist view of E. F. W. Pflüger, who held that "tissue not only consumes oxygen but controls its own appetite" (140); see also Lenoir, *Strategy of Life*, 217–19.
29. Rothschuh, *History of Physiology*, 254–56.
30. Rubner also worked briefly in Ludwig's laboratory; on his efforts to determine "which diet was appropriate for the optimal performance of the worker," see Rabinbach, *Human Motor*, 126–27. See also Corinna Treitel, "Max Rubner and the Biopolitics of Rational Nutrition," *Central European History* 41, no. 1 (2008): 1–25. For a recent discussion, see Richard D. Feinman and Eugene J. Fine, "'A Calorie Is a Calorie' Violates the Second Law of Thermodynamics," *Nutrition Journal* 3, no. 9 (2004): http://www.nutritionj.com/content/3/1/9.
31. Max Rubner, "Der Nahrungstrieb des Menschen," *Sitzungsberichte der Königlich-Preussischen Akademie der Wissenschaften zu Berlin* 11 (1920): 341–64, at 343.
32. Eduard F. W. Pflüger, "Die teleologische Mechanik der lebendigen Natur," *Archiv für die gesammte Physiologie des Menschen und der Thiere* 15 (1877): 57–101.
33. Pflüger admiringly invoked Aristotle as the originator of this insight about living bodies ("Die teleologische Mechanik," 58–60). On the varied character and rhythm of drives across the stages of human life, see Pflüger,

Ueber die Kunst der Verlängerung des menschlichen Lebens (Bonn: Emil Strauss, 1890).

34. Rudolf Heidenhain, *Handbuch der Physiologie der Absonderung und Aufsaugung*, vol. 5, pt. 1 of *Handbuch der Physiologie*, 5 vols, ed. Ludimar Hermann (Leipzig: F. C. W. Vogel, 1883), 117. For the judgment about oversimplification, see Vladislav Kruta, "Rudolf Peter Heinrich Heidenhain," *DSB* 6:224–27.
35. On Pavlov, see chapter 10.
36. On the origins of the concept of regulation, see Smith, *Inhibition*, esp. 20–21, 33–34, 43–45, 66–87; Georges Canguilhem, "The Development of the Concept of Biological Regulation in the Eighteenth and Nineteenth Centuries," in *Ideology and Rationality in the History of the Life Sciences*, trans. Arthur Goldhammer (Cambridge, MA: MIT Press, 1988), 81–102; E. F. Adolph, "Early Concepts of Physiological Regulation," *Physiological Reviews* 41, no. 4 (1961): 737–70.
37. See chapter 5, "Need and Desire in the Physiology of François Magendie."
38. Mirko D. Grmek, *Le legs de Claude Bernard* (Paris: Fayard, 1997), 121–79; Frederic L. Holmes, "Claude Bernard and the *milieu intérieur*," *Archives internationales d'histoire des sciences* 16 (1963): 369–76; Holmes, "Claude Bernard, the *Milieu intérieur*, and Regulatory Physiology," *History and Philosophy of the Life Sciences* 8, no. 1 (1986): 3–25.
39. Holmes, "Claude Bernard and the *milieu intérieur*," 369, 374.
40. Claude Bernard, "De l'origine du sucre dans l'économie animale," *AGM* 18 (1848): 303–19; Holmes, *Claude Bernard*, 426–44.
41. Holmes, "Claude Bernard and the *milieu intérieur*," 371–72, 374.
42. Claude Bernard, *Leçons sur les propriétés des tissus vivants*, ed. Émile Alglave (Paris: Germer Baillière, 1866), 376–80.
43. Bernard, *Leçons sur les propriétés des tissus vivants*, 381–82. On the "purely psychic or cerebral excitation" that could result from "the idea of succulent food," see also Claude Bernard, "Des fonctions du cerveau," *Revue des deux mondes* 98 (1872): 373–85, at 378.
44. Claude Bernard, *Leçons sur les phénomènes de la vie communs aux animaux et aux végétaux*, 2 vols. (Paris: J.-B. Baillière, 1878–79), 1:113–24.
45. Grmek notes the "cold" reception physiologists first gave Bernard's conception of the internal milieu; see Grmek, *Le legs de Claude Bernard*, 168–72. On later developments, see chapter 10.
46. Webb Haymaker and Francis Schiller, "Moritz Schiff (1823–1896)," in *The Founders of Neurology* (Springfield, IL: Thomas, 1970), 258–64. Schiff collaborated with Longet, who had experimented extensively on the pneumogastric nerves, on a publication treating the chemistry of digestion; see F.-A. Longet, *Mémoire sur la véritable nature des nerfs pneumo-gastriques et les usages de leurs anastomoses* (Paris: Rignoux, 1849); Longet and Schiff, *Fragments sur les phénomènes chimiques de la digestion* (Paris: n.p., 1857).

47. Helmut Heintel, "Moritz Schiffs gescheiterter Habilitationsversuch an der Universität Göttingen im Jahre 1855," *MhJ* 15, no. 4 (1980): 378–84; Haymaker and Schiller, "Moritz Schiff," 260.
48. Maurice Schiff, *Leçons sur la physiologie de la digestion, faites au Muséum d'Histoire naturelle de Florence*, 2 vols., ed. Émile Levier (Florence: Hermann Loescher, 1867). (Schiff's first name appeared as "Maurice" in French-language publications.)
49. Anticipating the later concern of psychologists about the reliability of reports of subjective feelings of hunger, Schiff noted that he sought out people who had no anatomical knowledge so that he did not obtain prejudiced results (*Leçons*, 1:31).
50. Schiff, *Leçons*, 1:36. For a parallel case that Schiff made against local nervous control of "inhibitory effects," see Smith, *Inhibition*, 86–87.
51. Rosenzweig, "Mechanisms of Hunger and Thirst," 74–76, 79, 82–89.
52. For a review of the results of animal experiments, see Heidenhain, *Handbuch der Physiologie der Absonderung und Aufsaugung*, 109–15. On the human patient described by the French physiologist Charles Richet, see chapter 8, "The 'New' Psychology in France."
53. Wilhelm Nicolai, *Ueber die Entstehung des Hungergefühls* (diss., Friedrich-Wilhelms- Universität Berlin, 1892).
54. Carlson, *Control of Hunger*, 19. Despite the hostility often evident between clinicians and experimentalists, this stomach-centered view of hunger both undergirded and drew on the increasingly important clinical work of German-language *Magenärzte*; see Edward Stainbrook, "Psychosomatic Medicine in the Nineteenth Century," *PSMed* 14, no. 3 (1952): 211–27, at 219.

CHAPTER EIGHT

1. Kurt Danziger, "Mid-Nineteenth-Century British Psycho-Physiology: A Neglected Chapter in the History of Psychology," in *The Problematic Science: Psychology in Nineteenth-Century Thought*, ed. William R. Woodward and Mitchell G. Ash (New York: Praeger, 1982), 119–46.
2. Robert M. Young, *Mind, Brain, and Adaptation in the Nineteenth Century: Cerebral Localization and Its Biological Context from Gall to Ferrier* (Oxford: Clarendon Press, 1970), 55–74, 88–94; Judith P. Swazey, "*Action propre* and *Action commune*: The Localization of Cerebral Function," *JHB* 3, no. 2 (Autumn 1970): 213–34, at 226–27, 232; Lorraine J. Daston, "British Responses to Psycho-physiology, 1860–1900," *Isis* 69, no. 2 (June 1978): 192–208, at 195–96.
3. Wilhelm Wundt, *Grundzüge der physiologischen Psychologie* (Leipzig: Wilhelm Engelmann, 1874); W. G. Bringmann, N. J. Bringmann, and G. A. Ungerer, "The Establishment of Wundt's Laboratory: An Archival and Documentary Study," in *Wundt Studies: A Centennial Collection*, ed. Wolfgang G. Bringmann and Ryan D. Tweney (Toronto: C. J. Hogrefe, 1980), 123–57.

4. Wundt largely steered clear of the psychology of ingestion and digestion, referring to hunger only briefly as a sensation under vegetative control; Wundt, *Grundzüge*, 276.
5. Stewart Wolf, *Brain, Mind, and Medicine: Charles Richet and the Origins of Physiological Psychology* (New Brunswick, NJ: Transaction, 1993), 61.
6. Charles Richet, *Du suc gastrique chez l'homme et les animaux: Ses propriétés chimiques et physiologiques* (Paris: Baillière, 1878), appendix, 158–65.
7. Richet, *Suc gastrique*, 159–60; Wolf, *Brain, Mind, and Medicine*, 29–31; see also Wolf, "The Psyche and the Stomach: A Historical Vignette," *Gastroenterology* 80, no. 3 (1981): 605–14, which recounts later work (much of it Wolf's own) on fistular patients.
8. See the discussion in chapter 7, "The Urge to Eat in Physicochemical Physiology."
9. Heidenhain, *Handbuch der Physiologie der Absonderung and Aufsaugung*, 117.
10. Wolf, *Brain, Mind, and Medicine*, 31.
11. John Warne Monroe, *Laboratories of Faith: Mesmerism, Spiritism, and Occultism in Modern France* (Ithaca, NY: Cornell University Press, 2008), 200, 207, 208, 211, 212, 218.
12. Charles Richet, ed., *Dictionnaire de physiologie* (Paris: Félix Alcan, 1895–1928); of the planned volumes of this dictionary, only the first ten appeared.
13. Charles Richet, "Dégoût," in *Dictionnaire de Physiologie*, 4:754–60, at 755.
14. Richet, "Dégoût," 756–57; this entry offered a précis of views Richet had put forward in *L'homme et l'intelligence: Fragments de physiologie et de psychologie* (Paris: Félix Alcan, 1884), 41–84.
15. The entry on hunger, by E. Bardier, is discussed in chapter 10.
16. Serge Nicolas, "Henry Beaunis (1830–1921), directeur-fondateur du laboratoire de Psychologie physiologique de la Sorbonne," *L'Année psychologique* 95, 2 (1995): 267–91.
17. H. Beaunis, *Les sensations internes* (Paris: Félix Alcan, 1889).
18. Nicolas, "Henry Beaunis," 281, 286; Martin S. Staum, *Nature and Nurture in French Social Sciences, 1859–1914 and Beyond* (Montreal: McGill-Queen's University Press, 2011), 126–41.
19. On Bernheim, see Mark S. Micale, *Approaching Hysteria: Disease and Its Interpretations* (Princeton, NJ: Princeton University Press, 1995), 26; Ruth Harris, "Murder under Hypnosis in the Case of Gabrielle Bompard: Psychiatry in the Courtroom in Belle Epoque Paris," in The *Anatomy of Madness: Essays in the History of Psychiatry*, 3 vols, ed. W. F. Bynum, Roy Porter, and Michael Shepherd, 2:197–241, esp. 203–4, 207–10.
20. "L'homme sain, qui meurt après plusieurs jours de jeûne, ne meurt pas d'inanition; . . . il *meurt de faim . . . la faim tue avant l'inanition*"; cited in Joanny Roux, "La faim, étude physio-psychologique," *Bulletin de la Société d'anthropologie de Lyon* 16 (1897): 409–55, at 433.
21. Hillel Schwartz, *Never Satisfied: A Cultural History of Diets, Fantasies and Fat* (New York: Free Press, 1986), 119–23.

22. Luigi Luciani, *Das Hungern: Studien und Experimente am Menschen*, foreword Jacob Moleschott, trans. M. O. Fraenkel (Hamburg: Leopold Voss, 1890), 1–2; Agustí Nieto-Galan, "Mr. Giovanni Succi Meets Dr Luigi Luciani in Florence: Hunger Artists and Experimental Physiology in the Late Nineteenth Century," *Social History of Medicine* 28, no. 1 (February 2015): 64–81.
23. Luciani later moved to Rome, where he succeeded Moleschott as professor of physiology; he died there in 1919; Nieto-Galan, "Mr. Giovanni Succi," 69–71; see also Bruno Zanobio and Gianluigi Porta, "Luigi Luciani," *DSB* 8:535–36.
24. Zanobio and Porta, "Luigi Luciani," 536; Soraya de Chadarevian, "Graphical Method and Discipline: Self-Recording Instruments in Nineteenth-Century Physiology," *Studies in History and Philosophy of Science* 24, no. 2 (1993): 267–91.
25. Nieto-Galan, "Mr. Giovanni Succi," 69–70.
26. Luciani, *Das Hungern*, ix.
27. Luciani, *Das Hungern*, 28–29. On the fast of his Paris rival, see E. Monin and Ph. Maréchal, *Stefano Merlatti: Histoire d'un jeûne célèbre précédée d'une étude anecdotique, physiologique et médicale sur le jeûne et les jeûneurs* (Paris: C. Marpon et E. Flammarion, 1887).
28. Luciani, *Das Hungern*, 6, 11–12; on instruments in use by the late nineteenth century, see Borell, "Instrumentation and the Rise of Modern Physiology."
29. Luciani, *Das Hungern*, 56–57; on autosuggestion in late nineteenth-century psychiatry, see Charles Baudouin and Edward B. Schlesinger, *Suggestion and Autosuggestion: A Psychological and Pedagogical Study Based upon the Investigations Made by the New Nancy School*, trans. Eden Paul and Cedar Paul (London: G. Allen and Unwin, 1920).
30. Luciani noted differing effects of the fast on specific tissues and organs; the spleen and liver suffered the most while the nervous system (victor in the "struggle for life") proved "the privileged part of the organism"; Luciani, *Das Hungern*, 78, 80–81, 90, 92, 234.
31. On his reputation and impact, see Nieto-Galan, "Mr. Giovanni Succi," 69–71; the judgment of his work on hunger is in Zanobio and Porta, "Luigi Luciani," 536.
32. Luciani, *Das Hungern*, translator's note, 1n1
33. Luciani, *Das Hungern*, 77, 229. Although Luciani's thinking on regulation owed a good deal to Bernard's concept of the internal milieu, he took issue with the French physiologist on the role of the blood, which, he held, should be regarded as itself a tissue, indeed the tissue that dominates all others, not just as "ein Ambient" (90).
34. Luciani rejected suggestions put forward by other authorities who commented on Succi, including the famed criminologist Cesare Lombroso, that the hunger artist's powers of resistance were typical of hysterics, epileptics, and paranoiacs; *Das Hungern* (216–17).
35. See the discussion in the introduction to part 2.

36. The classic study of the contest between Lamarckism and Darwinism in France is Yvette Conry, *L'introduction du darwinisme en France au XIXe siècle* (Paris: J. Vrin, 1974).
37. On early modern debates over animal intelligence, see Thomas, *Man and the Natural World*, 121–41. On this theme in French physiology of the 1820s–1840s, see Williams, *The Physical and the Moral*, 127–31, 170–73, 193–95.
38. "An action, which we ourselves require experience to enable us to perform, when performed by an animal, more especially by a very young one, without experience, and when performed by many individuals in the same way, without their knowing for what purpose it is performed, is usually said to be instinctive"; Charles Darwin, *The Origin of Species by Means of Natural Selection*, 6th ed. (London: John Murray, 1873), 205. See also Richards, *Darwin*, esp. 142–156.
39. Richards, *Darwin*, 387–88; see also Boakes, *From Darwin to Behaviourism*, 14.
40. On Spallanzani's observation of barnyard chicks, see chapter 3.
41. Richards, *Darwin*, 388.
42. Richards, *Darwin*, 388–89. Morgan discussed his results in "The Limits of Animal Intelligence," *Fortnightly Review* 60 (1893): 223–39, and later in *An Introduction to Comparative Psychology* (London: Walter Scott, 1894), 197–99, where he concluded that "these experiments and observations seem . . . to show that the skill in seizing is not perfect at birth, and that some practice is necessary" (198).
43. Richards, *Darwin*, 389–94; Boakes, *From Darwin to Behaviourism*, 32–44.
44. A typical account summarized a report by an official with the US Department of Agriculture that assessed the "injurious" or "beneficial" effects on agriculture of the feeding habits of horned larks and cedarbirds. See "J. A. A.," "Food Habits of Birds," *The Auk* 10, no. 4 (October 1893): 347.
45. On differences between amateur and scientific observers, see Mark V. Barrow, *A Passion for Birds: American Ornithology after Audubon* (Princeton, NJ: Princeton University Press, 1998), 75, 154–55, 172–80, 185.
46. This held true well past 1900. See "Instinct and Intelligence," *British Journal of Psychology* 3 (1909–1910): 209–70, cited by Richards, *Darwin*, who concludes: "Only after the Second World War did British psychologists really begin to think of their science in terms other than those fixed by the philosophers of mind" (404).
47. Marion Thomas, "Rethinking the History of Ethology: French Animal Behaviour Studies in the Third Republic (1870–1940)" (PhD diss., University of Manchester, 2003), 9–15.
48. On the perceived difficulties of tracking animal eating behavior in the wild, see "Appetite," in *The Oxford Companion to Animal Behavior*, ed. David McFarland (New York: Oxford University Press, 1982), 18–22, at 18–19.
49. On insect instincts as paradigmatic, see Charlotte Sleigh, *Six Legs Better: A Cultural History of Myrmecology* (Baltimore: Johns Hopkins University Press, 2007), 16, 39, 49–51, 60; A. J. Lustig, "Ants and the Nature of Nature in Au-

guste Forel, Erich Wasmann, and William Morton Wheeler," in *The Moral Authority of Nature*, ed. Lorraine Daston and Fernando Vidal (Chicago: University of Chicago Press, 2000), 282–307, at 282–83.

50. See chapter 3.
51. On "maternal instinct," see Sleigh, *Six Legs Better*, 59, 79.
52. Alfred Espinas, *Des sociétés animales: Étude de psychologie comparée* (Paris: Germer Baillière, 1877).
53. Espinas rejected the term "queen," saying that females who tended their young were purely "mothers"; Espinas, *Sociétés animales*, 188, 190.
54. Thomas, "Rethinking the History of Ethology," 154–55.
55. J. H. Fabre, *Fabre's Book of Insects*, retold from Alexander Teixeira de Mattos's translation of Fabre's "Souvenirs entomologiques" by Mrs. Rodolph Stawell, illustrated by E. J. Detmold (London: Hodder and Stoughton, [1921]), 82.
56. Thomas, "Rethinking the History of Ethology," 160.
57. On Fabre's embrace of a "religion of nature" promoted as a replacement for Catholicism by intellectual leaders of the French Third Republic, see Thomas, "Rethinking the History of Ethnology," 163.
58. Edmond Perrier, *La femme dans la nature, dans les moeurs, dans la légende, dans la société: Tableau de son évolution physique et psychique*, vol. 1 (Paris: Bong, 1908–1910) (volumes 2–4 of this work were by other authors); on Perrier's career, see Camille Limoges, "Edmond Perrier," in *DSB* 10:522–23.
59. Edmond Perrier, "L'instinct: Lecture faite devant les cinq Académies," *L'année psychologique* 8 (1901): 561–72, at 562; see also Thomas, "Rethinking the History of Ethology," 189–193.
60. E.-L. Bouvier, *Habitudes et métamorphoses des insectes* (Paris: Ernest Flammarion, 1921), 40–41. Compare Perrier's wasps, which, in the course of evolution, learned to kill prey with "surgical precision"; Perrier, *La femme*, 368.
61. Émile Roubaud, "Recherches biologiques sur les guêpes solitaires et sociales d'Afrique: La genèse de la vie sociale et l'évolution de l'instinct maternel chez les Vespides," *Annales des sciences naturelles: Zoologie et biologie animale*, ser. 10, 1 (1916): 1–160, at 122; Sleigh, *Six Legs Better*, 77–79.
62. On troubled gender relations during and after World War I, see Susan Grayzel, *Women's Identities at War: Gender, Motherhood, and Politics in Britain and France during the First World War* (Chapel Hill: University of North Carolina Press, 1999); Mary Louise Roberts, *Civilization without Sexes: Reconstructing Gender in Postwar France, 1917–1927* (Chicago: University of Chicago Press, 1994).
63. Alfred Giard, "Les origines de l'amour maternel," *Bulletin de l'Institut Général Psychologique* 5 (January–February 1905): 3–32.
64. Espinas, *Sociétés animales*, 124.
65. An instructive contrast to the florid entomological style of this era is found in the heavily quantitative account of Max Lafon, "Quelques documents sur l'appétit et la consommation alimentaire chez les insectes," in *Le comportement*

alimentaire et l'appétit, ed. H. Piéron et al. (Paris: Éditions du Centre National de la Recherche Scientifique, 1952), 103–22.

66. Alfred Giard, "Sur les parasites Bopyriens et la castration parasitaire," *Comptes rendus de la Société de Biologie* 39 (1887): 371–73, at 371; see also Raf De Bont, "Organisms in Their Milieu: Alfred Giard, His Pupils, and Early Ethology, 1870–1930," *Isis* 101, no. 1 (March 2010): 1–29.
67. On the Institut Général Psychologique as a shared venue for entomologists and physicians concerned with disordered eating, including Pierre Janet, Paul Sollier, and Charles Féré, see De Bont, "Organisms in Their Milieu," 19–21.
68. Gall and Spurzheim, *Anatomie et physiologie du système nerveux*, 3:209–17. Gall's sometime associate Spurzheim said more about the "desire to feed," locating it in the "anterior portion of the middle lobes," which he observed to be "very large in those who delight in an excellent dinner"; see J. G. Spurzheim, *Outlines of Phrenology*, 3rd ed. (Boston: Marsh, Capen and Lyon, 1834), 25.

CHAPTER NINE

1. "Anorexia, as a concomitant of phthisis, is the rule, its absence the exception"; J. P. Miller, "Phthisis—Its Successful Treatment," *JAMA* 2 (April 1884): 657–60, at 658 (cited in Brumberg, *Fasting Girls*, 308n3). See also Joseph Thomas, *Le cancer* (Paris: A. Maloine, 1906), 130.
2. Cited in Walter Vandereycken et al., "German Publications on Anorexia Nervosa in the Nineteenth Century," *International Journal of Eating Disorders* 10, no. 4 (1991): 473–90, at 484; on neurasthenia in the United States, see Charles E. Rosenberg, "The Place of George M. Beard in Nineteenth Century Psychiatry," *BHM* 36 (1962): 245–59.
3. Charles Bouchard, *Lectures on Auto-Intoxication in Disease; or, Self-poisoning of the Individual*, trans. and preface by Thomas Oliver (Philadelphia: F. A. Davis, 1894). For autointoxication in Germany, see Albert Albu, *Uber die Autointoxicationen des Intestinaltractus* (Berlin: Hirschwald, 1895), and for the United States and Britain, see James C. Whorton, *Inner Hygiene: Constipation and the Pursuit of Health in Modern Society* (Oxford: Oxford University Press, 2000), 22–27, 65–70, 77–79, 82–83, 241–42.
4. Whorton, *Inner Hygiene*, 25.
5. The literature on anorexia and obesity is vast; for synthetic discussions and titles of relevant primary sources, see Vandereycken and van Deth, *From Fasting Saints to Anorexic Girls*; Peter Stearns, *Fat History: Bodies and Beauty in the Modern West* (New York: New York University Press, 1997); Georges Vigarello, *Les métamorphoses du gras: Histoire de l'obésité du Moyen-Age au XXe siècle* (Paris: Seuil, 2010).
6. Andrew Pollack, "AMA Recognizes Obesity as a Disease," *New York Times*, June 18, 2013; Poulain, "Eléments pour une histoire de la médicalisation de l'obésité."

7. The phrase "symptom complex" was that of Charles Lasègue, who is often presented in the literature of eating disorders as the first to recognize anorexia nervosa as an independent disease despite the fact that he considered anorexia as part of hysteria. His famous article on "hysterical anorexia" situates it within the symptom complex of *troubles digestifs* that also included repeated vomiting, gastric pain, constipation, diarrhea, gastralgias, intestinal troubles, and "perversions bizarres de l'appétit"; hysterical anorexia was, thus, "une des formes de l'hystérie à foyer gastrique"; Charles Lasègue, "De l'anorexie hystérique," *AGM*, 6th ser., 21 (1873): 385–403, at 385–86.
8. Lasègue, "De l'anorexie hystérique," 386, 391, 393. On Lasègue's work on hysteria, including a course on mental and nervous diseases at the Paris medical faculty and management of patients at the Salpêtrière, see Christopher G. Goetz, Michel Bonduelle, and Toby Gelfand, *Charcot: Constructing Neurology* (New York: Oxford University Press, 1995), 175, 213n3. William Withey Gull, "Anorexia Nervosa (Apepsia Hysterica, Anorexia Hysterica)," *Transactions of the Clinical Society of London* 7 (1874): 22–28, at 22. Gull taught physiology and comparative anatomy at Guy's Hospital from 1846 to 1856 and served as Fullerian Professor of Physiology at the Royal Institution from 1847 to 1849; see Nick Hervey, "Gull, Sir William Withey, first baronet (1816–1890)," *Oxford Dictionary of National Biography*, https://doi.org/10.1093/ref:odnb/11730.
9. Lasègue, "De l'anorexie hystérique," 387. Some years later Lasègue again took up the problem of appetite, this time within the framework of medical semiotics, as one of the "subjective signs . . . for which objective controls are so entirely lacking that the physician plays with his patient the role of accoucheur in order to extract all the information that he needs"; Charles Lasègue, "Des appétits en général et de l'appétit digestif en particulier," *Gazette des hôpitaux civils et militaires* 54, no. 2 (1881): 10–11. In this piece Lasègue discusses the "digestive" appetite as an "aiguillon" similar to others—carnal, stomachal, cerebral—that spurred physiological function.
10. Lasègue, "De l'anorexie hystérique," 387. Lasègue stated that "anorexia" was a more appropriate term for the condition in question than "inanition" as the former involved a "less superficial, more delicate, and also more [genuinely] medical phenomenology" (386).
11. William A. Hammond, *Fasting Girls: Their Physiology and Pathology* (New York: G. P. Putnam's, 1879). Hammond credited Lasègue for recognizing that hysteria was "frequently marked, not only by the ability to endure lengthened periods of abstinence, but by the abolition of all desire for food" (69) and reported the case of a woman of twenty-three who "began to crave food, not, as she said, because her appetite had returned, but because she was afraid she would die." On the thirtieth day of her fast, she took "a little beef tea, and from that moment her appetite returned to her" (70).
12. Lucien Deniau, *De l'hystérie gastrique* (Paris: O. Doin, 1883), cited in Vandereycken and van Deth, *From Fasting Saints to Anorexic Girls*, 168. See also

the survey of competing views on the role of appetite in anorexia in Georges Noguès, "L'anorexie mentale, ses rapports avec la psychophysiologie de la faim" (thesis, Université de Toulouse, 1913), 57–70.

13. The particular shock Charcot described was the trauma the young girl suffered after acting as the medium in a prolonged séance organized by her mother. After initial recalcitrance, the patient accepted Charcot's program of "moral and physical hygiene" pursued in complete isolation from the family; see Jean-Martin Charcot, "De l'isolement dans le traitement de l'hystérie," *Progrès médical* 13, 2nd ser., no. 1 (February 28, 1885): 161–64.
14. Paul Sollier, "Anorexie hystérique (sitieirgie hystérique): Formes pathogéniques—Traitement moral," *Revue de médecine* 11 (1891): 625–50, at 629–30. Sollier later changed his view again, arguing that the essential point was the absence of hunger (not appetite) and that the disease should be called gastric psychopathy rather than anorexia, which, given its etymology, was misleading; see Vandereycken and van Deth, *From Fasting Saints to Anorexic Girls*, 170.
15. E. Brissaud and A. Souques, "'Délire de maigreur' chez une hystérique," *Nouvelle iconographie de la Salpêtrière* 7 (1894): 327–37.
16. Pierre Janet, *Les obsessions et la psychasthénie*, 2 vols. (Paris: Félix Alcan, 1903), 1:33–34, 36–40, quotations at 36–37.
17. Janet, *Les obsessions et la psychasthénie*, 1:413. On Janet's treatment of Pauline Lair Lamotte (known as "Madeleine Lebouc"), one of a number of Catholic women in France who engaged in prolonged abstinence from food in quest of bodily and spiritual purification, see Richard D. E. Burton, *Holy Tears, Holy Blood: Women, Catholicism, and the Culture of Suffering in France, 1840–1970* (Ithaca, NY: Cornell University Press, 2004), 123–27.
18. On the later history of anorexia nervosa in psychiatry, see chapter 12.
19. Tilmann Habermas, "The Role of Psychiatric and Medical Traditions in the Discovery and Description of Anorexia Nervosa in France, Germany, and Italy, 1873–1918," *Journal of Nervous and Mental Disease* 179, no. 6 (1991): 360–65, esp. 362; Tilmann Habermas et al., "Die Anorexia nervosa in der deutschsprachigen medizinischen Literatur des 19. Jahrhunderts," *Psychotherapie, Psychosomatik und medizinische Psychologie* 40 (1990): 466–73.
20. Moriz Rosenthal, *Magenneurosen und Magencatarrh sowie deren Behandlung* (Vienna: Urban and Schwarzenberg, 1886); Habermas et al., "Die Anorexia Nervosa," 468.
21. Tilmann Habermas, "Die Anorexia nervosa (Magersucht) in der deutschsprachigen medizinischen Literatur von 1900 bis 1945: Die Rolle der Magersucht in der Entstehung der Psychosomatik," *Sudhoffs Archiv: Zeitschrift für Wissenschaftsgeschichte* 76, 1 (1992): 37–62, at 41, 43–44. On "Simmonds's disease," see chapter 12.
22. The works on appetite of the physician Wilhelm Sternberg are discussed in chapter 12; others include Alois Pick, "Die Behandlung der Appetitlosigkeit mit besonderer Berücksichtigung ihrer nervösen Formen," *Wiener medizi-*

nische Wochenschrift 63, no. 2 (January 1913): 106–15; Berthold Stiller, "Die Pathologie des Appetits," *Archiv für Verdauungskrankheiten* 21 (1915): 23–34.

23. See chapter 5: "Digestive Experiments: A Place for Appetite?"
24. The existence of this large medical literature, attested to in the lengthy bibliographies provided in the encyclopedia entries discussed here, belies the claim of Peter Stearns that obesity became a major issue only in the 1890s. See Stearns, *Fat History*, vii–viii, 27, 53–54.
25. E. Demange, "Obésité," *DESM* (the article is signed "E.D." but the author is identified in the body of the article as E. Demange), 2nd ser., 14 (1880): 5–38, at 15.
26. Demange, "Obésité," 24–25. On Voit, see chapter 7, "The Urge to Eat in Physicochemical Physiology"; the experiments he conducted with Pettenkofer are discussed in Mani, "Die wissenschaftliche Ernährungslehre," 45–59.
27. William Banting, *Letter on Corpulence: addressed to the public*, 2nd ed. (London: Harrison and Sons, 1863); a significantly enlarged edition appeared in 1864.
28. Wilhelm Ebstein, *Corpulence and Its Treatment on Physiological Principles*, trans. A. H. Keane (London: H. Grevel, 1884), 37.
29. For a discussion that contextualizes the attack on fat within the German "life reform" movement, see Michael Hau, *The Cult of Health and Beauty in Germany: A Social History, 1890–1930* (Chicago: University of Chicago Press, 2003), esp. 35, 39, 73.
30. On the influential formulations of the philosopher Victor Cousin, who excluded adolescent girls from the "training" of the will promoted in the French lycée, see Jan Goldstein, "Mutations of the Self in Old Regime and Postrevolutionary France: From *Ame* to *Moi* to *Le Moi*," in Daston, *Biographies of Scientific Objects*, 86–116, esp. 111n63, 114.
31. Enoch Heinrich Kisch, "Fettsucht," *REGH*, 3rd ed., 7:551–70.
32. Kisch, "Fettsucht," 553. On Cohnheim, see Russell Maulitz, "Rudolf Virchow, Julius Cohnheim and the Program of Pathology," *BHM* 52, no. 2 (1978): 162–82.
33. Kisch, "Fettsucht," 555; Kisch attributed this phrase to Rudolf Virchow.
34. Demange, "Obésité," 6; Kisch, "Fettsucht," 552.
35. Demange, "Obésité," 7–10, Kisch, "Fettsucht," 555–60.
36. Kisch, "Fettsucht," 563; Kisch also implicitly acknowledged the difficulty of denying eating desires when he praised Ebstein for recognizing that a cure was achieved only when a sick person was able to maintain a dietetic regimen for life (563).
37. On these remedies, see Daniel P. Todes, *Pavlov's Physiology Factory: Experiment, Interpretation, Laboratory Enterprise* (Baltimore: Johns Hopkins University Press, 2002), 261.
38. S. Weir Mitchell, *Fat and Blood: And How to Make Them* (Philadelphia: J. B. Lippincott, 1877), 9–33, quotation at 25. Mitchell took the same approach to the widespread obesity seen in England, saying that the English were "larger

feeders" than Americans and big consumers of beer (15) but not questioning how or why such habits had taken hold.

39. T. Clifford Allbutt, *On Visceral Neuroses, being the Gulstonian Lectures on Neuralgia of the Stomach and Allied Disorders, Delivered at the Royal College of Physicians in March, 1884* (London: J. & A. Churchill, 1884), 4, 21, 31, 63.
40. Theodor Rosenheim, "Magenneurosen," *REGH* 14:392–426.
41. "On turning to consider the neuralgias of other viscera, we are met by the difficulty, as yet insuperable, of proving definitely in what nerves the pain may lie. We know that all essays to localise abdominal pains in the splanchnic nerves and their paths to the sensorium, in the sensory fibres around the large arteries, or in the deeper connections of the spinal nerves, are mere gymnastics, barren exercises which may stimulate the fancy, but which as yet can neither be proved nor disproved." This being the case, he asserted the primacy of clinical observation: "We can only find our way through those obscure gates and alleys of the body by patient clinical watchings, by grouping and contrasting our cases, and by noting the curious warps of disease"; Allbutt, *On Visceral Neuroses*, 76.

INTRODUCTION TO PART FOUR

1. Goody, "Industrial Food"; Petrick, "Industrial Food"; Oddy and Drouard, *Food Industries*; Levenstein, *Revolution at the Table*; Flandrin and Montanari, *Food*, pt. 7; Pierre Bourdieu, *Distinction: A Social Critique of the Judgement of Taste*, trans. Richard Nice (Cambridge, MA: Harvard University Press, 1984), 177–200.
2. Very recent developments have been treated by food journalists, scientists, and nutritionists; see Kessler, *End of Overeating*; Wilson, *First Bite*.
3. Galison and Hevly, *Big Science*; Capshew and Rader, "Big Science."
4. Richard W. Burkhardt Jr., *Patterns of Behavior: Konrad Lorenz, Niko Tinbergen, and the Founding of Ethology* (Chicago: University of Chicago Press, 2005), 61, 94, 118, 130–31, 176–77, 266–67; on complexities embedded in the terms "wild," "captive," and "domestic," see 250–54.
5. Bonner, *Becoming a Physician*, esp. 288–306; Kenneth M. Ludmerer, *Learning to Heal: The Development of American Medical Education* (Baltimore: Johns Hopkins University Press, 1985), 63, 75–79, 87, 93, 99, 102–3; Pickstone, *Ways of Knowing*, 180–81.
6. On these and related themes, see the essays in Kamminga and Cunningham, *Science and Culture of Nutrition*; see also Vernon, *Hunger*, esp. 91–95, 139–41, 163–80.
7. On the impact of World War I, see Brigitte Schroeder-Gudehus, *Les scientifiques et la paix: La communauté scientifique internationale au cour des années 20* (Montreal: Presses de l'Université de Montréal, 1978); Daniel J. Kevles, "'Into Hostile Political Camps': The Reorganization of International Science in World War I," *Isis* 62, no. 1 (Spring 1971): 47–60.

8. On the peculiarly American cast of behaviorism, see Mills, *Control*, esp. 1–2, 18–19, 23–24; on this division in ethology, see Burkhardt, *Patterns of Behavior*, esp. 390–98.
9. Daston, "Introduction: The Coming into Being of Scientific Objects," 5; I. P. Pavlov, *The Work of the Digestive Glands*, 2nd English ed., trans. W. H. Thompson (London: Charles Griffin, 1910), 94.
10. Todes, *Pavlov's Physiology Factory*, esp. 143–49, 293–94.
11. R. A. Gregory, "The Gastrointestinal Hormones," in *The Pursuit of Nature: Informal Essays on the History of Physiology*, ed. A. L. Hodgkin et al. (New York: Cambridge University Press, 1977), 109–10.
12. On subsequent debates over the differences between "reflex" and "conscious" action, see Nadine M. Weidman, *Constructing Scientific Psychology: Karl Lashley's Mind-Brain Debates* (New York: Cambridge University Press, 1999), 62–64.

CHAPTER TEN

1. Joanny Roux, "La faim, étude physio-psychologique," *Bulletin de la Société d'anthropologie de Lyon* 16 (1897): 409–55.
2. Roux, "La faim," 411–13. Rosenzweig points out in "Mechanisms of Hunger and Thirst" that both stomachal and cellular theories were "peripheral" in that they entailed "changes detected 'out in the body' rather than in the central nervous system" (74). Given the emphasis since Haller on activities or states of the stomach as the cause of hunger, however, the term "peripheral" often retained this meaning; see Carlson, *Control of Hunger*, 16–20.
3. Roux lamented that "documents positifs" were often lacking, given the impossibility of experimenting directly on human beings; Roux, "La faim," 426, 428.
4. See chapter 7, "The Urge to Eat in Physicochemical Physiology."
5. Carlson, *Control of Hunger*, 23.
6. E. Bardier, "Faim," in *Dictionnaire de Physiologie* 6:1–29; Bardier later authored or coauthored major works on eating and digestion, including *Les fonctions digestives* (Paris: O. Doin, 1911) and *Alimentation et digestion* (Paris: Masson, 1931).
7. On Richet's own article "Dégoût," which he discussed as the obverse of "appetition," see chapter 8, "The 'New' Psychology in France."
8. For objections to the view that appetite was the "first degree" of hunger, see, for example, the comments of Moritz Schiff discussed in chapter 7.
9. Todes, *Pavlov's Physiology Factory*, 125.
10. Pavlov, *Work of the Digestive Glands*, 89–90; noting the 1852 report by the physiologists Bidder and Schmidt of this "psychic" effect, Pavlov characterized its subsequent neglect as only one instance of physiology's disregard of facts that lay outside accepted bounds of interpretation (89–90). On Bidder and Schmidt, see chapter 7, "The Urge to Eat in Physicochemical Physiology."

11. Todes, *Pavlov's Physiology Factory*, 222.
12. Todes, *Pavlov's Physiology Factory*, 221.
13. Todes argues that despite such pronouncements Pavlov rejected clinical observations when they conflicted with results obtained in his laboratory or asserted that insufficient progress had yet been made to reconcile the two; *Pavlov's Physiology Factory*, 294–97.
14. Pavlov turned to the "systematic" study of temperament and personality from around 1910; see Jeffrey A. Gray, *Ivan Pavlov* (New York: Viking Press, 1979), 103–22, esp. 107.
15. Todes, *Pavlov's Physiology Factory*, 436n62.
16. See chapter 8: "Psychophysiology of the Hunger Artist."
17. Todes, *Pavlov's Physiology Factory*, 227, 236. On the confusion caused by the competing translations of Pavlov's terms as "conditional" or "conditioned" reflex, see Todes, 244–46.
18. For the Russian context, see Todes, *Pavlov's Physiology Factory*, 228–39. On the critical shift in thinking about reflexes, see Judith P. Swazey, *Reflexes and Motor Integration: Sherrington's Concept of Integrative Action* (Cambridge, MA: Harvard University Press, 1969). For an argument that Pavlov only partially assimilated this shift, see Gray, *Ivan Pavlov*, 102.
19. Ivan Petrovitch Pavlov, *Lectures on Conditioned Reflexes: Twenty-Five Years of Objective Study of the Higher Nervous Activity (Behaviour) of Animals*, vol. 1, with collaboration of G. Volborth, trans. and ed. W. Horsley Gantt, intro. Walter B. Cannon (New York: International, 1928), 147–55, at 148.
20. Pavlov acknowledged the prior work of the American psychologist E. L. Thorndike in moving away from "the subjective point of view," although he surmised that American researchers began not from physiological concerns but from "human interests," the conviction that "it is more important to be acquainted with the exact outward behaviour of man than with guesses about his internal states"; Pavlov, *Lectures*, 39–40.
21. Todes, *Pavlov's Physiology Factory*, 240–42.
22. J. P. Pawlow, *Die Arbeit der Verdauungsdrüsen*, trans. A. Walther (Wiesbaden: J. F. Bergmann, 1898); J. P. Pawlow, *Le travail des glandes digestives, leçons du professeur J. P. Pawlow*, trans. Victor Pachon and J. Sabrazès (Paris: Masson, 1901); Ivan Petrovich Pavlov, *The Work of the Digestive Glands*, trans. W. H. Thompson (London: Charles Griffin, 1902).
23. Todes, *Pavlov's Physiology Factory*, 305–16, 326–32.
24. W. M. Bayliss and E. H. Starling, "The Mechanism of Pancreatic Secretion," *Journal of Physiology* 28, 5 (1902): 325–53, at 327.
25. Ernest H. Starling, "The Croonian Lectures on the Chemical Correlation of the Functions of the Body," *Lancet* 2, no. 4275 (August 5, 1905): 339–41.
26. W. M. Bayliss, *Principles of General Physiology* (London: Longmans, Green, 1915); for this quotation and a discussion of Anrep and subsequent work, see Gregory, "Gastrointestinal Hormones," 109–10.

27. W. M. Bayliss, *The Physiology of Food and Economy in Diet* (London: Longman's, Green, 1917), 62.
28. Diana Long Hall, "The Critic and the Advocate: Contrasting British Views on the State of Endocrinology in the early 1920s," *JHB* 9, no. 2 (1976): 269–85, at 276–77. Hall cites this passage from Starling's Harveian Oration of 1923: "It seems almost a fairy tale that such widespread results, affecting every aspect of a man's life, should be conditioned by the presence or absence in the body of infinitesimal quantities of a substance which by its formula does not seem to stand out from the thousands of other substances with which organic chemistry has made us familiar" (277).
29. Cited in Hall, "The Critic and the Advocate," 278.
30. Hagner, "Scientific Medicine," 66–67; Pickstone, *Ways of Knowing*, 180–81.
31. C. F. Mohr, "Hypertrophie (markschwammige Entartung?) der *Hypophysis cerebri* und dadurch bedingter Druck auf die Hirngrundfläche," *Wochenschrift für die gesammte Heilkunde* 6 (1840): 565–71.
32. Joseph Babinski, "Tumeur du corps pituitaire sans acromégalie et avec arrêt de développement des organes génitaux," *Revue neurologique* 8 (1900): 531–33; Alfred Fröhlich, "Ein Fall von Tumor der Hypophysis cerebri ohne Akromegalie," *Wiener klinische Rundschau* 15 (1901): 883–86, 906–8.
33. Artur Biedl, *Innere Sekretion: ihre physiologischen Grundlagen und ihre Bedeutung für die Pathologie* (Berlin: Urban and Schwarzenberg, 1910); citations are to the English translation, *The Internal Secretory Organs: Their Physiology and Pathology* (New York: William Wood, 1913). This work includes a bibliography of nearly 150 pages.
34. For an overview of Cushing's work on the pituitary, see Michael Bliss, *Harvey Cushing: A Life in Surgery* (New York: Oxford University Press, 2005), 204–20.
35. Harvey Cushing, *The Pituitary Body and its Disorders* (Philadelphia: J. B. Lippincott, 1912), 190.
36. Cited in Elin L. Wolfe, A. Clifford Barger, and Saul Benison, *Walter B. Cannon, Science and Society* (Boston Medical Library in the Francis A. Countway Library of Medicine; distributed by Harvard University Press, 2000), 421 (a letter from Cushing to T. R. Elliott, December 1930). For a laudatory account of Cushing's "rare combination of clinical and experimental work" as well as his surgical ingenuity in developing "a safe technique for the operation of the almost inaccessible pituitary," see Erwin H. Ackerknecht, "The History of Metabolic Diseases," *Ciba symposia* 6 (June–July 1944): 1834–44. Ackerknecht referred to the pituitary not as a "monarch" but as the "leader of the endocrine orchestra" (1842).
37. David Ferrier, *The Functions of the Brain* (New York: G. P. Putnam's, 1876), 193–97. Although convinced "that there is a causal relation between the removal of the occipital lobes and the annihilation of the appetite for food" (195), Ferrier did not pursue the matter further; see Ferrier, *The Functions of the Brain*, 2nd ed. (London: Smith, Elder, 1886).

38. Stephen Paget, "On Cases of Voracious Hunger and Thirst from Injury or Disease of the Brain," *Transactions of the Clinical Society of London* 30 (1897): 113–19, at 113.
39. Rosenzweig, "Mechanisms of Hunger and Thirst," 105–6.
40. Percival Bailey and Frédéric Bremer, "Experimental Diabetes Insipidus," *Archives of Internal Medicine* 28 (1921): 773–803, at 774.
41. Bailey and Bremer, "Experimental Diabetes Insipidus," 798, 801.
42. Jean Camus and Gustave Roussy, "Les syndromes hypophysaires: Anatomie et physiologie pathologiques," *Revue neurologique* 38 (1922): 622–37, at 623.
43. J. F. Fulton, C. F. Jacobsen, and Margaret A. Kennard, "A Note Concerning the Relation of the Frontal Lobes to Posture and Forced Grasping in Monkeys," *Brain* 55, no. 4 (1932): 524–36; Fulton, "Some Functions of the Cerebral Cortex," 178; "John Brobeck," in *History of the American Physiological Society: The First Century, 1887–1987*, ed. John R. Brobeck, Orr E. Reynolds, and Toby Appel (Bethesda, MD: American Physiological Society, 1987), 209–12.
44. John R. Brobeck, Jay Tepperman, and C. N. H. Long, "Experimental Hypothalamic Hyperphagia in the Albino Rat," *YJBM* 15, no. 6 (July 1943): 831–53, at 831–32.
45. Brobeck, Tepperman, and Long, "Experimental Hypothalamic Hyperphagia," 832; Randy L. Jensen, James L. Stone, and Robert A. Hayne, "Introduction of the Human Horsley-Clarke Stereotactic Frame," *Neurosurgery* 38, no. 3 (1996): 563–67.
46. Brobeck, Tepperman, and Long, "Experimental Hypothalamic Hyperphagia," 832.
47. See also John R. Brobeck, "Food Intake as a Mechanism of Temperature Regulation," *YJBM* 20, no. 6 (1948): 545–52; Brobeck, "Neural Control of Hunger, Appetite, and Satiety," *YJBM* 29, 6 (1957): 565–74; Brobeck, "Neural Regulation of Food Intake," in *Regulation*, 44–55.
48. Brobeck, "Neural Regulation of Food Intake," 44.
49. Brobeck cited as the principal source of his thinking about regulation the Rochester physiologist Edward F. Adolph (American Physiological Society website, John Brobeck Papers, 1935–1993, Wheaton College Archives and Special Collections, and "John R. Brobeck," American Physiological Society, http://www.the-aps.org/fm/presidents/introjrb.html); see Adolph, *Physiological Regulations* (Lancaster, PA: Jacques Cattell, 1943): "Regulations mean . . . in physiological science, those self-managements of organisms which result in constancy of function" (1).
50. Walter B. Cannon, *The Mechanical Factors of Digestion* (New York: Longmans, Green, 1911).
51. Walter B. Cannon, *Digestion and Health* (New York: W. W. Norton, 1936), 16. Cannon's biographers do not speculate on the matter, but they provide ample evidence that his interest in the influence of emotions on appetite and eating was, at least to some extent, grounded in personal experience. See

their discussions of the strange diets Cannon's father imposed on his children, of Cannon's own stomach ills and attempts at dietary treatment, and of the advice given by his mother-in-law to his wife to cure his "blues" with good food. Saul Benison, A. Clifford Barger, and Elin L. Wolfe, *Walter B. Cannon: The Life and Times of a Young Scientist* (Cambridge, MA: Harvard University Press, 1987); 10–11, 85, 137; Wolfe, Barger, and Benison, *Walter B. Cannon*, 292, 298.

52. Cannon hoped at this point to integrate his work with that of Pavlov and of Bayliss and Starling, who then saw peristaltic contraction as "a coordinated local reflex carried out by the neural network in the wall of the gut"; see Benison, Barger, and Wolfe, *Walter B. Cannon*, 161 and, on this phase of Cannon's work overall, 52–56, 60–65.
53. Wolfe, Barger, and Benison, *Walter B. Cannon*, 160–61; Walter B. Cannon, "Die Notfallsfunktionen des sympathico-adrenalen Systems," *Ergebnisse der Physiologie* 27 (1928): 380–406.
54. Walter B. Cannon, "A Consideration of the Nature of Hunger" (New York: [Harvey Society], 1911–1912), 130–52.
55. Cannon, "Consideration," 136–40. Cannon had espoused a "general" view after work he did in 1899 on hunger in children, reported in an unpublished manuscript entitled "An X-Ray Investigation of the Human Stomach"; Benison, Barger, and Wolfe, *Walter B. Cannon*, 64–65, 417n42.
56. W. B. Cannon and A. L. Washburn, "An Explanation of Hunger," *AJP* 29, no. 5 (1912): 441–54.
57. Cannon, "Die Notfallsfunktionen"; Wolfe, Barger, and Benison, *Walter B. Cannon*, 83–84, 93, 149, 151, 161, 170, 175.
58. Cross and Albury, "Walter B. Cannon, L. J. Henderson, and the Organic Analogy," 170, 172n18.
59. Cannon's theory extended well beyond nutritional needs, encompassing all "conditions which must be maintained constant in the fluid matrix," including "water, oxygen, temperature and nutriment"; see Cannon, "Organization for Physiological Homeostasis," *Physiological Reviews* 9, no. 3 (1929): 399–431, at 401; Wolfe, Barger, and Benison, *Walter B. Cannon*, 163.
60. The published version was Cannon, *Digestion and Health.*
61. Cannon, *Digestion and Health*, 81. When Cannon delivered the Beaumont lecture, he had just published his acclaimed work *The Wisdom of the Body*, in which he argued that hunger alerted the body to the need to replenish essential supplies but that the drive to eat was not purely a matter of local pangs in the stomach or, as a new theory had it, of low levels of sugar in the blood, but had to do with the "psychic elements" of "delight" and "pleasure"; see Cannon, *The Wisdom of the Body* (New York: W. W. Norton, 1932), 70–76.
62. On Cannon's friendship with Pavlov, see Wolfe, Barger, and Benison, *Walter B. Cannon*, 125–26, 128–29, 131–33, 251–52, 333–34, 335, 347, 348–49.

63. On Cannon's unease at treading the ground of psychology, see Wolfe, Barger, and Benison, *Walter B. Cannon*, 185–87.
64. Wolfe, Barger, and Benison, *Walter B. Cannon*, 170.
65. Cannon, *Digestion and Health*, 19, 119–20.
66. On Cannon's role in the rise of psychosomatic medicine, see chapter 12.
67. Miner, *Regulation*.
68. Jean Mayer, "Regulation of Energy Intake and the Body Weight: The Glucostatic Theory and the Lipostatic Hypothesis," in *Regulation*, 15–43, at 30; Morton I. Grossman, "Integration of Current Views on the Regulation of Hunger and Appetite," in *Regulation*, 76–91, at 79; J. P. Quigley, "The Role of the Digestive Tract in Regulating the Ingestion of Food," in *Regulation*, 6–14, at 6–8; Isidore Snapper, "Food Preferences in Man: Special Cravings and Aversions," in *Regulation*, 92–106.
69. Henry D. Janowitz and Franklin Hollander, "The Time Factor in the Adjustment of Food Intake to Varied Caloric Requirement in the Dog: A Study of the Precision of Appetite Regulation," in *Regulation*, 56–67, at 56.
70. Hilde Bruch, "Role of the Emotions in Hunger and Appetite," in *Regulation*, 68–75; Snapper, "Food Preferences," 106.
71. Brobeck, "Neural Regulation of Food Intake," 45, 50.
72. Brobeck, "Neural Regulation of Food Intake," 50; Brobeck cited this conclusion from Jack L. Strominger and John R. Brobeck, "A Mechanism of Regulation of Food Intake," *YJBM* 25, no. 5 (1953): 383–90, at 383.
73. Brobeck, "Neural Regulation of Food Intake," 44.
74. Pavlov, "Food Centre," 151.
75. Franklin Hollander, introduction to Miner, *Regulation*, 3.

CHAPTER ELEVEN

1. Danziger, "History of Introspection."
2. On behaviorism in Europe, see George Mandler, *A History of Modern Experimental Psychology: From James and Wundt to Cognitive Science* (Cambridge, MA: MIT Press, 2007), 176–77.
3. Alexandra W. Logue, "The Origins of Behaviorism: Antecedents and Proclamation" and "The Growth of Behaviorism: Controversy and Diversity" in *Points of View in the Modern History of Psychology*, ed. Claude E. Buxton (New York: Academic Press, 1985), 141–67 and 169–96; on varieties of behaviorism, see Mills, *Control*, 2–4.
4. For an early usage of this term, see *Current Trends in Psychology and the Behavioral Sciences: Eighth Annual Conference on Current Trends in Psychology Held at the University of Pittsburgh, 1954* (Pittsburgh: University of Pittsburgh Press, 1955).
5. On the problem of causation and the emergence of "neo-behaviorism," see Mitchell Ash, "Psychology," in *The Cambridge History of Science*, vol. 7, *The*

Modern Social Sciences, ed. Theodore M. Porter and Dorothy Ross (Cambridge: Cambridge University Press, 2003), 251–74, at 265.

6. Logue, "Growth of Behaviorism," 186.
7. On late nineteenth-century theories of inheritance, see Richards, *Darwin*; on "degeneration," see Nye, *Crime, Madness and Politics in Modern France*; Mendelism is discussed in Garland Allen, *Life Science in the Twentieth Century* (New York: Cambridge University Press, 1978), 41–72.
8. On comparative psychology, see O'Donnell, *Origins of Behaviorism*, 163–66, 180–98; Cheryl A. Logan, "The Altered Rationale for the Choice of a Standard Animal in Experimental Psychology: Henry H. Donaldson, Adolf Meyer, and 'the' Albino Rat," *History of Psychology* 2, 1 (1999): 3–24.
9. On these themes, see Burkhardt, *Patterns of Behavior*.
10. John B. Watson, "Psychology as the Behaviorist Views It," *Psychological Review* 20, no. 2 (1913): 158–77; O'Donnell, *Origins*, 187–91, 200–207; Kerry W. Buckley, *Mechanical Man: John Broadus Watson and the Beginnings of Behaviorism* (New York: Guilford Press, 1989), 37–44, 52–54, 56–58, 74–81.
11. Mills, *Control*, 56.
12. Edward L. Thorndike, *Animal Intelligence: Experimental Studies*, intro. Darryl Bruce (New Brunswick, NJ: Transaction, 2000); on Romanes, see chapter 8, "Feeding in Insects and Other Animals."
13. For the quotation, see O'Donnell, *Origins*, 166.
14. Thorndike's assumption that the experimentalist could easily recognize an animal's internal state of hunger belied his claim that what was being observed was purely external movements or actions of the animal in question. Mills notes that the assumption that food deprivation produced a state of hunger rather than some other response (frustration, increased tolerance of hunger) also marked later work by "drive theorists"; see Mills, *Control*, 15.
15. Thorndike came to be associated with "operant" or "instrumental" conditioning (based on rewards and punishments) as opposed to "classical" or "Pavlovian" conditioning; see Gray, *Ivan Pavlov*, 28, 58, 126.
16. Buckley notes that Karl Lashley, a partisan of innatism, charged Watson with seeking to use the conditioned reflex "as a basis for a systematic psychology" without concerning himself with "'the nature of the reaction itself'"; Buckley, *Mechanical Man*, 204n40.
17. "The classic behaviorists related all actions to the solving of a few basic biological needs: feeding and drinking, fleeing predators, fighting, and reproducing"; Charles T. Snowdon, "Genetic Influences, Learning, and Instinct," in *Foundations of Animal Behavior: Classic Papers with Commentaries*, ed. Lynne D. Houck and Lee C. Drickamer (Chicago: University of Chicago Press, 1996), 261–73, at 272.
18. John B. Watson, *Behaviorism* (New York: W. W. Norton, 1925), 12.

19. "By swallowing a small balloon and attaching it to a recording instrument we can easily register the response of the stomach to lack of food and note the lack of response when food is present"; Watson, *Behaviorism*, 12.
20. B. F. Skinner sought to account for differing strengths of ingestive reflexes by reference to a hunger "drive," which he characterized as a "hypothetical state interpolated between operation and behavior"; Skinner, *The Behavior of Organisms: An Experimental Analysis* (New York: D. Appleton-Century, 1938), 368. On Skinner's use of "drive" in his early work on eating, see Herrnstein, "Nature as Nurture," 45–49.
21. On Watson's turn away from "animal work" and his subsequent career, see Buckley, *Mechanical Man*, 88–176; on behaviorists' embrace of "applied psychology," see O'Donnell, *Origins*, 206, 210, 226–29, 237.
22. Jay Schulkin, *Curt Richter: A Life in the Laboratory* (Baltimore: Johns Hopkins University Press, 2005), x.
23. Schulkin, *Curt Richter*, 7–21; Logan, "Altered Rationale."
24. Curt P. Richter, "The Behavior of the Rat: A Study of General and Specific Activities" (PhD diss., Johns Hopkins University, 1921); see also Richter, "A Behavioristic Study of the Activity of the Rat," *Comparative Psychology Monographs* 1 (1922): 1–55. This move distanced Richter decisively from Watson, something he implicitly acknowledged when he stated that the "emphasis is usually placed on . . . the training element. . . . Interest has only recently begun to be directed toward the spontaneous activities of humans and animals" (3–4).
25. The cyclic pattern of eating in the rat was the starting point for Richter's work on biological clocks, but his investigations came to include a wide variety of "spontaneous" activities ranging from basic biological functions such as urination and defecation to complex behaviors such as nest building and, in humans, "periodic illnesses"; see Curt P. Richter, *Biological Clocks in Medicine and Psychiatry* (Springfield, IL: Charles C. Thomas, 1965).
26. Curt P. Richter, "Animal Behavior and Internal Drives," *Quarterly Review of Biology* 2, no. 3 (September 1927): 307–43, at 307.
27. Schulkin, *Curt Richter*, 4.
28. "The simultaneous records of activity and feeding suggest . . . a close relationship . . . between the periods observed in the simple activity cage and the hunger 'drive' of the animal"; Richter, "Animal Behavior and Internal Drives," 313. Perhaps to preclude any note of doubt at the outset, this device was not used for "drives" in the article title (307).
29. Richter, "Animal Behavior and Internal Drives," 314; A. J. Carlson and H. Ginsburg, "Contributions to the Physiology of the Stomach. XXIV. The Tonus and Hunger Contractions of the Stomach of the New-Born," *AJP* 38, no. 1 (July 1, 1915): 29–32. Richter and his team did not have the same success and abandoned the method after the babies either "would not go to sleep or else the tube was completely blocked by strong spasms of the cardiac sphincter"; Richter, "Animal Behavior and Internal Drives," 315.

30. Curt P. Richter, "Biology of Drives," *PSMed* 3 (1941): 105–10, at 105. On the prior interest of the Johns Hopkins biologist Herbert Spencer Jennings in animal behavior as "regulatory" in character, see Weidman, *Constructing Scientific Psychology*, 25–26.
31. The first in this series of publications was Curt P. Richter, "Increased Salt Appetite in Adrenalectomized Rats," *AJP* 115 (1936): 155–61; for a fuller discussion, see Schulkin, *Curt Richter*, 47–76.
32. On the "appetite method," see Schulkin, *Curt Richter*, 55.
33. Richter, "Biology of Drives," 107–8.
34. Arguing that heredity also endowed the rat with specific mechanisms that guided food choices—in the case of salt appetite, the taste nerves—Richter devised experiments showing that a rat with its adrenal glands removed and its taste nerves sectioned "no longer showed an increased appetite for salt and, as a result, died within a few days"; Richter, "Biology of Drives," 108. On Richter's lifelong interest in "innate structure, genetically endowed," see Schulkin, *Curt Richter*, 25.
35. Richter, "Animal Behavior and Internal Drives," 318–19.
36. Richter, "Biology of Drives," 109; Clara Davis, "Self Selection of Diet by Newly Weaned Infants: An Experimental Study," *American Journal of Diseases of Children* 36, no. 4 (1928): 651–79. Davis's work is differently interpreted in recent discussions, some commentators arguing that she proved the children made innately "wise" choices, others that she showed simply that they thrived when presented only healthy foods to choose from. See Benjamin Scheindlin, "'Take One More Bite for Me': Clara Davis and the Feeding of Young Children," *Gastronomica* 5, no. 1 (Winter 2005): 65–69; Wilson, *First Bite*, 32–38, 42–43.
37. Richter, "Biology of Drives," 109.
38. On this debate and on experimental techniques indebted to Richter, see Paul Thomas Young, "The Experimental Analysis of Appetite," *Psychological Bulletin* 38, no. 3 (March 1941): 129–64, quotation at 136.
39. Alexandra W. Logue, "Watson's Behaviorist Manifesto: Past Positive and Current Negative Consequences," in *Modern Perspectives on John B. Watson and Classical Behaviorism*, ed. James T. Todd and Edward K. Morris (Westport, CT: Greenwood, 1994), 109–23, at 111–12, 113.
40. Cited in Logan, "Altered Rationale," 3–4; on Skinner, see Frank A. Beach, "The Snark Was a Boojum," *American Psychologist* 5 (1950): 115–24.
41. Burkhardt, *Patterns of Behavior*.
42. Wallace Craig, "Appetites and Aversions as Constituents of Instincts," *Biological Bulletin of the Marine Biological Laboratory* 34 (1918): 91–107. Craig's position in the nature-nurture debate is still disputed. Burkhardt emphasizes the impact Craig had on Konrad Lorenz's instinctivism, while present-day animal behaviorists assert that Craig's work challenged the tendency to attribute complex behavior to instinct; see Burkhardt, *Patterns of Behavior*, 33, 58, 173–74, 178, 224, 469, 472, 480; Snowdon, "Genetic Influences," 262.

43. Burkhardt, *Patterns of Behavior*, 36–37, 39–45, 53–55.
44. Craig, "Appetites and Aversions," 95; here Craig reversed physiological descriptions that presented appetite as the "first phase" of hunger, urging that "hunger is clearly one kind of appetite" (95n1).
45. Craig, "Appetites and Aversions," 96–101; see also Craig, "Observations on Doves Learning to Drink," *Journal of Animal Behavior* 2, 4 (1912): 273–79. In "Appetites and Aversions" Craig stated that his view of instinct held for human behavior as well, seeing "a difference of degree only" (104) between humans and animals in this regard. Curiously, however, when he turned to human beings, he emphasized innate over acquired patterns. Leaving behind basic "appetites," he discerned in human emotional and aesthetic states (homesickness, enjoyment of music) cycles of restlessness, searching, satisfaction, and rest, which showed that "we, like the birds, are but little able to alter the course of our behavior cycles" (107).
46. Others dispute a marked European-American division of this kind; see Lee C. Drickamer and Charles T. Snowdon, "Defining the Goals, Approaches, and Methods," in *Foundations of Animal Behavior: Classic Papers with Commentaries*, ed. Lynne D. Houck and Lee C. Drickamer (Chicago: University of Chicago Press, 1996), 71–86, at 71.
47. Burkhardt, *Patterns of Behavior*, 163–72, quotation at 164.
48. Konrad Lorenz, "Die angeborenen Formen möglicher Erfahrung," *Zeitschrift für Tierpsychologie* 5 (1943): 235–409.
49. Burkhardt, *Patterns of Behavior*, 271, 141–42.
50. Burkhardt, *Patterns of Behavior*, 271–72.
51. Niko Tinbergen, "Ueber die Ernährung einer Waldohreulenbrut (Asio otus otus (L.))," *Beiträge zur Fortpflanzungsbiologie der Vögel* 8 (1932): 54–55.
52. Harrington, *Reenchanted Science*, xx, 30–33.
53. Harrington, *Reenchanted Science*, 48–52.
54. Fred C. Dyer and H. Jane Brockmann, "Biology of the *Umwelt*," in *Foundations of Animal Behavior: Classic Papers with Commentaries*, ed. Lynne D. Houck and Lee C. Drickamer (Chicago: University of Chicago Press, 1996), 529–38, at 530.
55. Harrington, *Reenchanted Science*, 43.
56. On Wernicke, see Engstrom, *Clinical Psychiatry in Imperial Germany*, 90, 101–2.
57. Goldstein's special area of observation was the process of compensation by which brain-damaged patients came to enjoy restored functions supposedly made impossible by destruction of specific brain-sites; Harrington, *Reenchanted Science*, 145–51.
58. Harrington, *Reenchanted Science*, 140–74.
59. Harrington, *Reenchanted Science*, 106–9.
60. Mitchell G. Ash, *Gestalt Psychology in German Culture, 1890–1967: Holism and the Quest for Objectivity* (New York: Cambridge University Press, 1995), 2.
61. David Katz, *Hunger und Appetit: Untersuchungen zur medizinischen Psychologie* (Leipzig: Johann Ambrosius Barth, 1932); Katz's related works include:

"Psychologische Probleme des Hungers und Appetits," *Der Nervenarzt* 1, no. 6 (1928): 345–51; "Psychologische Probleme des Hungers und Appetits, insbesondere beim Kinde," *Zeitschrift für Kinderforschung* 34 (1928): 15–97; "Psychologische Untersuchungen über Hunger und Appetit," *Archiv für die gesamte Psychologie* 65 (1928): 269–320; "La faim et l'appétit en psychologie générale et en biotypologie," *Biotypologie: Bulletin de la Société de Biotypologie* 6, no. 4 (December 1938): 233–51.

62. Katz later said of Husserl that except for Müller no teacher had "more deeply influenced my method of work and my attitude in psychological matters"; Katz, "David Katz," in *A History of Psychology in Autobiography*, vol. 4, ed. Edwin G. Boring et al. (New York: Russell and Russell, 1952), 189–211, at 194; see also Wilhelm Arnold, "David Katz," *DB*; Robert B. MacLeod, "David Katz, 1884–1953," *Psychological Review* 61, no. 1 (January 1954): 1–4; Ulfried Geuter, *The Professionalization of Psychology in Nazi Germany*, trans. Richard J. Holmes (Cambridge: Cambridge University Press, 1992), 9, 42, 48, 51, 53, 79.
63. Katz's appointment was part of the expansion of psychology into pedagogy; Geuter, *Professionalization*, 48–49.
64. Kurt Lewin, *Vorsatz, Wille und Bedürfnis, mit Vorbemerkungen über die psychischen Kräfte und Energien und die Struktur der Seele* (Berlin: Julius Springer, 1926). On Lewin's career, see Ash, *Gestalt Psychology*, 263–75; Alexandre Métraux, "Kurt Lewin: Philosopher-Psychologist," *Science in Context* 5, 2 (1992): 373–84.
65. Katz attributed neglect of such topics as hunger and appetite to the emphasis, ever since Descartes, on phenomena of consciousness to the detriment of processes "tied to the vegetative nervous system that entered consciousness . . . only here and there"; "Psychologische Probleme des Hungers und Appetits," 345.
66. Katz, "David Katz," 204.
67. Katz argued, for example, against an absolute state of satiety, referring to "partial satieties" that developed in stages in reference to specific foodstuffs; Katz, *Hunger und Appetit*, 15–16.
68. Katz, *Hunger und Appetit*, 14, 21; Katz also argued that animals in captivity lost their sex drive but not the drive to eat. He did not mention the very different claim of Freud for the primacy of the sexual instinct; on Freud's view of the instincts, see chapter 12, "'The Mysterious Leap from Mental to Bodily': Appetite in Psychoanalysis."
69. On "activity-readiness" ("Tätigkeitsbereitschaft"), compare Lewin, *Wille*, 69.
70. Katz, *Hunger and Appetit*, 56. Katz, like Richter, lauded the American researcher Clara M. Davis for illuminating children's instinctive capacity to select appropriate foods; Katz, "La faim et l'appétit," 237; on the importance of Davis's work to Richter, see "'Salt Appetite': Curt P. Richter and the Instinctive Search for Nutrients."
71. For prior statements of "two-component" theory, see Erwin Bayer, "Beiträge zur Zweikomponententheorie des Hungers," *Zeitschrift für Psychologie* 112

(1929): 1–54; Gallus Beck, "Neue Beiträge zur Zweikomponententheorie des Hungers," *Zeitschrift für Psychologie* 118 (1929): 283–349.

72. Katz, *Hunger und Appetit*, 27; on Lewin's opposition to classifying objects of study as "normal" or "pathological," see Ash, *Gestalt Psychology*, 267.
73. Katz, *Hunger und Appetit*, 57–64; see also "La faim et l'appétit," where Katz discusses further the influence of social factors such as eating alone or in the presence of another (239).
74. Katz, *Hunger und Appetit*, 64; in this connection Katz cited the work of Max Rubner, discussed in chapter 7, "The Urge to Eat in Physicochemical Physiology."
75. Katz, *Hunger und Appetit*, 28–31; Katz adapted these labels from the typology of the psychiatrist Ernst Kretschmer. On the relation between Kretschmer's thinking and Nazi "race psychology," see Geuter, *Professionalization*, 111, 121, 123.
76. Katz's article "La faim et l'appétit" (1938) was a published version of an address he gave to the French Société de Biotypologie in which he admitted that it was not yet clear whether "psychophysical constitution" could be altered by diet while nonetheless stressing how essential the question was to practical medicine (251).
77. "With their expulsion and emigration, certain theoretical developments in Germany came to an abrupt end"; Geuter, *Professionalization*, 9. On the later careers of some Gestalt psychologists, see Mandler, *History of Modern Experimental Psychology*, 139–64. One who made an important contribution to the study of food choice was Karl Duncker, who worked briefly at Swarthmore before committing suicide in 1940; see Duncker, "Experimental Modification of Children's Food Preferences through Social Suggestion," *Journal of Abnormal and Social Psychology* 33, no. 4 (1938): 489–507.
78. On the study of hunger by physicians in the Warsaw Ghetto, see Myron Winick, ed., *Hunger Disease: Studies by the Jewish Physicians in the Warsaw Ghetto* (New York: John Wiley, 1979). On his famed wartime study of food-deprived conscientious objectors, see Ancel Keys, *The Biology of Human Starvation*, 2 vols. (Minneapolis: University of Minnesota Press, 1950).
79. [Mario] Autuori et al., *L'instinct dans le comportement des animaux et de l'homme* (Paris: Masson, 1956); Burkhardt, *Patterns of Behavior*, 390–98. On an earlier American meeting, see W. S. Hunter, "Summary Comments on the Heredity-Environment Symposium," *Psychological Review* 54, no. 6 (November 1947): 348–52. Hunter saw the renewed emphasis on heredity at this symposium as a major departure from the "anti-instinct movement" of the previous twenty-five years.
80. Gaston Viaud, "Taxies et tropismes dans le comportement instinctif," in *L'instinct*, 5–49, at 36–37; Marc Klein, "Aspects biologiques de l'instinct reproducteur dans le comportement des mammifères," in *L'instinct*, 287–339, at 326–27.

81. Curt P. Richter, "Salt Appetite of Mammals: Its Dependence on Instinct and Metabolism," in *L'instinct*, 577–632, at 578.
82. T. C. Schneirla, "Interrelationships of the 'Innate' and the 'Acquired' in Instinctive Behavior," in *L'instinct*, 387–452, at 392.
83. Daniel S. Lehrman, "On the Organization of Maternal Behavior and the Problem of Instinct," in *L'instinct*, 475–520, at 487.
84. Lehrman, in discussion of Richter, "Salt Appetite," 630. In response to Richter's claim about appetite and the "taste-bud mechanism," Lehrman stated that other data "suggest that the effect of depletion may not be to change the sensitivity of the taste buds, but to change the role of salt as a reward" (630).
85. Marcel Mauss, *The Gift: Forms and Functions of Exchange in Archaic Societies*, trans. Ian Cunnison, intro. E. E. Evans-Pritchard (New York: W. W. Norton, 1967), 76–77.
86. Young, "Experimental Analysis of Appetite," 160.

CHAPTER TWELVE

1. Nicholas J. Talley, "Dyspepsia and Non-Ulcer Dyspepsia: An Historical Perspective," *Medical Journal of Australia* 145 (December 1 and 15, 1986): 614–18; John C. Chatel and Roger Peele, "A Centennial Review of Neurasthenia," *American Journal of Psychiatry* 126, 10 (1970): 1404–13.
2. Frederic L. Holmes, "The Transformation of the Science of Nutrition," *JHB* 8, no. 1 (Spring 1975): 135–44.
3. Mani, "Die wissenschaftliche Ernährungslehre," esp. 65–68; Kamminga and Cunningham, "Introduction," 7–8; Kenneth J. Carpenter, "The Life and Times of W. O. Atwater (1844–1907)," *Journal of Nutrition* 124, supplement 9 (September 1994): 1707S–1714S; Marion Nestle and Malden Nesheim, *Why Calories Count: From Science to Politics* (Berkeley: University of California Press, 2012), 5, 32–39.
4. On the emergence of biochemistry, see Robert E. Kohler Jr., "The Enzyme Theory and the Origin of Biochemistry," *Isis* 64, no. 2 (June 1973): 181–96.
5. Mark Weatherall, "Bread and Newspapers: The Making of 'A Revolution in the Science of Food,'" in *Science and Culture of Nutrition*, 179–212, at 191–92.
6. Rima D. Apple, *Vitamania: Vitamins in American Culture* (New Brunswick, NJ: Rutgers University Press, 1996).
7. David Smith and Malcolm Nicolson, "The 'Glasgow School' of Paton, Findlay and Cathcart: Conservative Thought in Chemical Physiology, Nutrition and Public Health," *Social Studies of Science* 19, no. 2 (May 1989): 195–238; F. Gowland Hopkins, "Cameron Prize Lectures on the Present Position of the Vitamin Problem," *BMJ* 2, no. 3277 (October 20, 1923): 691–93; 2, 3278 (October 27, 1923): 748–50.
8. Quoted in Weatherall, "Bread and Newspapers," 189.

9. Aside from the most prominent deficiency diseases (rickets, beriberi, scurvy, and pellagra), vitamin-related conditions discussed in the pages of *BMJ* between 1920 and 1950 included dental caries, respiratory infections, diarrhea, eye diseases, hepatitis, tuberculosis, diphtheria, senility, menopausal flushes, whooping cough, and many others. On vitamin deficiencies as the cause of "psychoneuroses," see A. Grey Clarke and F. Prescott, "Studies in Vitamin B Deficiency with Special Reference to Mental and Oral Manifestations," *BMJ* 2, no. 4320 (October 23, 1943): 503–5.
10. E. V. McCollum, *The Newer Knowledge of Nutrition: The Use of Food for the Preservation of Vitality and Health* (New York: Macmillan, 1918), 63–64.
11. Frederick Gowland Hopkins, "The Study of Human Nutrition: The Outlook To-Day," *BMJ* 1, no. 3872 (March 23, 1935): 571–77, at 571.
12. L. S. P. Davidson, "Refresher Course for General Practitioners: The Use and Abuse of Vitamins," *BMJ* 2, no. 4635 (November 5, 1949): 1036–38, at 1037.
13. For developments in the United States, see Levenstein, *Fear of Food*, 79–106.
14. T. B. Wood and F. G. Hopkins, *Food Economy in War Time* (Cambridge: Cambridge University Press, 1916); Bayliss, *Physiology of Food and Economy in Diet*. For a comparison of British and German approaches to wartime nutrition, see Mikuláš Teich, "Science and Food during the Great War: Britain and Germany," in *Science and Culture of Nutrition*, 213–34.
15. Weatherall, "Bread and Newspapers," 204; Robert E. Kohler, "Walter Fletcher, F. G. Hopkins, and the Dunn Institute of Biochemistry: A Case Study in the Patronage of Science," *Isis* 69, no. 3 (1978): 331–55, esp. 343, 346n73, 347.
16. Bayliss, *Physiology of Food*, 62.
17. For discussion of some of the pertinent issues, see the epilogue.
18. A striking irony is that women, so often maligned for perverse or undisciplined eating, assumed leading roles in this process, thus helping to give credence to scientific claims on the once private and domestic management of food; see Apple, "Science Gendered."
19. The phrase is in Louis H. Nahum and H. E. Himwich, "Insulin and Appetite. I. A Method for Increasing Weight in Thin Patients," *American Journal of the Medical Sciences* 183, no. 5 (May 1932): 608–13, at 609.
20. On traditional herbal remedies, see Roger Hyvert, *Précis de pathologie interne et de diagnostic*, 5th ed. (Paris: Maloine, 1921), 396; Todes, *Pavlov's Physiology Factory*, 265–66.
21. Todes, *Pavlov's Physiology Factory*, 259–88.
22. Todes, *Pavlov's Physiology Factory*, 275–76, 284; on the effectiveness of a French variant, see P. Le Gendre, "L'opothérapie gastrique au moyen du suc gastrique de chien à estomac isolé (*gastérine de Frémont*)," *Bulletins et mémoires de la Société médicale des hôpitaux de Paris*, 3rd ser., 17 (1900): 61–70.
23. On nineteenth-century practice, see Merriley Borell, "Brown-Séquard's Organotherapy and Its Appearance in America at the End of the Nineteenth Century," *BHM* 50, no. 3 (1976): 309–20.

24. On Falta's claims, see Nahum and Himwich, "Insulin and Appetite," 609.
25. Nahum and Himwich, "Insulin and Appetite," 609–10.
26. Ben Harvey Gottesfeld and Amando Caiuby Novaes, "Narco-Analysis and Sub-Shock Insulin in the Treatment of Anorexia Nervosa," *Digest of Neurology and Psychiatry* 13 (1945): 486–94.
27. G. von Bergmann, "Magerkeit und Magersucht," *DMW* 60, nos. 4–5 (January–February 1934): 123–26, 159–63.
28. G. von Bergmann, "Das Streben nach Synthese wissenschaftlicher Medizin und natürlicher Heilweisen," *DMW* 64, nos. 13–14 (March–April 1938): 455–57, 501–5; for an argument that Bergmann's "contradictory and inconclusive" approach to Magersucht helped prevent recognition of anorexia nervosa in Germany, see Habermas, "Die Anorexia nervosa (Magersucht)," 49.
29. John M. Berkman, "Anorexia Nervosa, Anorexia, Inanition and Low Basal Metabolic Rate," *American Journal of Medical Science* 180 (July 1930): 411–24.
30. Berkman, "Anorexia Nervosa," 419, 423; on Berkman, see Brumberg, *Fasting Girls*, 208, 210, 211.
31. On work that targeted the hypothalamus as the brain site of appetite, see chapter 10: "Nerves versus Hormones."
32. Bailey and Bremer, "Experimental Diabetes Insipidus," 802.
33. Brobeck, "Neural Regulation of Food Intake," 44, 51; Norman Jolliffe, *Reduce and Stay Reduced* (New York: Simon & Schuster, 1952).
34. Brobeck, "Neural Control of Hunger, Appetite, and Satiety."
35. Mark F. Lesses and Abraham Myerson, "Human Autonomic Pharmacology: XVI. Benzedrine Sulfate as an Aid in the Treatment of Obesity," *New England Journal of Medicine* 218, no. 3 (1938): 119–24; David Adlersberg and Martin E. Mayer, "Results of Prolonged Medical Treatment of Obesity with Diet Alone, Diet and Thyroid Preparations, and Diet and Amphetamine," *Journal of Clinical Endocrinology* 9, no. 3 (1949): 275–84; see also Laura Dawes, *Childhood Obesity in America: Biography of an Epidemic* (Cambridge, MA: Harvard University Press, 2014), 101–9.
36. The rationale for pyloroplasty was that appetite failed when the pylorus was too narrow for the ready passage of food into the stomach.
37. John A. Ryle, "Anorexia Nervosa," *Lancet* 228, no. 5903 (October 17, 1936): 893–99, quotation at 897.
38. On Ryle as one of London's patrician old guard, see Lawrence, "Still Incommunicable," 96, 103.
39. Cited in Hall, "The Critic and the Advocate," 273.
40. Jean Carrier, "L'anorexie mentale: Trouble instinctivo-affectif" (thesis, Faculté de Médecine et de Pharmacie de Lyon, 1939), 1.
41. Morris Simmonds, "Über embolische Prozesse in der Hypophysis," *Virchows Archiv für pathologische Anatomie* 217 (1914): 226–39; Simmonds, "Ueber Hypophysisschwund mit tödlichem Ausgang," *DMW* 40, no. 7 (February 1914): 322–23.

42. For a discussion of nearly fifty publications on pituitary cachexia and anorexia nervosa published between 1914 and 1939 (mostly German to around 1930, and then English and American as well), see Heinz Magendantz and Samuel Proger, "Anorexia Nervosa or Hypopituitarism?" *JAMA* 114, no. 20 (May 18, 1940): 1973–83; see also Vandereycken and van Deth, *From Fasting Saints to Anorexic Girls*, 179–80; Habermas, "Die Anorexia nervosa (Magersucht)," 39, 45, 48, 49, 50, 51–52.
43. Roberto F. Escamilla and H. Lisser, "Simmonds' Disease: A Clinical Study with Review of the Literature; Differentiation from Anorexia Nervosa by Statistical Analysis of 595 Cases, 101 of Which Were Proved Pathologically," *Journal of Clinical Endocrinology* 2, no. 2 (February 1942): 65–96.
44. Alexander R. Lucas, "Toward the Understanding of Anorexia Nervosa as a Disease Entity," *Mayo Clinic Proceedings* 56, no. 4 (1981): 254–64. Brumberg follows Lucas on this point; see Brumberg, *Fasting Girls*, 208 and 337n7.
45. For analyses decrying what the authors saw as persistent confusion of the two conditions, see P. Lôo, "Anorexie mentale, maladie de Simmonds, maladie de Sheehan," *Annales médico-psychologiques* 113, no. 1 (March 1955): 369–75; H. Frahm, "Simmonds'sche Krankheit, Sheehan-Syndrom und nervöse Magersucht. Ätiologie und Differentialdiagnose," *Die medizinische Welt* 5 (January 29, 1966): 235–38.
46. The contrast with Ryle's judgment of 1936 was striking: whereas Escamilla and Lisser emphasized "the truly difficult problem in differential diagnosis" (83), Ryle had stated: "It is only necessary to know of the existence of the malady to recognise it"; Ryle, "Anorexia Nervosa," 899.
47. Hagendantz and Proger, "Anorexia Nervosa or Hypopituitarism?," 1982.
48. Review of 1910 translation of *Work of the Digestive Glands* in *Lancet* 176, no. 4543 (September 24, 1910): 949.
49. Oskar Fischer, "Der Einfluss des Appetits auf die Magentätigkeit und seine Bedeutung für die funktionelle Magendiagnostik," *Münchener medizinische Wochenschrift* 58, no. 7 (February 14, 1911): 345–48.
50. J. Sim Wallace, "Some Principles of Dietetics," *Lancet* 185, no. 4789 (June 12, 1915): 1229–32.
51. On these issues, see chapter 11, "Appetite of 'the Organism': The Holism of David Katz."
52. Wilhelm Sternberg, "Der Appetit in der exakten Medizin," *Zeitschrift für Sinnesphysiologie* 45 (1911): 433–59; this article was reproduced with only minor changes in Sternberg, *Das Nahrungs-Bedürfnis, der Appetit und der Hunger: Eine diätetische Studie* (Leipzig: Johann Ambrosius Barth, 1913) (unless otherwise indicated, all citations are to the latter work).
53. Sternberg, *Das Nahrungs-Bedürfnis*, vi–x; Sternberg took Freud to task for insisting on the primacy of the sex drive, arguing that "the drive that serves to maintain the type must be weaker than that which first of all serves to maintain the individual. And this is appetite"; Sternberg, *Nahrungs-Bedürfnis*, x.

54. Arnold Durig, *Appetit: Vortrag, gehalten in der Gesellschaft der Aerzte in Wien am 9. Januar 1925* (Vienna: Julius Springer, 1925). On Durig's biography, including his wartime service at a military hospital in Sarajevo, see Rudolf W. Soukup, "Arnold Durig (1872–1961): Einer der letzten groβen Physiologen, die das gesamte Fach noch souverän überblickten," in *Die wissenschaftliche Welt von gestern*, ed. Rudolf W. Soukup and W. Gerhard Pohl (Vienna: Böhlau, 2004), 128–32. Durig published other works on nutrition, including, just after World War I, *Zum Ernährungsproblem Österreichs* (Vienna: Braumüller, 1920).
55. Durig, *Appetit*, 20, 21, 23, 39, 49; urging that Pavlov had overstated his case, Durig pointed to subsequent research showing that gastric secretion could occur in the absence of appetite and, conversely, that appetite could be present in the absence of secretion (49).
56. For an argument that psychosomaticists mischaracterized nineteenth-century medicine as restricted to "narrow-minded philistines," neglecting those who posited close interrelationships between psychic and somatic, see Erwin H. Ackerknecht, "Das Märchen vom verlorenen Psychosomatismus," *Gesnerus* 25, nos. 1–2 (1968): 113–15.
57. Michael Langenbach, "Phenomenology, Intentionality, and Mental Experiences: Edmund Husserl's *Logische Untersuchungen* and the First Edition of Karl Jaspers's *Allgemeine Psychopathologie*," *History of Psychiatry* 6 (1995): 209–24, at 212.
58. Christoph Mundt, "Impact of Karl Jaspers' *General Psychopathology*: The Range of Appraisal," in *One Century of Karl Jaspers'* General Psychopathology, ed. Giovanni Stanghellini and Thomas Fuchs (Oxford: Oxford University Press, 2013), 42–54, at 42. Jaspers left an important legacy to psychosomatic medicine, although he proved hostile to the form it took in Germany; see Thomas Henkelmann, "Zur Geschichte der Psychosomatik in Heidelberg: V. v. Weizsäcker und A. Mitscherlich als Klinikgründer," *Psychotherapie, Psychosomatik, medizinische Psychologie* 42 (1992): 175–86, esp. 179, 180–81.
59. In an "important programmatic statement" of 1931, the neurologist Kurt Goldstein declared that "'mind' and 'body' did not point to genuine entities but were just 'symbols' (human abstractions) of a holistic organismic reality'"; Harrington, *Reenchanted Science*, 161.
60. Walter B. Cannon, *Bodily Changes in Pain, Hunger, Fear and Rage: An Account of Recent Researches into the Function of Emotional Excitement* (rpt., New York: D. Appleton and Co., 1918), vii.
61. Walter B. Cannon, "The Interrelations of Emotions as Suggested by Recent Physiological Researches," *American Journal of Psychology* 25, no. 2 (1914): 252–82, cited in Benison, Barger, and Wolfe, *Walter B. Cannon*, 315.
62. Quotation in anonymous review of [Helen] Flanders Dunbar, *Emotions and Bodily Changes: A Survey of Literature on Psychosomatic Interrelationships, 1910–1933* (1935) in *BMJ* 1, no. 3925 (March 28, 1936): 642; see also Carl Binger,

"Doctor Walter B. Cannon," *PSMed* 8, no. 1 (1946): 60; Benison, Barger, and Wolfe, *Walter B. Cannon*, 320, 484nn24–28.

63. Wolfe, Barger, and Benison, *Walter B. Cannon*, 161–62, 168–69, 184.
64. James L. Halliday, "The Rising Incidence of Psycho-somatic Illness," *BMJ* 2, no. 4043 (July 2, 1938): 11–14. On British clinicians who argued that "we must not go germ-mad," see Lawrence, "Still Incommunicable," 102; see also Rhodri Hayward, "Enduring Emotions: James L. Halliday and the Invention of the Psychosocial," *Isis* 100, no. 4 (December 2009): 827–38.
65. Dowbiggin, *Inheriting Madness*, 2–7, 11–37, 104, 116–60; Engstrom, *Clinical Psychiatry in Imperial Germany*, 26–35, 99–101, 175–76; Nye, *Crime, Madness and Politics in Modern France*; Paul Lerner, *Hysterical Men: War, Psychiatry, and the Politics of Trauma in Germany, 1890–1930* (Ithaca, NY: Cornell University Press, 2003), at 1.
66. Sigmund Freud, "Three Contributions to the Theory of Sex," in *The Basic Writings of Sigmund Freud*, trans., ed., intro. A. A. Brill (New York: Modern Library, 1938), 551–629.
67. Freud's phrase, cited in Sándor Ferenczi, "The Phenomena of Hysterical Materialization: Thoughts on the Conception of Hysterical Conversion and Symbolism," in *Further Contributions to the Theory and Technique of Psycho-analysis*, comp. John Rickman, trans. Jane Isabel Suttie (New York: Boni and Liveright, 1927), 89–103, at 90–91.
68. Freud, "Three Contributions," 612; see also C. G. Jung, *Psychology of the Unconscious: A Study of the Transformations and Symbolisms of the Libido*, trans. Beatrice M. Hinkle, intro. William McGuire (1916; Princeton, NJ: Princeton University Press, 1991).
69. Sigmund Freud, "Draft G—Melancholia," in *The Standard Edition of the Complete Psychological Works of Sigmund Freud*, 24 vols., ed. J. Strachey (London: Hogarth Press, 1966), 1:200–206, at 200.
70. Sigmund Freud, "From the History of an Infantile Neurosis," in *Standard Edition*, 17:7–122, at 106.
71. Ferenczi, "Hysterical Materialization," 93.
72. Lazslo Antonio Avila, "Georg Groddeck: Originality and Exclusion," *History of Psychiatry* 14, no. 1 (March 2003): 83–101, at 85.
73. "Introductory Statement," *PSMed* 1, no. 1 (January 1939): 3–5, at 3.
74. Stress was a concept first advanced by the Viennese-born physician Hans Selye, who, though standing apart from explicit formulations of psychosomatic medicine, had considerable influence on its practitioners; see Russell Viner, "Putting Stress in Life: Hans Selye and the Making of Stress Theory," *Social Studies of Science* 29, no. 3 (June 1999): 391–410, esp. 395–96.
75. Max Hamilton, *Psychosomatics*, foreword by D. R. MacAlman (New York: John Wiley, 1955), 1; see also Halliday, "Rising Incidence," 13.
76. On the connections between psychosomatics and psychoanalysis in the United States, see Nathan G. Hale Jr., *The Rise and Crisis of Psychoanalysis*

in the United States: Freud and the Americans, 1917–1985 (New York: Oxford University Press, 1995), 178–84, 258–63, 322–27, 383–84.

77. Brown, "Alan Gregg," quotation at 180. On Alexander's role in American psychosomatic medicine, see also Anne Harrington, *The Cure Within: A History of Mind-Body Medicine* (New York: W. W. Norton, 2008), 91–94; Hale, *Rise and Crisis*, 180–82.
78. Edward Weiss, "Psychosomatic Medicine," *American Journal of Nursing* 45, no. 3 (March 1945): 189–93, at 191, emphasis in the original.
79. Franz Alexander et al., "The Influence of Psychologic Factors upon Gastro-intestinal Disturbances: A Symposium," *Psychoanalytic Quarterly* 3, no. 4 (1934): 501–88; Alexander stated that the symposium was "part of a more comprehensive study" under way at the Chicago Institute on "the influence of psychic factors upon the different vegetative systems" (501).
80. Alexander, "Influence," 502–3; Ferenczi, "Hysterical Materialization"; Ferenczi, "Disgust for Breakfast," in *Further Contributions*, 326; Karl Abraham, "Contributions to the Theory of the Anal Character," in *Essential Papers on Obsessive-Compulsive Disorder*, ed. Dan J. Stein and Michael H. Stone (New York: New York University Press, 1997), 73–90; Ernest Jones, "Anal-Erotic Character Traits," in *Papers on Psycho-Analysis*, 3rd ed. (London: Baillière, Tindall and Cox, 1923), 680–704.
81. John V. Waller, M. Ralph Kaufman, and Felix Deutsch, "Anorexia Nervosa: A Psychosomatic Entity," *PSMed* 11, no. 1 (January 1940): 3–16; see also Lincoln Rahman, Henry B. Richardson, and Herbert S. Ripley, "Anorexia Nervosa with Psychiatric Observations," *PSMed* 1, no. 3 (July 1939): 335–65.
82. Viktor von Weizsäcker, "Über Traume bei sogenannter endogener Magersucht," *DMW* 63, nos. 7–8 (February 1937): 253–57, 294–97; on Weizsäcker's orientation within psychosomatic medicine, see Harrington, *Cure Within*, 84–86.
83. Grace Nicolle, "Pre-Psychotic Anorexia," *Proceedings of the Royal Society of Medicine* 32 (1938): 153–62.
84. Jules H. Masserman, "Psychodynamisms in Anorexia Nervosa and Neurotic Vomiting," *Psychoanalytic Quarterly* 10, no. 2 (1941): 211–42; on the "heavy-handed" character of American psychoanalytic theorizing about anorexia, see Brumberg, *Fasting Girls*, 225.
85. Bernard C. Meyer and Leonard A. Weinroth, "Observations on Psychological Aspects of Anorexia Nervosa," *PSMed* 19, no. 5 (September–October 1957): 389–98.
86. Hilde Bruch, "Obesity and Anorexia Nervosa," *Psychosomatics* 19, no. 4 (April 1978): 208–12. Habermas traces a different path away from the previously strong emphasis on lost appetite among German physicians of the 1940s; see Habermas, "Die Anorexia nervosa (Magersucht)," 50–51.
87. Habermas, "Die Anorexia nervosa (Magersucht)," 38.
88. On recent approaches to eating disorders that emphasize "the eating itself," see Wilson, *First Bite*, 263–302. The emergence of these frameworks, which

are one-sided in their own way, represents yet another sharp zigzag in the medical history of troubled eating. For further discussion, see the epilogue.

89. Carl von Noorden, *Die Fettsucht*, 2nd rev. ed. (Vienna: Hölder, 1910).
90. In 1930 the British physician William F. Christie estimated the proportion of endogenous cases at 10 percent; Christie, "Corpulence: Its Causes and Cure," *Lancet* 215, no. 5564 (April 19, 1930): 894–96, at 896. "Maladies of nutrition" also included arthritis, gout, and diabetes; see Hyvert, *Précis de pathologie interne*, 89, 223, 292; Henri-Alphonse Richardière, *Maladies de la nutrition: Goutte, obésité, diabète*, 6th ed. (Paris: J.-B. Baillière, 1920); see also Erwin H. Ackerknecht, "The History of Metabolic Diseases," *Ciba symposia* 6 (1944–1945): 1834–1844.
91. Bergmann, "Magerkeit und Magersucht," 123–24.
92. Hyvert stated that 50 percent of cases were inherited; *Précis de pathologie interne*, 394.
93. Jolliffe, *Reduce and Stay Reduced*, 3, 4.
94. Christie, "Corpulence," 894.
95. Hyvert, *Précis de pathologie interne*, 396.
96. Harry Gottesfeld, "Body and Self-Cathexis of Super-Obese Patients," *Journal of Psychosomatic Research* 6, no. 3 (July–September 1962): 177–83.
97. Hilde Bruch and Grace Touraine, "Obesity in Childhood: V. The Family Frame of Obese Children," *PSMed* 2, no. 2 (April 1940): 141–206, at 141. Bruch appeared as sole author of the first four articles in the series.
98. Bruch discussed psychiatric interpretations in "Obesity in Childhood: III. Physiologic and Psychologic Aspects of the Food Intake of Obese Children," *American Journal of Diseases of Children* 59, no. 4 (1940): 739–81, at 760–70.
99. Bruch and Touraine, "Obesity in Childhood: V," 157–58, 175–76.
100. Bruch, "Obesity in Childhood: III," 740.
101. Myer Mendelson, "The Treatment of Obesity," *Psychosomatics* 1, no. 3 (May–June 1960): 158–60. Bruch later changed her views, arguing that the obese exhibited no single psychological profile; see Hilde Bruch, "Psychiatric Aspects of Obesity," *Metabolism* 6 (1957): 461–65. For further discussion, including Bruch's move toward behaviorism, see Dawes, *Childhood Obesity in America*, 121–22.
102. Wilhelm Falta, *The Ductless Glandular Diseases*, trans. Milton K. Meyers, foreword by Archibald E. Garrod (1913; Philadelphia: P. Blakiston's, 1915), 571–72.
103. On the origins of "personality" studies, see Danziger, *Constructing the Subject*, 158–65.
104. Whether obesity has definite psychological components and should be ranged among eating disorders is currently a subject of debate; see J. Day, A. Ternouth, and D. A. Collier, "Eating Disorders and Obesity: Two Sides of the Same Coin?" *Epidemiologia e psichiatria sociale* 18, no. 2 (April–June 2009): 96–100.

105. On treatment programs that dominated in the United States from the 1970s to 2000, especially successive versions of behavior modification, see Dawes, *Childhood Obesity*, 122–26.
106. Quigley, "Role of the Digestive Tract," 6.
107. Hollander, introduction to Miner, *Regulation*, 3.
108. Edward C. Klein Jr., "The Pernicious Type of Nervous Dyspepsia," *Annals of Internal Medicine* 7, no. 8 (1934): 960–65, at 961.
109. Meyer and Weinroth, "Observations," 397.
110. On the "reducing" industry around 1950, see Schwartz, *Never Satisfied*, esp. 242–68.
111. Swale Vincent, "The Arris and Gale Lecture on a Critical Examination of Current Views on Internal Secretion," *Lancet* 200, no. 5163 (August 12, 1922): 313–20, at 319.

EPILOGUE

1. Kremer, "Physiology," esp. 358–66.
2. The first use of the phrase that I have found is Hilde Bruch, "Psychotherapeutic Problems in Eating Disorders," *Psychoanalytic Review* 50 (1963): 573–87.
3. Francesca Brambilla, "1969–1989: Twenty Years of Life of the International Society of Psychoneuroendocrinology," *Psychoneuroendocrinology* 14, no. 4 (1989): 247–49; only twenty years into the existence of the group, the author speculated that the field might have "exhaust[ed] its possibilities and perspectives" as its investigations shed light on the hypothalamus but not on "higher centers which are more deeply involved in cognition, affect, and behavior" (249). A point of interest about this society is that it gives an annual Curt Richter Award; on Richter, see chapter 11.
4. Bal K. Anand, "Nervous Regulation of Food Intake," *Physiological Reviews* 41, no. 4 (1961): 677–708; Robert L. Isaacson, *The Limbic System* (New York: Plenum, 1974), 75–79, 88–98; Eliot Stellar, "The Biological Nature of Appetite," *Proceedings of the American Philosophical Society* 136, no. 4 (December 1992): 507–20.
5. A. S. Levine and R. L. Atkinson, "Opioids in the Regulation of Food Intake and Energy Expenditure: Introduction," *Symposium Presented by the American Institute of Nutrition at the 70th Annual Meeting of the Federation of American Societies for Experimental Biology, St. Louis, Missouri, April 16, 1986*, in *Federation Proceedings* 46, no. 1 (January 1987): 159–62.
6. Richard J. Bodnar, "Endogenous Opioids and Feeding Behavior: A 30-Year Historical Perspective," *Peptides* 25 (2004): 697–725.
7. Allen, *Life Science in the Twentieth Century*, 187–228.
8. Gregory S. Barsh, I. Sadaf Farooqi, and Stephen O'Rahilly, "Genetics of Body-Weight Regulation," *Nature* 404, no. 6778 (April 6, 2000): 644–51; John R.

Speakman, "Gene Environment Interactions and the Origin of the Modern Obesity Epidemic: A Novel 'Nonadaptive Drift' Scenario," in *Appetite and Body Weight: Integrative Systems and the Development of Anti-Obesity Drugs*, ed. Tim C. Kirkham and Steven J. Cooper (Amsterdam: Elsevier, 2007), 301–22.

9. Rosenzweig, "Mechanisms of Hunger and Thirst," 128–33.
10. "For many years it was thought that eating behavior is driven directly by hunger drives, and that food was a reward because it reduces hunger drive. But actually, physiological hunger works primarily by modulating food incentive values, not driving behavior directly through deficit or drive signals"; Kent C. Berridge, "Brain Reward Systems for Food Incentives and Hedonics in Normal Appetite and Eating Disorders," in Kirkham and Cooper, *Appetite and Body Weight*, 191–215, at 208–9.
11. For a study that drew on ethology but cited classic work of the 1930s-50s rather than recent investigations, see John E. Morley et al., "Opioid Modulation of Appetite," *Neuroscience and Biobehavioral Reviews* 7 (1983): 281–305, at 286.
12. On the shift toward interest in the "unique, species-specific perceptual worlds" of animals," see Dyer and Brockmann, "Biology of the *Umwelt*," 529, 530.
13. Vernon identifies as a key figure Audrey I. Richards, a student of Bronislaw Malinowski, who argued against the universalizing claims of nutrition science and for the importance of "'social heritage'" in food selection; see Vernon, *Hunger*, 110–11, 113–15, 134, 136, 138; Audrey I. Richards, *Hunger and Work in a Savage Tribe: A Functional Study of Nutrition among the Southern Bantu*, foreword by Bronislaw Malinowski (London: Routledge, 1932).
14. A notable case in the United States was the work of the anthropologist Margaret Mead as executive secretary of the Committee on Food Habits of the National Research Council; Margaret Mead, "The Factor of Food Habits," *Annals of the American Academy of Political and Social Science* 225, no. 1 (January 1943): 136–41. On related developments in Britain, see Vernon, *Hunger*, 139–46.
15. Ellen Messer, "Anthropological Perspectives on Diet," *Annual Review of Anthropology* 13 (1984): 205–49; Sidney W. Mintz and Christine M. Du Bois, "The Anthropology of Food and Eating," *Annual Review of Anthropology* 31 (2002): 99–119. Messer noted the emergence of a new subfield of "nutritional anthropology" (206).
16. For background, see Mintz and Du Bois, "Anthropology," 109–10; for a recent example, see Judith Farquhar, *Appetites: Food and Sex in Postsocialist China* (Durham, NC: Duke University Press, 2002).
17. Rebecca J. Lester, "Critical Therapeutics: Cultural Politics and Clinical Reality in Two Eating Disorder Treatment Centers," *Medical Anthropology Quarterly*, new series, 21, no. 4 (December 2007): 369–87; Helen Gremillion, *Feeding Anorexia: Gender and Power at a Treatment Center* (Durham, NC: Duke University Press, 2003).

18. Ron van Deth and Walter Vandereycken, "Eating Disorders: Clinical Section," in *A History of Clinical Psychiatry: The Origin and History of Psychiatric Disorders*, ed. G. E. Berrios and Roy Porter (New York: New York University Press, 1995), 593–601.
19. These included refusal to maintain body weight over a minimal normal; intense fear of gaining weight or becoming fat; disturbance in the way body weight, size, or shape is experienced; and, in females, amenorrhea; see *Diagnostic and Statistical Manual of Mental Disorders* (*DSM-III*) (3rd rev. ed.) (Washington, DC: American Psychiatric Association, 1987), 67.
20. *DSM III*, 69–71; *DSM-III-R*, 67–69. "Binge-eating disorder" appeared in appendix B among "Criteria Sets and Axes Provided for Further Study" in *Diagnostic and Statistical Manual of Mental Disorders* (4th ed., text rev.) (Washington, DC: American Psychiatric Association, 2000), 785–87. On arguments over the classification of bulimia, see Van Deth and Vandereycken, "Eating Disorders," 597; Habermas, *Heißhunger*, 13–21.
21. English-language journals included *Eating Disorders Review* (1990), *Eating Disorders* (1993), *European Eating Disorders Review* (1993), *Eating Behaviors* (2000), *Advances in Eating Disorders* (2013), and *Journal of Eating Disorders* (2013).
22. Claire Thompson et al., "What Does It Mean to Be a 'Picky Eater'?: A Qualitative Study of Food-Related Identities and Practices," *Appetite* 84 (2015): 235–39; Rachel Bryant-Waugh, "Avoidant Restrictive Food Intake Disorder: An Illustrative Case Example," *International Journal of Eating Disorders* 46, no. 5 (2013): 420–23. For a claim that the most common childhood syndrome is "EDNOS: Eating Disorder Not Otherwise Specified" and a distinction between the eating disorders of adults and the "feeding disorders" of small children, see Wilson, *First Bite*, 267–68.
23. Quotation in Nestle and Nesheim, *Why Calories Count*, 137. Nutritional authorities give widely differing dates, however, for the beginning of the obesity crisis: Nestle and Nesheim specify the 1980s for the United States (7), as does Kessler, *End of Overeating*, 3–6, although Kessler cites a study of overweight, not obesity: R. J. Kuczmarski et al., "Increasing Prevalence of Overweight among US Adults: The National Health and Nutrition Examination Surveys, 1960 to 1991," *JAMA* 272, no. 3 (1994): 205–11. The date of origin is important as these authors tie increased obesity and overweight to changes in the "food environment" in the United States dating to the 1970s. Yet other researchers pinpoint the 1950s: "Since the 1950s there has been a major epidemic of obesity . . . throughout the Western world"; Speakman, "Gene Environment Interactions," 301.
24. Poulain, "Eléments pour une histoire de la médicalisation de l'obésité," 15; on the ideological implications of the language of "epidemic," see Michael Gard and Jan Wright, *The Obesity Epidemic: Science, Morality and Ideology* (London: Routledge, 2005), 173–74.

25. Avner Offer, "Body Weight and Self-Control in the United States and Britain since the 1950s," *Social History of Medicine* 14, no. 1 (2001): 79–106; Derek J. Oddy and Peter J. Atkins, introduction to *The Rise of Obesity in Europe: A Twentieth Century Food History*, ed. Derek J. Oddy, Peter J. Atkins, and Virginie Amilien (Farnham, UK: Ashgate, 2009), 1–11; Stanley J. Ulijaszek and Hayley Lofink, "Obesity in Biocultural Perspective," *Annual Review of Anthropology* 35 (2006): 337–60.
26. Gard and Wright, *Obesity Epidemic*, esp. 168–86; Michael Gard, *The End of the Obesity Epidemic* (London: Routledge, 2011). Historical skeptics include Peter Stearns, who sees not a health crisis but hostility to fat as a broad cultural phenomenon that set in in the 1890s; Stearns, *Fat History*; see also Sander L. Gilman, introduction to *Diets and Dieting: A Cultural Encyclopedia* (New York: Routledge, 2008), ix–xii. Laura Dawes, though not contesting the language of "epidemic," emphasizes that the issue has a long history; Dawes, *Childhood Obesity in America*, 10–18. For a claim that obesity is a threat equaled only by global climate change, see Kirkham and Cooper, "Introduction and Overview," in *Appetite and Body Weight*, 1.
27. Hollander, introduction to Miner, *Regulation*, 3.
28. See, for example, in the volume *Appetite and Body Weight*, Morten L. Kringelbach, "Cortical Systems Involved in Appetite and Food Consumption," 5–26; Dianne Figlewicz Lattemann et al., "Integration of Peripheral Adiposity Signals and Psychological Controls of Appetite," 167–90; Berridge, "Brain Reward Systems for Food Incentives and Hedonics in Normal Appetite and Eating Disorders."
29. Teitelbaum, "Appetite," 470; on early work on taste in physiological psychology, see Carl Pfaffmann, "Carl Pfaffmann," in *A History of Psychology in Autobiography*, ed. Gardner Lindzey (Stanford, CA: Stanford University Press, 1989), 8:421–47, esp. 426–35.
30. Catherine Goodman, "Monell Chemical Senses Center," *Nature Chemical Biology* 5, no. 1 (January 2009): 2; Goodman discusses interdisciplinary research by investigators from biophysics, biopsychology, bioengineering, neuroscience, physiological psychology, genetics, and neurobiology.
31. Bodnar, "Endogenous Opioids."
32. Keith W. Sehnert, "Development of Phentermine, an Appetite-Control Drug," *Clinical Medicine* 70 (February 1963): 400–403; "Clinical Case No. 442, Fever and Loss of Appetite," *Case Reports of the Children's Memorial Hospital, Chicago* 6, no. 12 (1947): 1049–54; Charles Upton Lowe et al., "Fibrosis of the Pancreas: Enterogastrone and the 'Duodenal Mechanism' in Relation to Increased Appetite," *American Journal of Diseases of Children* 79, no. 1 (1950): 91–98.
33. Hilde Bruch, *The Golden Cage: The Enigma of Anorexia Nervosa* (Cambridge, MA: Harvard University Press, 1978), ix–x; Gremillion, *Feeding Anorexia*, 226n21.

34. For a succinct summary of varieties of family therapy, see Gremillion, *Feeding Anorexia*, 227–28n2.; see also Johan Vanderlinden and Walter Vandereycken, "Guidelines for the Family Therapeutic Approach to Eating Disorders," *Psychotherapy and Psychosomatics* 56, nos. 1–2 (1991): 36–42.
35. Berridge, "Brain Reward Systems," 192–93; on recent views of anorexia nervosa as a biological illness, see Carrie Arnold, *Decoding Anorexia: How Breakthroughs in Science Offer Hope for Eating Disorders* (New York: Routledge, 2012).
36. Judith Rodin, Gordon Mogenson, and David Booth, "Editorial," *Appetite* 1, no. 1 (March 1980): n.p.
37. David Booth, "The Next Twenty Years: An Editorial Perspective," *Appetite* 34, no. 1 (2000): 1–3.
38. Booth, "The Next Twenty Years," 1; Booth discusses efforts at integration made in the 1970s, citing in particular the requirement by funding entities in the United Kingdom that "research on food intake . . . be extended to food choice (as understood in psychological science)" as well as a German book series, Appetite and Food Intake, that was interdisciplinary and intercontinental (Europe and the United States) in coverage, and a Swiss conference and publication, *Criteria of Food Acceptance*, with "a more applied orientation" (1).
39. These failures were also operational: "Research . . . will have to become much better integrated," he observed, for use by business, policy-makers, practitioners, and educators; Booth, "The Next Twenty Years," 2.
40. David A. Booth, "Thirty Years of the Journal *Appetite*: The Citation Record," *Appetite* 54, no. 1 (February 2010): 1–4.
41. Arezzo di Trifiletti et al., "Comparison of the Performance of Four Different Tools in Diagnosing Disease-Associated Anorexia and Their Relationship with Nutritional, Functional and Clinical Outcome Measures in Hospitalized Patients," *Clinical Nutrition* 32, no. 4 (August 2013): 527–32.
42. For a discussion of imaging techniques in psychology, see the "Special Section" in *Perspectives on Psychological Science* 5, no. 6 (November 2010): 714–75.
43. Nestle and Nesheim, *Why Calories Count*, 6–7, 143, 177–85.
44. Nestle and Nesheim, *Why Calories Count*, 106–9.
45. Discussing the applicability to overeating and other disorders of his widely accepted distinction between "liking" and "wanting," Berridge states: "Different eating disorders may require different answers. Perhaps even different individuals with the 'same' disorder will involve different answers"; Berridge, "Brain Reward Systems," 193.
46. Nutritionists concerned with overweight and obesity assume an innate tendency to overeat when food is ample while simultaneously asserting that "for thousands of years human body weight stayed remarkably stable"; Kessler, *End of Overeating*, 3; Nestle and Nesheim, *Why Calories Count*: "The

urge to eat is so fundamental to life that multiple overlapping and powerful regulatory factors evolved to maintain that urge and to prompt you to eat as much as you can at every opportunity" (108).

47. For a helpful summary of the role of "fast food" that recognizes the importance of "standardized products" yet urges attention to "different levels of analysis" (political economy and culture, "rational systems and social practice"), see Steve Penfold, "Fast Food," in *Oxford Handbook of Food History*, ed. Pilcher, 279–301.
48. Kessler, *End of Overeating*, 12–28.
49. Perhaps out of concern for being seen to dictate, Nestle and Nesheim state the rules as a "personal preference": "Our personal preference is for diets that provide plenty of fruits, vegetables, fish or lean meats, and whole grains and that avoid sugary drinks and minimize high-calorie snacks"; Nestle and Nesheim, *Why Calories Count*, 172.
50. David Julian McClements et al., "In Defense of Food Science," *Gastronomica* 11, no. 2 (Summer 2011): 76–84.
51. Nick Cullather, "The Foreign Policy of the Calorie," *American Historical Review* 112, no. 2 (April 2007): 337–64; Nestle and Nesheim, *Why Calories Count*, 28–29.
52. Robert N. Proctor, *Cancer Wars: How Politics Shapes What We Know and Don't Know about Cancer* (New York: Basic Books, 1995).
53. "Despite our increasingly tailored treatments, we will not offer the best care to our cancer patients unless we invest the time and energy to discuss, to listen, to talk with, and to care for each patient as an individual"; Jessica Grubman and Afsaneh Barzi, "Lost in Translation: The Patient-Physician Relationship in the Molecular Era," *Journal of Palliative Medicine* 18, no. 11 (2015): 987–88.
54. This is not to discount critiques of the environmental impact of modern food-producing techniques; see, on this cluster of problems, Harvey Blatt, *America's Food: What You Don't Know about What You Eat* (Cambridge, MA: MIT Press, 2008), 25–48.
55. Scrinis, "On the Ideology of Nutritionism."
56. On stigmatization, see Gard and Wright, *Obesity Epidemic*, 14, 148, 161, 174, 179.
57. In recent years the discourse of "deportation" in the United States has included calls to deport "ugly fat people"; "Milo Yiannopoulos Rails against Liberals, Muslims, Feminism and 'Fat Acceptance' at CU Boulder," *CUIndependent*, January 26, 2017, http://cuindependent.com/2017/10/26.
58. Anna Kirkland, "The Environmental Account of Obesity: A Case for Feminist Skepticism," *Signs* 36, no. 2 (Winter 2011): 463–85.
59. The focus of most work on the "contagion" of anorexia nervosa has been the dangers of group treatment in which ways "to be really, really good at anorexia" are shared; Walter Vandereycken, "Can Eating Disorders Become

'Contagious' in Group Therapy and Specialized Inpatient Care?" *European Eating Disorders Review* 19 (2011): 289–95.

60. Gremillion, *Feeding Anorexia*, 3–4, 8–10, 14–15, 30, 33, 60–69.
61. On rhetorical strategies shared by the war on terror and the "war on obesity," see Charlotte Biltekoff, "The Terror Within: Obesity in Post 9/11 U.S. Life," *American Studies* 48, no. 3 (Fall 2007): 29–48.
62. Wilson, *First Bite*, 9.
63. Jörg Niewöhner et al., "Cardiovascular Disease and Obesity Prevention in Germany: An Investigation into a Heterogeneous Engineering Project," *Science, Technology, & Human Values* 36, no. 5 (September 2011): 723–51.
64. For a very different approach to harms done by nutritionism, see Landecker, "Food as Exposure." Working from the perspective of "nutritional epigenetics," Landecker explores unintended consequences of health policies developed on the assumption that what is good for one eater is good for all. One example involves legislation of 1998 that mandated folic acid fortification of all wheat products based on findings of benefits for pregnant women and recent research suggesting that this step may be involved in the increase in colon cancer in the aged (182).
65. Canguilhem, "La santé," English translation by Todd Meyers and Stefanos Geroulanos: "Health: Crude Concept and Philosophical Question," *Public Culture* 20, no. 3 (2008): 467–77. Canguilhem contrasts the "sociopolitical-medical ambition of regulating the lives of individuals" to a "free health [that] is not an object for those who believe themselves its specialists" (474): "I go well insofar as I feel able to take responsibility for my acts" (477).
66. Gard, *End of the Obesity Epidemic*, 117; Wilson, *First Bite*, 351–56; Nestle and Nesheim, *Why Calories Count*, 172 (their policy recommendations involve not individuals but the "food environment").
67. Jerome Groopman, *How Doctors Think* (New York: Houghton Mifflin, 2008), 169.

Bibliography

Abraham, Karl. "Contributions to the Theory of the Anal Character." In *Essential Papers on Obsessive-Compulsive Disorder*, edited by Dan J. Stein and Michael H. Stone, 73–90. New York: New York University Press, 1997.

Académie des Sciences (Paris). "Rapport fait à l'Académie des Sciences au nom de la Commission dite *de la gélatine*." *Comptes-rendus hebdomadaires des séances de l'Académie des Sciences*, 2nd semester, 13 (1841): 237–96.

Ackerknecht, Erwin H. "The History of the Discovery of the Vegetative (Autonomic) Nervous System." *Medical History* 18 (1974): 1–8.

———. "The History of Metabolic Diseases." *Ciba symposia* 6 (1944–1945): 1834–44.

———. "Das Märchen vom verlorenen Psychosomatismus." *Gesnerus* 25, nos. 1–2 (1968): 113–15.

———. *Medicine at the Paris Hospital, 1794–1848*. Baltimore: Johns Hopkins University Press, 1967.

Adelon, N., P.-A. Béclard, L.-T. Biett, G. Breschet, A.-F. Chomel, H. Cloquet, J. Cloquet, et al. *Dictionnaire de médecine*. 21 vols. Paris: Béchet, 1821–1828.

Adelon, N., and F. V. Mérat, eds. *Dictionnaire des sciences médicales*. 60 vols. Paris: C. L. F. Panckoucke, 1812–1822.

Adlersberg, David, and Martin E. Mayer. "Results of Prolonged Medical Treatment of Obesity with Diet Alone, Diet and Thyroid Preparations, and Diet and Amphetamine." *Journal of Clinical Endocrinology* 9, no. 3 (1949): 275–84.

Adolph, E. F. "Early Concepts of Physiological Regulation." *Physiological Reviews* 41, no. 4 (1961): 737–70.

———. *Physiological Regulations*. Lancaster, PA: Jaques Cattell, 1943.

Albala, Kenneth. *Eating Right in the Renaissance*. Berkeley: University of California Press, 2002.

———. *Food in Early Modern Europe*. Westport, CT: Greenwood Press, 2003.

———. "Weight Loss in the Age of Reason." In Forth and Carden-Coyne, *Cultures of the Abdomen*, 169–83.

Albu, Albert. *Ueber die Autointoxicationen des Intestinaltractus*. Berlin: August Hirschwald, 1895.

Albury, William R. "Experiment and Explanation in the Physiology of Bichat and Magendie." In *Studies in History of Biology*, edited by William R. Coleman and Camille Limoges, 1:47–131. Baltimore: Johns Hopkins University Press, 1977.

Alexander, Franz, Catherine Bacon, George W. Wilson, Harry B. Levey, and Maurice Levine. "The Influence of Psychologic Factors upon Gastro-intestinal Disturbances: A Symposium." *Psychoanalytic Quarterly* 3, no. 4 (1934): 501–88.

Alexander, William S. "An Inaugural Dissertation on Dyspepsia." Thesis, University of Pennsylvania, 1811.

Allbutt, T. Clifford. *On Visceral Neuroses, being the Gulstonian Lectures on Neuralgia of the Stomach and Allied Disorders, Delivered at the Royal College of Physicians in March, 1884*. London: J. & A. Churchill, 1884.

Allen, Garland E. *Life Science in the Twentieth Century*. New York: Cambridge University Press, 1978.

Allen, Richard C. *David Hartley on Human Nature*. Albany: State University of New York Press, 1999.

American Psychiatric Association. *Diagnostic and Statistical Manual of Mental Disorders*. 3rd ed. Washington, DC: American Psychiatric Association, 1980.

American Psychiatric Association. *Diagnostic and Statistical Manual of Mental Disorders*. 3rd rev. ed. Washington, DC: American Psychiatric Association, 1987.

American Psychiatric Association. *Diagnostic and Statistical Manual of Mental Disorders*. 4th ed. Washington, DC: American Psychiatric Association, 1994.

American Psychiatric Association. *Diagnostic and Statistical Manual of Mental Disorders*. 4th ed., text revision. Washington, DC: American Psychiatric Association, 2000.

Anand, Bal K. "Nervous Regulation of Food Intake." *Physiological Reviews* 41, no. 4 (1961): 677–708.

"Appetit." In Zedler, *Grosses vollständiges Universal-Lexicon aller Wissenschafften und Künste*. Supplement 2:58.

Apple, Rima D. "Science Gendered: Nutrition in the United States, 1840–1940." In Kamminga and Cunningham, *Science and Culture of Nutrition*, 129–54.

———. *Vitamania: Vitamins in American Culture*. New Brunswick, NJ: Rutgers University Press, 1996.

Aristotle. *The Complete Works of Aristotle: The Revised Oxford Translation*. Edited by Jonathan Barnes. 2 vols. Princeton, NJ: Princeton University Press, 1984.

Arnaud, Sabine. *L'invention de l'hystérie au temps des Lumières (1670–1820)*. Paris: École des Hautes Études en Sciences Sociales, 2014.

Arnold, Carrie. *Decoding Anorexia: How Breakthroughs in Science Offer Hope for Eating Disorders*. New York: Routledge, 2012.

Arnold, Wilhelm. "David Katz." In *Neue Deutsche Biographie* 11 (1977): https://www.deutsche-biographie.de/sfz40136.html#ndbcontent.

Ash, Mitchell G. *Gestalt Psychology in German Culture, 1890–1967: Holism and the Quest for Objectivity*. New York: Cambridge University Press, 1995.

———. "Psychology." In *Modern Social Sciences*. Vol. 7 of *The Cambridge History of Science*, edited by Dorothy Ross and Theodore M. Porter, 251–74. Cambridge: Cambridge University Press, 2003.

Astruc, Jean. *Traité des maladies des femmes*. 6 vols. Paris: Cavelier, 1761–1765.

Aumont, Arnulphe d'. "Faim (*Séméiotique*)." In Diderot and d'Alembert, *Encyclopédie*, 6:376.

Autuori, M., M.-P. Bénassy, J. Benoît, R. Courrier, E.-P. Deleurance, M. Fontaine, K. von Frisch, et al., eds. *L'instinct dans le comportement des animaux et de l'homme*. Paris: Masson, 1956.

Avila, Lazslo Antonio. "Georg Groddeck: Originality and Exclusion." *History of Psychiatry* 14, no. 1 (March 2003): 83–101.

Azouvi, François. "Quelques jalons dans la préhistoire des sensations internes." *Revue de synthèse*, 3rd ser., nos. 113–14 (January–June 1984): 113–33.

Babinski, Joseph. "Tumeur du corps pituitaire sans acromégalie et avec arrêt de développement des organes génitaux." *Revue neurologique* 8 (1900): 531–33.

Bailey, Percival, and Frédéric Bremer. "Experimental Diabetes Insipidus." *Archives of Internal Medicine* 28 (1921): 773–803.

Banting, William. *Letter on Corpulence: Addressed to the Public*. 2nd ed. London: Harrison and Sons, 1863.

Bardier, É[mile]. "Faim." In Richet, *Dictionnaire de Physiologie* 6:1–29.

Baron, Jeremy Hugh. *The Stomach: A Biography. Four Thousand Years of Stomach Pains: Literature, Symptoms and Epidemiology*. North Charleston, SC: CreateSpace, 2013.

Baron, J. H., and A. Sonnenberg. "Alimentary Diseases in the Poor and Middle Class in London 1773–1815, and in New York Poor 1797–1818." *Alimentary Pharmacology and Therapeutics* 16 (2002): 1709–14.

Barras, J.-P.-T. *Traité sur les gastralgies et les entéralgies ou maladies nerveuses de l'estomac et des intestins*. 2nd ed. Paris: Béchet, 1827.

Barrow, Mark V. *A Passion for Birds: American Ornithology after Audubon*. Princeton, NJ: Princeton University Press, 1998.

Barsh, Gregory S., I. Sadaf Farooqi, and Stephen O'Rahilly. "Genetics of Body-Weight Regulation." *Nature* 404, no. 6778 (April 6, 2000): 644–51.

Baudouin, Charles, and Edward B. Schlesinger. *Suggestion and Autosuggestion: A Psychological and Pedagogical Study Based upon the Investigations Made by the New Nancy School*. Translated by Eden Paul and Cedar Paul. London: G. Allen and Unwin, 1920.

Bayer, Erwin. "Beiträge zur Zweikomponententheorie des Hungers." *Zeitschrift für Psychologie* 112 (1929): 1–54.

Bayle, A.-L.-J. *Traité des maladies du cerveau et de ses membranes* (1826). New York: Arno Press, 1976.

Bayle, François-Antoine. "Considérations physiologiques et pathologiques sur la faim, Thèse présentée et soutenue à la Faculté de Médecine de Paris, le 26 avril 1816." Thesis, Faculté de Médecine de Paris, 1816.

Bayliss, William M. *The Physiology of Food and Economy in Diet*. London: Longmans, Green, 1917.

———. *Principles of General Physiology*. London: Longmans, Green, 1915.

Bayliss, W. M., and E. H. Starling. "The Mechanism of Pancreatic Secretion." *Journal of Physiology* 28, no. 5 (1902): 325–53.

Beach, Frank A. "The Snark Was a Boojum." *American Psychologist* 5 (1950): 115–24.

Beauchêne, E.-P. Chauvot de. *De l'influence des affections de l'âme dans les maladies nerveuses des femmes*. Paris: Méquignon, 1781.

Beaumont, William. *Experiments and Observations on the Gastric Juice and the Physiology of Digestion*. 1833. Biographical essay by William Osler. New York: Dover, 1959.

Beaunis, H[enry]. *Les sensations internes*. Paris: Félix Alcan, 1889.

Beck, Gallus. "Neue Beiträge zur Zweikomponententheorie des Hungers." *Zeitschrift fur Psychologie* 118 (1929): 283–349.

Bell, Whitfield J., Jr. "Philadelphia Medical Students in Europe, 1750–1800." In *The Colonial Physician and Other Essays*, 41–69. New York: Science History Publications, 1975.

Benison, Saul, A. Clifford Barger, and Elin L. Wolfe. *Walter B. Cannon: The Life and Times of a Young Scientist*. Cambridge, MA: Belknap Press of Harvard University Press, 1987.

Berg, Maxine, and Helen Clifford, eds. *Consumers and Luxury: Consumer Culture in Europe, 1650–1850*. Manchester: Manchester University Press, 1999.

Berg, Maxine, and Elizabeth Eger, eds. *Luxury in the Eighteenth Century: Debates, Desires and Delectable Goods*. Basingstoke, UK: Palgrave Macmillan, 2003.

Berg, Maxine, and Elizabeth Eger. "The Rise and Fall of the Luxury Debates." In *Luxury in the Eighteenth Century: Debates, Desires and Delectable Goods*, edited by Maxine Berg and Elizabeth Eger, 7–27. Basingstoke, UK: Palgrave Macmillan, 2003.

Berg, Maxine, and Elizabeth Eger. Introduction to *Luxury in the Eighteenth Century: Debates, Desires and Delectable Goods*, edited by Maxine Berg and Elizabeth Eger, 1–4. Basingstoke, UK: Palgrave Macmillan, 2003.

Bergmann, G. von. "Magerkeit und Magersucht." *Deutsche medizinische Wochenschrift* 60, no. 4–5 (January–February 1934): 123–26, 159–63.

———. "Das Streben nach Synthese wissenschaftlicher Medizin und naturlicher Heilweisen." *Deutsche medizinische Wochenschrift* 64, nos. 13–14 (March–April 1938): 455–57, 501–5.

Berkman, John M. "Anorexia Nervosa, Anorexia, Inanition and Low Basal Metabolic Rate." *American Journal of Medical Science* 180 (July 1930): 411–24.

Bernard, Claude. "De l'origine du sucre dans l'économie animale." *Archives générales de médecine* 18 (1848): 303–19.

———. "Des fonctions du cerveau." *Revue des deux mondes* 98 (1872): 373–85.

———. *Leçons de physiologie expérimentale appliquée à la médecine*. Paris: Baillière, 1855.

———. *Leçons sur les phénomènes de la vie communs aux animaux et aux végétaux*. 2 vols. Paris: J.-B. Baillière, 1878–79.

———. *Leçons sur les propriétés des tissus vivants*. Edited by Émile Alglave. Paris: Germer Baillière, 1866.

Berridge, Kent C. "Brain Reward Systems for Food Incentives and Hedonics in Normal Appetite and Eating Disorders." In Kirkham and Cooper, *Appetite and Body Weight*, 191–215.

Berry, Wendell. "The Pleasures of Eating." Center for Ecoliteracy. https://www.ecoliteracy.org/article/wendell-berry-pleasures-eating.

Bichat, Xavier. *Anatomie générale, appliquée à la physiologie et à la médecine*. 4 vols. Paris: Brosson, Gabon, 1801.

———. *Physiological Researches upon Life and Death*. Translated by Tobias Watkins. Philadelphia: Smith and Maxwell, 1809.

Bichat, F. X. [completed by M. F. R. Buisson and P.-J. Roux]. *Traité d'anatomie descriptive*. 5 vols. Paris: Brosson, Gabon, 1801–1803.

Bidder, Friedrich H., and Carl Schmidt. *Die Verdauungssaefte und der Stoffwechsel: Eine physiologisch-chemische Untersuchung*. Mitau: Reyher, 1852.

Bidder, F. H., and A. W. Volkmann. *Die Selbständigkeit des sympathischen Nervensystems durch anatomische Untersuchungen nachgewiesen*. Leipzig: Breitkopf und Härtel, 1842.

Biedl, Artur. *The Internal Secretory Organs: Their Physiology and Pathology*. Translated by Linda Forster. New York: William Wood, 1913.

Biltekoff, Charlotte. "The Terror Within: Obesity in Post 9/11 U.S. Life." *American Studies* 48, no. 3 (Fall 2007): 29–48.

Bing, Franklin C. "Friedrich Bidder (1810–1894) and Carl Schmidt (1822–1894)—A Biographical Sketch." *Journal of Nutrition* 103, no. 5 (May 1973): 637–48.

Binger, Carl. "Doctor Walter B. Cannon." *Psychosomatic Medicine* 8, no. 1 (1946): 60.

Binswanger, Otto. *Die Pathologie und Therapie der Neurasthenie. Vorlesungen für Studierende und Aerzte*. Jena: Gustav Fischer, 1896.

Bischoff, T. L. W., and Carl von Voit. *Die Gesetze der Ernährung des Fleischfressers durch neue Untersuchungen*. Leipzig: C. F. Winter, 1860.

Blanche, Antoine-Émile. "Cathétérisme oesaphagien chez les aliénés." Thesis, Faculté de Médecine de Paris, 1848.

Blatt, Harvey. *America's Food: What You Don't Know about What You Eat*. Cambridge, MA: MIT Press, 2008.

Bliss, Michael. *Harvey Cushing: A Life in Surgery*. New York: Oxford University Press, 2005.

Blondlot, Nicolas. *Traité analytique de la digestion considérée particulièrement dans l'homme et dans les animaux vertébrés*. Paris: Fortin, Masson, 1843.

Blumenbach, J. F. *The Institutions of Physiology*. Translated by John Elliotson. 3rd ed. London: Longman, Hurst, Rees, Orme, and Brown, 1820.

Boakes, Robert. *From Darwin to Behaviourism: Psychology and the Mind of Animals.* Cambridge: Cambridge University Press, 1984.

Bodnar, Richard J. "Endogenous Opioids and Feeding Behavior: A 30-Year Historical Perspective." *Peptides* 25 (2004): 697–725.

Bonner, Thomas Neville. *Becoming a Physician: Medical Education in Britain, France, Germany, and the United States, 1750–1945.* Baltimore: Johns Hopkins University Press, 1995.

Bont, Raf De. "Organisms in Their Milieu: Alfred Giard, His Pupils, and Early Ethology, 1870–1930." *Isis* 101, no. 1 (March 2010): 1–29.

Booth, David. "The Next Twenty Years: An Editorial Perspective." *Appetite* 34, no. 1 (2000): 1–3.

———. "Thirty Years of the Journal *Appetite*: The Citation Record." *Appetite* 54, no. 1 (February 2010): 1–4.

Bordeu, Théophile. "Recherches anatomiques sur les glandes, et sur leur action." In *Oeuvres complètes de Bordeu,* edited by A.-B. Richerand, 1:45–208. Paris: Caille, 1818.

———. "Recherches sur les maladies chroniques." In *Oeuvres complètes de Bordeu,* edited by A.-B. Richerand, 2:797–929. Paris: Caille, 1818.

Borell, Merriley. "Brown-Séquard's Organotherapy and Its Appearance in America at the End of the Nineteenth Century." *Bulletin of the History of Medicine* 50, no. 3 (1976): 309–20.

———. "Instrumentation and the Rise of Modern Physiology." *Science and Technology Studies* 5, no. 2 (1987): 53–62.

Bouchard, Charles. *Lectures on Auto-Intoxication in Disease; or Self-poisoning of the Individual.* Translated by Thomas Oliver. Philadelphia: F. A. Davis, 1894.

Bourdieu, Pierre. *Distinction: A Social Critique of the Judgement of Taste.* Translated by Richard Nice. Cambridge, MA: Harvard University Press, 1984.

Boussingault, J. B. *Rural Economy, in Its Relations with Chemistry, Physics, and Meteorology, or, Chemistry Applied to Agriculture.* Introduction and translation by George Law. New York: Orange Judd, 1865.

Bouvier, E.-L. *Habitudes et métamorphoses des insectes.* Paris: Ernest Flammarion, 1921.

Bowler, Peter J., and John Pickstone, eds. *The Modern Biological and Earth Sciences.* Vol. 6 of *The Cambridge History of Science.* Cambridge: Cambridge University Press, 2009.

———. Introduction to *Modern Biological and Earth Sciences,* edited by Peter. J. Bowler and John Pickstone, 1–12. Vol. 6 of *The Cambridge History of Science.* Cambridge: Cambridge University Press, 2009.

Boyd, John T. "An Inaugural Dissertation on Dyspepsia." Thesis, University of Pennsylvania, 1810.

Boylan, Michael. "The Digestive and 'Circulatory' Systems in Aristotle's Biology." *Journal of the History of Biology* 15, no. 1 (Spring 1982): 89–118.

Brachet, J.-L. *Recherches expérimentales sur les fonctions du système nerveux ganglionaire, et sur leur application à la pathologie.* Paris: Gabon, 1830.

Brambilla, Francesca. "1969–1989: Twenty Years of Life of the International Society of Psychoneuroendocrinology." *Psychoneuroendocrinology* 14, no. 4 (1989): 247–49.

Braunstein, Jean-François. *Broussais et le matérialisme: Médecine et philosophie au XIXe siècle*. Paris: Méridiens Klincksieck, 1986.

Bray, George A. "Commentary on Classics in Obesity 6. Science and Politics of Hunger." *Obesity Research* 1, no. 6 (1993): 489–93.

———. "Obesity: Historical Development of Scientific and Cultural Ideas." *International Journal of Obesity* 14 (1990): 909–26.

Breschet, Gilbert, Henri Milne-Edwards, and Pierre Vavasseur. *Recherches expérimentales sur les fonctions du système nerveux: 1er mémoire, De l'influence du système nerveux sur la digestion stomacale*. Paris: Migneret, 1823.

Brieger, Gert. "Dyspepsia: The American Disease? Needs and Opportunities for Research." In *Healing and History: Essays for George Rosen*, edited by Charles E. Rosenberg, 179–90. Folkestone, UK: Dawson Science History Publications, 1979.

Bringmann, W. G., N. J. Bringmann, and G. A. Ungerer. "The Establishment of Wundt's Laboratory: An Archival and Documentary Study." In *Wundt Studies: A Centennial Collection*, edited by W. G. Bringmann and Ryan D. Tweney, 123–57. Toronto: C. J. Hogrefe, 1980.

Brissaud, E., and A. Souques. "'Délire de maigreur' chez une hystérique." *Nouvelle iconographie de la Salpêtrière* 7 (1894): 327–37.

Brobeck, John R. "Neural Control of Hunger, Appetite, and Satiety." *Yale Journal of Biology and Medicine* 29, no. 6 (1957): 565–74.

———. "Neural Regulation of Food Intake." In Miner, *Regulation of Hunger and Appetite*, 44–55.

Brobeck, John R., Orr E. Reynolds, and Toby Appel, eds. *History of the American Physiological Society: The First Century, 1887–1987*. Bethesda, MD: American Physiological Society, 1987.

Brobeck, John R., Jay Tepperman, and C. N. H. Long. "Experimental Hypothalamic Hyperphagia in the Albino Rat." *Yale Journal of Biology and Medicine* 15, no. 6 (July 1943): 831–53.

Brockliss, L. W. B. "The Medico-Religious Universe of an Early Eighteenth Century Parisian Doctor: The Case of Philippe Hecquet." In *The Medical Revolution of the Seventeenth Century*, edited by R. K. French and Andrew Wear, 191–221. Cambridge: Cambridge University Press, 1989.

———. "The Moment of No Return: The University of Paris and the Death of Aristotelianism." *Science & Education* 15 (2006): 259–78.

Brockliss, Laurence, and Colin Jones. *The Medical World of Early Modern France*. Oxford, UK: Clarendon Press, 1997.

Broussais, F.-J.-V. *De l'irritation et de la folie*. Paris: Delaunay, 1828.

Brown, Theodore M. "Alan Gregg and the Rockefeller Foundation's Support of Franz Alexander's Psychosomatic Research." *Bulletin of the History of Medicine* 61, no. 2 (1987): 155–82.

———. *The Mechanical Philosophy and the "Animal Oeconomy."* New York: Arno Press, 1981.

———. "From Mechanism to Vitalism in Eighteenth-Century English Physiology." *Journal of the History of Biology* 7, no. 2 (Autumn 1974): 179–216.

———. "Medicine in the Shadow of the *Principia*." *Journal of the History of Ideas* 48, no. 4 (1987): 629–48.

Bruch, Hilde. *The Golden Cage: The Enigma of Anorexia Nervosa.* Cambridge, MA: Harvard University Press, 1978.

———. "Hunger and Instinct." *Journal of Nervous and Mental Diseases* 149, no. 2 (1969): 91–114.

———. *The Importance of Overweight.* New York: W. W. Norton, 1957.

———. "Obesity and Anorexia Nervosa." *Psychosomatics* 19, no. 4 (April 1978): 208–12.

———. "Obesity in Childhood: III. Physiologic and Psychologic Aspects of the Food Intake of Obese Children." *American Journal of Diseases of Children* 59, no. 4 (1940): 739–81.

———. "Psychiatric Aspects of Obesity." *Metabolism* 6 (1957): 461–65.

———. "Psychotherapeutic Problems in Eating Disorders." *Psychoanalytic Review* 50, no. 1 (1963): 573–87.

———. "Role of the Emotions in Hunger and Appetite." In Miner, *Regulation of Hunger and Appetite,* 68–75.

Bruch, Hilde, and Grace Touraine. "Obesity in Childhood: V. The Family Frame of Obese Children." *Psychosomatic Medicine* 2, no. 2 (April 1940): 141–206.

Brügelmann, Jan. "Observations on the Process of Medicalization in Germany, 1770–1830, Based on Medical Topographies." *Historical Reflections/Réflexions historiques* 9 (1982): 131–49.

Brumberg, Joan Jacobs. *Fasting Girls: The History of Anorexia Nervosa.* New York: Vintage/Random House, 2000.

Brush, Louise A. "Recent Literature Relative to the Psychiatric Aspects of Gastrointestinal Disorders–A Review." *Psychosomatic Medicine* 1, no. 3 (July 1939): 423–28.

Bryant-Waugh, Rachel. "Avoidant Restrictive Food Intake Disorder: An Illustrative Case Example." *International Journal of Eating Disorders* 46, no. 5 (2013): 420–23.

Buckley, Kerry W. *Mechanical Man: John Broadus Watson and the Beginnings of Behaviorism.* New York: Guilford Press, 1989.

Buffon, Georges Louis Leclerc de. *Histoire naturelle, générale et particulière: avec la description du Cabinet du Roy,* 15 vols. Paris: Imprimerie Royale, 1749–1767. http://www.buffon.cnrs.fr/.

Burkhardt, Richard W., Jr. "Le comportement animal et la biologie française 1920–1950." In *Les sciences biologiques et médicales en France 1920–1950: Actes du colloque de Dijon, 25–27 juin 1992,* edited by Claude Debru, Jean Gayon, and Jean-François Picard, 99–111. Paris: Centre national de la recherche scientifique, 1994.

———. "Darwin on Animal Behavior and Evolution." In *The Darwinian Heritage*, edited by David Kohn and Malcolm J. Kottler, 327–65. Princeton, NJ: Princeton University Press, 1985.

———. "Ethology, Natural History, the Life Sciences, and the Problem of Place." *Journal of the History of Biology* 32, no. 3 (Winter 1999): 489–508.

———. *Patterns of Behavior: Konrad Lorenz, Niko Tinbergen, and the Founding of Ethology*. Chicago: University of Chicago Press, 2005.

———. "The Zoological Philosophy of J. B. Lamarck." In J. B. Lamarck, *Zoological Philosophy: An Exposition with Regard to the Natural History of Animals*, translated by Hugh Elliot, xv–xxxix. Chicago: University of Chicago Press, 1984.

Burton, Richard D. E. *Holy Tears, Holy Blood: Women, Catholicism, and the Culture of Suffering in France, 1840–1970*. Ithaca, NY: Cornell University Press, 2004.

Buxton, Claude E., ed. *Points of View in the Modern History of Psychology*. New York: Academic Press, 1985.

Bynum, William F. "Nosology." In *Companion Encyclopedia of the History of Medicine*, edited by W. F. Bynum and Roy Porter, 2 vols., 1:335–56. London: Routledge, 1993.

Bynum, William F. "William Cullen (1710–1790), Chemist and Physician." *Oxford Dictionary of National Biography*. https://doi.org/10.1093/ref:odnb/6874.

Bynum, William F., and Roy Porter, eds. *Medicine and the Five Senses*. Cambridge: Cambridge University Press, 1993.

Bynum, W. F., Roy Porter, and Michael Shepherd, eds. *The Anatomy of Madness: Essays in the History of Psychiatry*. 3 vols. London: Tavistock, 1985.

Cabanis, P.-J.-G. *Oeuvres philosophiques de Cabanis*. 2 vols. Edited by Claude Lehec and Jean Cazeneuve. Paris: Presses Universitaires de France, 1956.

Camporesi, Pietro. *Bread of Dreams: Food and Fantasy in Early Modern Europe*. Translated by David Gentilcore. Chicago: University of Chicago Press, 1989.

Camus, Jean, and Gustave Roussy. "Les syndromes hypophysaires: Anatomie et physiologie pathologiques." *Revue neurologique* 38 (1922): 622–37.

Canguilhem, Georges. "La constitution de la physiologie comme science." In *Études d'histoire et de philosophie des sciences*, 226–73. Paris: J. Vrin, 1979.

———. "The Development of the Concept of Biological Regulation in the Eighteenth and Nineteenth Centuries." In *Ideology and Rationality in the History of the Life Sciences*, translated by Arthur Goldhammer, 81–102. Cambridge, MA: MIT Press, 1988.

———. *La formation du concept de réflexe aux XVIIe et XVIIIe siècles*. 2nd rev. ed. Paris: J. Vrin, 1977.

———. *Le normal et le pathologique*. Paris: Presses Universitaires de France, 1966.

———. "La santé, concept vulgaire et question philosophique." *Cahiers du séminaire de philosophie* 8 (1988): 119–33.

———. "La santé, concept vulgaire et question philosophique." Translated by Todd Meyers and Stefanos Geroulanos as "Health: Crude Concept and Philosophical Question." *Public Culture* 20, no. 3 (2008): 467–77.

Cannon, Walter B. *Bodily Changes in Pain, Hunger, Fear and Rage: An Account of Recent Researches into the Function of Emotional Excitement.* New York: D. Appleton and Co., 1915.

———. *A Consideration of the Nature of Hunger.* New York: [Harvey Society], 1911–1912.

———. *Digestion and Health.* New York: W. W. Norton, 1936.

———. "The Interrelations of Emotions as Suggested by Recent Physiological Researches." *American Journal of Psychology* 25, no. 2 (1914): 252–82.

———. *The Mechanical Factors of Digestion.* New York: Longmans, Green, 1911.

———. "Die Notfallsfunktionen des sympathico-adrenalen Systems." *Ergebnisse der Physiologie* 27 (1928): 380–406.

———. "Organization for Physiological Homeostasis." *Physiological Reviews* 9, no. 3 (1929): 399–431.

———. *The Wisdom of the Body.* New York: W. W. Norton, 1932.

Cannon, Walter B., and A. L. Washburn. "An Explanation of Hunger." *American Journal of Physiology* 29, no. 5 (1912): 441–54.

Cantor, David, ed. *Reinventing Hippocrates.* Aldershot, UK: Ashgate, 2002.

Capshew, James H., and Karen A. Rader. "Big Science: Price to the Present." *Osiris* 7 (1992): 2–25.

Carlson, A. J. "Contributions to the Physiology of the Stomach.—I. The Character of the Movements of the Empty Stomach in Man." *American Journal of Physiology* 31, no. 3 (1912): 151–68.

———. "Contributions to the Physiology of the Stomach.—II. The Relation between the Contractions of the Empty Stomach and the Sensation of Hunger." *American Journal of Physiology* 31 (1913): 175–92.

———. *The Control of Hunger in Health and Disease.* Chicago: University of Chicago Press, 1916.

Carlson, A. J., and H. Ginsburg. "Contributions to the Physiology of the Stomach. XXIV. The Tonus and Hunger Contractions of the Stomach of the New-Born." *American Journal of Physiology* 38, no. 1 (July 1, 1915): 29–32.

Carpenter, Kenneth J. "The Life and Times of W. O. Atwater (1844–1907)." *Journal of Nutrition* 124, supplement 9 (September 1994): 1707S–1714S.

Carrier, Jean. "L'anorexie mentale: Trouble instinctivo-affectif." Thesis, Faculté de médecine et de pharmacie de Lyon, 1939.

Carson, Anne. "Putting Her in Her Place: Woman, Dirt, and Desire." In *Before Sexuality: The Construction of Erotic Experience in the Ancient Greek World,* edited by David M. Halperin, John J. Winkler, and Froma I. Zeitlin, 135–69. Princeton, NJ: Princeton University Press, 1990.

Chadarevian, Soraya de. "Graphical Method and Discipline: Self-Recording Instruments in Nineteenth-Century Physiology." *Studies in History and Philosophy of Science* 24, no. 2 (1993): 267–91.

Charcot, Jean-Martin. "De l'isolement dans le traitement de l'hystérie." *Le Progrès médical* 13, 2nd ser., 1 (February 28, 1885): 161–64.

Chatel, John C., and Roger Peele. "A Centennial Review of Neurasthenia." *American Journal of Psychiatry* 126, no. 10 (1970): 1404–13.

Cheung, Tobias. "Hermann Samuel Reimarus' Theorie der 'Lebensarten' und 'Triebe.'" *Sudhoffs Archiv* 90, no. 2 (2006): 143–60.

Cheyne, George. *The English Malady; or, A Treatise of Nervous Diseases of All Kinds, as Spleen, Vapours, Lowness of Spirits, Hypochondriacal, and Hystrial Distempers, etc.* London: George Strahan; Bath: James Leake, 1733.

Chomel, A.-F. *Des dyspepsies*. Paris: Victor Masson, 1857.

Christie, William F. "Corpulence: Its Causes and Cure." *Lancet* 215, no. 5564 (April 19, 1930): 894–96.

Clarke, A. Grey, and F. Prescott. "Studies in Vitamin B Deficiency with Special Reference to Mental and Oral Manifestations." *British Medical Journal* 2, no. 4320 (October 23, 1943): 503–5.

Clarke, Edwin, and L. S. Jacyna. *Nineteenth-Century Origins of Neuroscientific Concepts*. Berkeley: University of California Press, 1987.

Clarke, Edwin, and C. D. O'Malley. *The Human Brain and Spinal Cord: A Historical Study Illustrated by Writings from Antiquity to the Twentieth Century*. Berkeley: University of California Press, 1968.

"Clinical Case No. 442, Fever and Loss of Appetite." *Case Reports of the Children's Memorial Hospital, Chicago* 6, no. 12 (1947): 1049–54.

Coleman, William. *Biology in the Nineteenth Century: Problems of Form, Function, and Transformation*. New York: Cambridge University Press, 1977.

———. "Health and Hygiene in the *Encyclopédie*: A Medical Doctrine for the Bourgeoisie." *Journal of the History of Medicine and Allied Sciences* 29 (1974): 399–421.

Coleman, William B., and Frederic L. Holmes, eds. *The Investigative Enterprise: Experimental Physiology in Nineteenth-Century Medicine*. Berkeley: University of California Press, 1988.

Colman, Eric. "Anorectics on Trial: A Half Century of Federal Regulation of Prescription Appetite Suppressants." *Annals of Internal Medicine* 143, no. 5 (September 2005): 380–85.

Conry, Yvette. *L'introduction du darwinisme en France au XIXe siècle*. Paris: J. Vrin, 1974.

Cooter, Roger, and Claudia Stein. *Writing History in the Age of Biomedicine*. New Haven, CT: Yale University Press, 2013.

Counihan, Carole, and Penny Van Esterik, eds. *Food and Culture: A Reader*. New York: Routledge, 1997.

Craig, Wallace. "Appetites and Aversions as Constituents of Instincts." *Biological Bulletin of the Marine Biological Laboratory* 34 (1918): 91–107.

———. "Observations on Doves Learning to Drink." *Journal of Animal Behavior* 2, no. 4 (1912): 273–79.

Cross, Stephen J., and William R. Albury. "Walter B. Cannon, L. J. Henderson, and the Organic Analogy." *Osiris*, 2nd ser., 3, no. 1 (1987): 165–92.

Cullather, Nick. "The Foreign Policy of the Calorie." *American Historical Review* 112, no. 2 (2007): 337–64.

Cullen, William. *Nosology, or, A Systematic Arrangement of Diseases, by classes, orders, genera, and species; with the Distinguishing Characters of Each, and outlines of the systems of Sauvages, Linnaeus, Vogel, Sagar, and Macbride.* Translated from Latin of William Cullen. Edinburgh: Printed by C. Stewart for William Creech, 1800.

"Cullen Project: The Consultation Letters of Dr William Cullen (1710–1790) at the Royal College of Physicians of Edinburgh." http://www.cullenproject.ac.uk.

Culotta, Charles A. "Tissue Oxidation and Theoretical Physiology: Bernard, Ludwig, and Pflüger." *Bulletin of the History of Medicine* 44 (1970): 109–40.

Cunningham, Andrew, and Perry Williams, eds. *The Laboratory Revolution in Medicine.* Cambridge: Cambridge University Press, 1992.

Current Trends in Psychology and the Behavioral Sciences: Eighth Annual Conference on Current Trends in Psychology Held at the University of Pittsburgh, 1954. Pittsburgh: University of Pittsburgh Press, 1954.

Cushing, Harvey. *The Pituitary Body and Its Disorders: Clinical States Produced by Disorders of the Hypophysis Cerebri.* Philadelphia: J. B. Lippincott, 1912.

Dacome, Lucia. "Living with the Chair: Private Excreta, Collective Health and Medical Authority in the Eighteenth Century." *History of Science* 39 (2001): 467–500.

———. "Useless and Pernicious Matter: Corpulence in Eighteenth-Century England." In Forth and Carden-Coyne, *Cultures of the Abdomen*, 185–204.

Dalby, Andrew. *Siren Feasts: A History of Food and Gastronomy in Greece.* London: Routledge, 1996.

Danziger, Kurt. *Constructing the Subject: Historical Origins of Psychological Research.* Cambridge: Cambridge University Press, 1990.

———. "The History of Introspection Reconsidered." *Journal of the History of the Behavioral Sciences* 16 (1980): 241–62.

———. "Mid-Nineteenth-Century British Psycho-Physiology: A Neglected Chapter in the History of Psychology." In *The Problematic Science: Psychology in Nineteenth-Century Thought*, edited by William R. Woodward and Mitchell G. Ash, 119–46. New York: Praeger, 1982.

Darwin, Charles. *The Origin of Species by Means of Natural Selection.* 6th ed. London: John Murray, 1873.

Darwin, Erasmus. *Phytologia; or, The philosophy of agriculture and gardening. With the theory of draining morasses, and with an improved construction of the drill plow.* London: Printed for J. Johnson, St. Paul's Churchyard, 1800.

———. *Zoonomia, or The Laws of Organic Life.* 2nd ed., corrected. 2 vols. London: Printed for J. Johnson, in St. Paul's Church-Yard, 1796.

Daston, Lorraine. "Attention and the Values of Nature in the Enlightenment." In *Moral Authority of Nature*, edited by Daston and Vidal, 100–126.

———, ed. *Biographies of Scientific Objects.* Chicago: University of Chicago Press, 2000.

———. "British Responses to Psycho-physiology, 1860–1900." *Isis* 69, no. 2 (June 1978): 192–208.

———. "Introduction: The Coming into Being of Scientific Objects." In *Biographies of Scientific Objects*, edited by Daston, 1–14.

Daston, Lorraine, and Fernando Vidal, eds. *The Moral Authority of Nature*. Chicago: University of Chicago Press, 2004.

Davidovici, Salomon. *Obésité et gestation*. Paris: Jouve, 1933.

Davidson, L. S. P. "Refresher Course for General Practitioners: The Use and Abuse of Vitamins." *British Medical Journal* 2, no. 4635 (November 5, 1949): 1036–38.

Davis, Clara M. "Self Selection of Diet by Newly Weaned Infants: An Experimental Study." *American Journal of Diseases of Children* 36, no. 4 (1928): 651–79.

Dawes, Laura. *Childhood Obesity in America: Biography of an Epidemic*. Cambridge, MA: Harvard University Press, 2014.

Day, J., A. Ternouth, and D. A. Collier. "Eating Disorders and Obesity: Two Sides of the Same Coin?" *Epidemiologia e psichiatria sociale* 18, no. 2 (April–June 2009): 96–100.

Debus, Allen. *The French Paracelsians: The Chemical Challenge to Medical and Scientific Tradition in Early Modern France*. New York: Cambridge University Press, 1991.

Delaporte, François. *Nature's Second Kingdom: Explorations of Vegetality in the Eighteenth Century*. Cambridge, MA: MIT Press, 1982.

Demange, E. "Obésité." In Raige-Delorme and Dechambre, *Dictionnaire encyclopédique des sciences médicales*, 2nd ser., 14:5–38.

Deniau, Lucien. *De l'hystérie gastrique*. Paris: O. Doin, 1883.

Descartes, René. *Treatise of Man*. Translation and commentary by Thomas Steele Hall. Cambridge, MA: Harvard University Press, 1972.

Desmond, Adrian. *The Politics of Evolution: Morphology, Medicine and Reform in Radical London*. Chicago: University of Chicago Press, 1989.

Desportes, Émile. "Du refus de manger chez les aliénés." Thesis, Faculté de Médecine de Paris, 1864.

Diderot, Denis, and Jean Le Rond d'Alembert, eds. *Encyclopédie, ou Dictionnaire raisonné des sciences, des arts et des métiers*. 17 vols. Paris: Briasson, 1751–1777.

Dowbiggin, Ian. *Inheriting Madness: Professionalization and Psychiatric Knowledge in Nineteenth-Century France*. Berkeley: University of California Press, 1991.

Drickamer, Lee C., and Charles T. Snowdon. "Defining the Goals, Approaches, and Methods." In *Foundations of Animal Behavior*, edited by Houck and Drickamer, 71–86.

Duchesneau, François. *La physiologie des Lumières: Empirisme, modèles et théories*. The Hague: Martinus Nijhoff, 1982.

———. "Vitalism in Late Eighteenth-Century Physiology: The Cases of Barthez, Blumenbach and John Hunter." In *William Hunter and the Eighteenth-Century Medical World*, edited by W. F. Bynum and Roy Porter, 259–95. Cambridge: Cambridge University Press, 1985.

Duchet, Michèle. *Anthropologie et histoire au siècle des lumières: Buffon, Voltaire, Rousseau, Helvétius, Diderot*. Paris: Francois Maspéro, 1971.

Duffin, Jacalyn. "Jodocus Lommius's Little Golden Book and the History of Diagnostic Semeiology." *Journal of the History of Medicine and Allied Sciences* 61, no. 3 (2006): 249–87.

Dumas, C.-L. *Principes de physiologie, ou introduction à la science expérimentale, philosophique et médicale de l'homme vivant*. 4 vols. Paris: Crapelet, Déterville, 1800–1803.

———. *Principes de physiologie, ou introduction à la science expérimentale, philosophique et médicale de l'homme vivant*. 2nd ed. 4 vols. Paris: Déterville, 1806.

Duncker, Karl. "Experimental Modification of Children's Food Preferences through Social Suggestion." *Journal of Abnormal and Social Psychology* 33, no. 4 (1938): 489–507.

Durig, Arnold. *Appetit: Vortrag, gehalten in der Gesellschaft der Aerzte in Wien am 9. Januar 1925*. Vienna: Julius Springer, 1925.

Dyer, Fred C., and H. Jane Brockmann. "Biology of the *Umwelt*." In *Foundations of Animal Behavior*, edited by Houck and Drickamer, 529–38.

Ebstein, Wilhelm. *Corpulence and Its Treatment on Physiological Principles*. Translated by A. H. Keane. London: H. Grevel, 1884.

Eckart, Wolfgang U. "'Und setzet eure Worte nicht auf Schrauben': Medizinische Semiotik vom Ende des 18. bis zum Beginn des 20. Jahrhunderts—Gegenstand und Forschung." *Berichte zur Wissenschaftsgeschichte* 19 (1996): 1–18.

Elliott, Paul. "Erasmus Darwin, Herbert Spencer, and the Origins of the Evolutionary Worldview in British Provincial Scientific Culture, 1770–1850." *Isis* 94, no. 1 (March 2003): 1–29.

Engelhardt, Dietrich von. "Hunger und Appetit: Essen und Trinken im System der Diätetik–Kulturhistorische Perspektiven." In *Kulturthema Essen: Ansichten und Problemfelder*, edited by Alois Wierlacher, Gerhard Neumann, and Hans Jürgen Teuteberg, 137–49. Berlin: Akademie Verlag, 1993.

Engstrom, Eric J. *Clinical Psychiatry in Imperial Germany: A History of Psychiatric Practice*. Ithaca, NY: Cornell University Press, 2003.

Engstrom, Eric J., and Volker Roelcke, eds. *Psychiatrie im 19. Jahrhundert: Forschungen zur Geschichte von psychiatrischen Institutionen, Debatten und Praktiken im deutschen Sprachraum*. Basel: Schwabe, 2003.

Escamilla, Roberto F., and H. Lisser. "Simmonds' Disease: A Clinical Study with Review of the Literature; Differentiation from Anorexia Nervosa by Statistical Analysis of 595 Cases, 101 of Which Were Proved Pathologically." *Journal of Clinical Endocrinology* 2, no. 2 (February 1942): 65–96.

Espinas, Alfred. *Des sociétés animales: Étude de psychologie comparée*. Paris: Germer Baillière, 1877.

Esquirol, J.-E.-D. "Folie." In Adelon and Mérat, *Dictionnaire des sciences médicales*, 16:151–240.

———. *Des maladies mentales, considérées sous les rapports médical, hygiénique et médico-légal* (1838). 2 vols. New York: Arno Press, 1976.

———. *Des passions considérées comme causes, symptômes et moyens curatifs de l'aliénation mentale* (1805). Paris: Librairie des deux-mondes, 1980.

Ettmüller, Michael. *Etmullerus abridg'd: or, A compleat system of the theory and practice of physic. Being a description of all diseases incident to men, women and children. With an account of their causes, symptoms, and most approved methods of cure, physical and chirurgical. To which is prefix'd a short view of the animal and vital functions; and the several vertues and classes of med'cines.* London: Printed for E. Harris, 1699.

Eulenberg, Albert, ed. *Real-Encyclopädie der gesammten Heilkunde: Medicinisch-chirurgisches Handwörterbuch für praktische Aerzte.* 3rd ed. 26 vols. Vienna: Urban and Schwarzenberg, 1894–1901.

Eulner, Hans Heinz. "Die Lehre von der Ernährung im Universitätsunterricht." In Heischkel-Artelt, *Ernährung und Ernährungslehre im 19. Jahrhundert*, 76–97.

Fabre, J. H. *Fabre's Book of Insects.* Retold from Alexander Teixeira de Mattos's translation of Fabre's "Souvenirs entomologiques" by Mrs. Rodolph Stawell. Illustrated by E. J. Detmold. London: Hodder and Stoughton, [1921].

"Faim, appétit (Gram. Syn.)." In Diderot and d'Alembert, *Encyclopédie*, 6:373.

Falta, Wilhelm. *The Ductless Glandular Diseases.* Translated by Milton K. Meyers. Foreword by Archibald E. Garrod. Philadelphia: P. Blakiston's, 1915.

Farber, Paul L. "Discussion Paper: The Transformation of Natural History in the Nineteenth Century." *Journal of the History of Biology* 15, no. 1 (Spring 1982): 145–52.

Farquhar, Judith. *Appetites: Food and Sex in Postsocialist China.* Durham, NC: Duke University Press, 2002.

Feinman, Richard D., and Eugene J. Fine. "'A Calorie Is a Calorie' Violates the Second Law of Thermodynamics." *Nutrition Journal* 3, no. 9 (2004): https://doi.org/10.1186/1475-2891-3-9.

Ferenczi, Sándor. "Disgust for Breakfast." In *Further Contributions to the Theory and Technique of Psycho-analysis*, compiled by John Rickman, translated by Jane Isabel Suttie, 26. New York: Boni and Liveright, 1927.

———. "The Phenomena of Hysterical Materialization: Thoughts on the Conception of Hysterical Conversion and Symbolism." In *Further Contributions to the Theory and Technique of Psycho-analysis*, compiled by John Rickman, translated by Jane Isabel Suttie, 89–103. New York: Boni and Liveright, 1927.

Ferrier, David. *The Functions of the Brain.* New York: Putnam's, 1876.

Finlay, Mark R. "Early Marketing of the Theory of Nutrition: The Science and Culture of Liebig's Extract of Meat." In Kamminga and Cunningham, *Science and Culture of Nutrition*, 48–74.

Fischer, Oskar. "Der Einfluss des Appetits auf die Magentätigkeit und seine Bedeutung für die funktionelle Magendiagnostik." *Münchener medizinische Wochenschrift* 58, no. 7 (February 14, 1911): 345–48.

Fissell, Mary E. "The Disappearance of the Patient's Narrative and the Invention of Hospital Medicine." In *British Medicine in an Age of Reform*, edited by Roger French and Andrew Wear, 92–109. London: Routledge, 1991.

Flandrin, Jean-Louis. "Dietary Choices and Culinary Technique, 1500–1800." In Flandrin and Montanari, *Food: A Culinary History from Antiquity to the Present*, 403–17.

———. "Introduction: The Early Modern Period." In Flandrin and Montanari, *Food: A Culinary History from Antiquity to the Present*, 357.

Flandrin, Jean-Louis, and Massimo Montanari, eds. *Food: A Culinary History from Antiquity to the Present*. Translated by Albert Sonnenfeld. New York: Columbia University Press, 1999.

Flemming, Rebecca, and Ann Ellis Hanson. "Hippocrates' *Peri Partheniôn* ('Diseases of Young Girls'): Text and Translation." *Early Science and Medicine* 3, no. 3 (1998): 241–52.

Forth, Christopher E., and Ana Carden-Coyne, eds. *Cultures of the Abdomen: Diet, Digestion, and Fat in the Modern World*. New York: Palgrave Macmillan, 2005.

Foucault, Michel. *The History of Sexuality*. Vol. 1, *An Introduction*. Translated by Robert Hurley. New York: Vintage, 1990.

———. *The History of Sexuality*. Vol. 2, *The Use of Pleasure*. Translated by Robert Hurley. New York: Vintage Books, 1990.

Frahm, H. "Simmonds'sche Krankheit, Sheehan-Syndrom und nervöse Magersucht. Ätiologie und Differentialdiagnose." *Die medizinische Welt* 5 (January 29, 1966): 235–38, 243.

Freedman, Paul, Joyce E. Chaplin, and Ken Albala, eds. *Food in Time and Place: The American Historical Association Companion to Food History*. Oakland: University of California Press, 2014.

French, R. K. "Sauvages, Whytt and the Motion of the Heart: Aspects of Eighteenth-Century Animism." *Clio medica* 7 (1972): 35–54.

Freud, Sigmund. "Draft G—Melancholia." In *Standard Edition*, 1:200–206.

———. "From the History of An Infantile Neurosis." In *Standard Edition*, 17:7–122.

———. *The Standard Edition of the Complete Psychological Works of Sigmund Freud*. 24 vols. Edited by J. Strachey. London: Hogarth Press, 1953–1974.

———. "Three Contributions to the Theory of Sex." In *The Basic Writings of Sigmund Freud*. Translated, edited, and introduced by A. A. Brill, 551–629. New York: Modern Library, 1938.

Fröhlich, Alfred. "Ein Fall von Tumor der Hypophysis cerebri ohne Akromegalie." *Wiener klinische Rundschau* 15 (1901): 883–86, 906–8.

Fulton, J. F. "Some Functions of the Cerebral Cortex. Lecture I. Autonomic Representation in the Cerebral Cortex." *Journal of the Michigan State Medical Society* 33, no. 4 (April 1934): 175–82.

Fulton, J. F., C. F. Jacobsen, and Margaret A. Kennard. "A Note Concerning the Relation of the Frontal Lobes to Posture and Forced Grasping in Monkeys." *Brain* 55, no. 4 (1932): 524–36.

Galen. *Galen on Food and Diet*. Edited and translated by Mark Grant. London: Routledge, 2000.

———. *On the Usefulness of the Parts of the Body*. 2 vols. Translated and introduced by Margaret Tallmadge May. Ithaca, NY: Cornell University Press, 1968.

———. *A Translation of Galen's Hygiene (De sanitate tuenda)*. Translated by Robert Montraville Green. Introduction by Henry E. Sigerist. Springfield, IL: Charles C. Thomas, 1951.

Galison, Peter, and Bruce Hevly, eds. *Big Science: The Growth of Large-Scale Research*. Stanford, CA: Stanford University Press, 1992.

Gall, F.-J., and G. Spurzheim, *Anatomie et physiologie du système nerveux en général, et du cerveau en particulier*. 5 vols. Paris: F. Schoell, 1810–1819.

Gallouin, F., and J. Le Magnen. "Evolution historique des concepts de faim, satiété et appétits." *Reproduction, Nutrition, Développement* 27, no. 1B (1987): 109–28.

Gard, Michael. *The End of the Obesity Epidemic*. London: Routledge, 2011.

Gard, Michael, and Jan Wright. *The Obesity Epidemic: Science, Morality and Ideology*. London: Routledge, 2005.

Garfinkle, Norton. "Science and Religion in England, 1790–1800: The Critical Response to the Work of Erasmus Darwin." *Journal of the History of Ideas* 16, no. 3 (June 1955): 376–88.

Geist-Hofman, A. M. "Jacob Moleschott." In Gillispie, *Dictionary of Scientific Biography*, 9:456–57.

Georget, E.-J.-G. *De la folie: Considérations sur cette maladie* (1820). New York: Arno Press, 1976.

———. *De la physiologie du système nerveux, et spécialement du cerveau: Recherches sur les maladies nerveuses en général, et en particulier sur le siège, la nature et le traitement de l'hystérie, de l'hypochondrie, de l'épilepsie et de l'asthme convulsive*. 2 vols. Paris: J.-B. Baillière, 1821.

Geuter, Ulfried. *The Professionalization of Psychology in Nazi Germany*. Translated by Richard J. Holmes. Cambridge: Cambridge University Press, 1992.

Geyer-Kordesch, Johanna. "Georg Ernst Stahl's Radical Pietist Medicine and Its Influence on the German Enlightenment." In *The Medical Enlightenment of the Eighteenth Century*, edited by Andrew Cunningham and R. K. French, 67–87. New York: Cambridge University Press, 1990.

Giard, Alfred. "Les origines de l'amour maternel." *Bulletin de l'Institut Général Psychologique* 5 (January–February 1905): 3–32.

———. "Sur les parasites Bopyriens et la castration parasitaire." *Comptes rendus de la Société de Biologie* 39 (1887): 371–73.

Gibson, Mary E. "Alfred Fröhlich and his Correspondence with Sir Ronald Ross." *Journal of Medical Biography* 11 (2003): 41–46.

Gillispie, Charles C., ed. *Dictionary of Scientific Biography*. 18 vols. New York: Scribner, 1970–1980.

Gilman, Sander L. Introduction to *Diets and Dieting: A Cultural Encyclopedia*, ix–xii. New York: Routledge, 2008.

Goetz, Christopher G., Michel Bonduelle, and Toby Gelfand. *Charcot: Constructing Neurology*. New York: Oxford University Press, 1995.

Goldstein, Jan. *Console and Classify: The French Psychiatric Profession in the Nineteenth Century*. New York: Cambridge University Press, 1987.

———. "Mutations of the Self in Old Regime and Postrevolutionary France: From *Ame* to *Moi* to *Le Moi*." In Daston, *Biographies of Scientific Objects*, 86–116.

———. *The Post-Revolutionary Self: Politics and Psyche in France, 1750–1850*. Cambridge, MA: Harvard University Press, 2005.

Goodman, Catherine. "Monell Chemical Senses Center." *Nature Chemical Biology* 5, no. 1 (January 2009): 2.

Goody, Jack. "Industrial Food: Towards the Development of a World Cuisine." In Counihan and Van Esterik, *Food and Culture: A Reader*, 72–90.

Gottesfeld, Ben Harvey, and Amando Caiuby Novaes. "Narco-Analysis and Sub-Shock Insulin in the Treatment of Anorexia Nervosa." *Digest of Neurology and Psychiatry* 13 (1945): 486–94.

Gottesfeld, Harry. "Body and Self-Cathexis of Super-Obese Patients." *Journal of Psychosomatic Research* 6, no. 3 (July–September 1962): 177–83.

Grange, Jules. "Essai sur les névroses de l'estomac." Thesis, Faculté de Médecine de Paris, 1845.

Granger, Benjamin. "Account of the Second Watch of the Reputed Fasting-Woman." *Edinburgh Medical and Surgical Journal* 9 (1813): 323–26.

———. "On Unusual Cases of Anorexy." *Edinburgh Medical and Surgical Journal* 9 (1813): 157–59.

———. "Some Account of the Fasting Woman at Titbury, Who Has at Present Lived above Two Years without Food." *Edinburgh Medical and Surgical Journal* 5 (1809): 319–26.

Gray, Jeffrey A. *Ivan Pavlov*. New York: Viking Press, 1979.

Grayzel, Susan. *Women's Identities at War: Gender, Motherhood, and Politics in Britain and France during the First World War*. Chapel Hill: University of North Carolina Press, 1999.

Gregory, Frederick. *Scientific Materialism in Nineteenth Century Germany*. Dordrecht, The Netherlands: D. Reidel, 1977.

Gregory, R. A. "The Gastrointestinal Hormones." In *The Pursuit of Nature: Informal Essays on the History of Physiology*, edited by A. L. Hodgkin, A. F. Huxley, W. Feldberg, W. A. H. Rushton, R. A. Gregory, and R. A. McCance, 105–32. Cambridge: Cambridge University Press, 1977.

Gremillion, Helen. *Feeding Anorexia: Gender and Power at a Treatment Center*. Durham, NC: Duke University Press, 2003.

Grimaud, J.-C.-M.-G. *Cours complet de physiologie*. 2nd rev. ed. 2 vols. Edited by Étienne Lanthois. Paris: Egron, 1824.

———. *Mémoire sur la nutrition*. Montpellier: J. Martel, 1787.

———. *Second mémoire sur la nutrition: Supplément au Mémoire françois sur la nutrition*. Montpellier: Jean Martel, 1789.

Grmek, Mirko Drazen. "Examen critique de la genèse d'une grande découverte: 'La piqûre diabétique' de Claude Bernard." *Clio medica* 1 (1966): 341–50.

———. *Le legs de Claude Bernard*. Paris: Fayard, 1997.

Grondona, Felice. "Des expériments de Lazzaro Spallanzani sur la digestion aux applications thérapeutiques du suc gastrique." *Clio medica* 8, no. 4 (1973): 285–94.

Groopman, Jerome. *How Doctors Think*. New York: Houghton Mifflin, 2008.

Gross, Michael. "The Lessened Locus of Feelings: A Transformation in French Physiology in the Early Nineteenth Century." *Journal of the History of Biology* 12, no. 2 (1979): 231–71.

Grossman, Morton I. "Integration of Current Views on the Regulation of Hunger and Appetite." In Miner, *Regulation of Hunger and Appetite*, 76–91.

Grossman, M. I., G. M. Cummins, and A. C. Ivy. "The Effect of Insulin on Food Intake after Vagotomy and Sympathectomy." *American Journal of Physiology* 149 (April 1, 1947): 100–102.

Grossman, M. I., and I. F. Stein. "Vagotomy and the Hunger-Producing Action of Insulin in Man." *Journal of Applied Physiology* 1, no. 4 (October 1948): 263–69.

Grubman, Jessica, and Afsaneh Barzi. "Lost in Translation: The Patient-Physician Relationship in the Molecular Era." *Journal of Palliative Medicine* 18, no. 11 (2015): 987–88.

Gruman, Gerald J. *A History of Ideas about the Prolongation of Life*. Philadelphia: American Philosophical Society, 1966.

Guerrini, Anita. "Archibald Pitcairne and Newtonian Medicine." *Medical History* 31, no. 1 (1987): 70–83.

———. *Obesity and Depression in the Enlightenment: The Life and Times of George Cheyne*. Norman: University of Oklahoma Press, 2000.

Gull, William Withey. "Anorexia Nervosa (Apepsia Hysterica, Anorexia Hysterica)." *Transactions of the Clinical Society of London* 7 (1874): 22–28.

Gusdorf, Georges. *Les sciences humaines et la pensée occidentale*. Vol. 7, *Naissance de la conscience romantique au siècle des Lumières*. Paris: Payot, 1976.

Habermas, Tilmann. "Die Anorexia nervosa (Magersucht) in der deutschsprachigen medizinischen Literatur von 1900 bis 1945: Die Rolle der Magersucht in der Entstehung der Psychosomatik." *Sudhoffs Archiv: Zeitschrift für Wissenschaftsgeschichte* 76, no. 1 (1992): 37–62.

———. *Heißhunger: Historische Bedingungen der Bulimia nervosa*. Frankfurt: Fischer, 1990.

———. "Historical Continuities and Discontinuities between Religious and Medical Interpretations of Extreme Fasting: The Background to Giovanni Brugnoli's Description of Two Cases of Anorexia Nervosa in 1875." *History of Psychiatry* 3 (1992): 431–55.

———. "The Role of Psychiatric and Medical Traditions in the Discovery and Description of Anorexia Nervosa in France, Germany, and Italy, 1873–1918." *Journal of Nervous and Mental Disease* 179, no. 6 (1991): 360–65.

Habermas, Tilmann, W. Vandereycken, R. van Deth, and R. Meermann. "Die Anorexia nervosa in der deutschsprachigen Literatur des 19. Jahrhunderts." *Psychotherapie, Psychosomatik, medizinische Psychologie* 40 (1990): 466–73.

Haffter, Carl. "Die Entstehung des Begriffs der Zivilisationskrankheiten." *Gesnerus* 36, nos. 3–4 (1979): 228–37.

Hagendantz, Heinz, and S. Proger. "Anorexia Nervosa or Hypopituitarism?" *Journal of the American Medical Association* 114, no. 20 (May 18, 1940): 1973–83.

Hagner, Michael. "Scientific Medicine." In *From Natural Philosophy to the Sciences: Writing the History of Nineteenth-Century Science*, edited by David Cahan, 49–87. Chicago: University of Chicago Press, 2003.

Hale, Nathan G., Jr. *The Rise and Crisis of Psychoanalysis in the United States: Freud and the Americans, 1917–1985*. New York: Oxford University Press, 1995.

Hall, Diana Long. "The Critic and the Advocate: Contrasting British Views on the State of Endocrinology in the Early 1920s." *Journal of the History of Biology* 9, no. 2 (1976): 269–85.

Hall, Marshall. *Commentaries on Some of the More Important of the Diseases of Females*. London: Longman, Rees, Orme, Brown, and Green, 1827.

Hall, Thomas S. *History of General Physiology, 600 B.C. to A.D. 1900*. Vol. 1, *From Pre-Socratic Times to the Enlightenment*. Chicago: University of Chicago Press, 1975.

Haller, Albrecht von. *Anfangsgründe der Phisiologie des menschlichen Körpers*. 8 vols. Edited by Johann Samuel Haller. Berlin: Christian Friedrich Voss, 1759–1776.

———. *First Lines of Physiology*. 1786. 2 vols. Introduction to reprint by Lester S. King. New York: Johnson, 1966.

Halliday, James L. "The Rising Incidence of Psycho-somatic Illness." *British Medical Journal* 2, no. 4043 (July 2, 1938): 11–14.

Hamilton, Max. *Psychosomatics*. Foreword by D. R. MacAlman. New York: John Wiley, 1955.

Hammond, William A. *Fasting Girls: Their Physiology and Pathology*. New York: G. P. Putnam's, 1879.

Hannaway, Caroline, and Ann La Berge, eds. *Constructing Paris Medicine*. Amsterdam: Rodopi, 1998.

Harrington, Anne. *The Cure Within: A History of Mind-Body Medicine*. New York: W. W. Norton, 2008.

———. *Reenchanted Science: Holism in German Culture from Wilhelm II to Hitler*. Princeton, NJ: Princeton University Press, 1996.

Harris, John. "An Essay on Dyspepsia." Thesis, University of Pennsylvania, 1816.

Harris, Ruth. "Murder under Hypnosis in the Case of Gabrielle Bompard: Psychiatry in the Courtroom in Belle Epoque Paris." In *The Anatomy of Madness: Essays in the History of Psychiatry*, vol. 2, *Institutions and Society*, edited by W. F. Bynum, Roy Porter, and Michael Shepherd, 197–241. London: Tavistock, 1985.

Hartley, David. *Observations on Man, His Frame, His Duty, and His Expectations*. 1749. 3 vols. London: J. Johnson, 1791.

Hatfield, Gary. "Remaking the Science of Mind: Psychology as Natural Science." In *Inventing Human Science: Eighteenth-Century Domains*, edited by Christopher

Fox, Roy Porter, and Robert Wokler, 184–231. Berkeley: University of California Press, 1995.

Hau, Michael. *The Cult of Health and Beauty in Germany: A Social History, 1890–1930*. Chicago: University of Chicago Press, 2003.

Haymaker, Webb, and Francis Schiller. "Moritz Schiff (1823–1896)." In *The Founders of Neurology: One Hundred and Forty-Six Biographical Sketches by Eighty-Eight Authors*, edited by Francis Schiller and Webb Haymaker, 258–64. Springfield, IL: Thomas, 1970.

Hayward, Rhodri. "Enduring Emotions: James L. Halliday and the Invention of the Psychosocial." *Isis* 100, no. 4 (December 2009): 827–38.

Heidenhain, Rudolf. *Handbuch der Physiologie der Absonderung und Aufsaugung*. Vol. 5, pt. 1. *Handbuch der Physiologie*. 5 vols. Edited by Ludimar Hermann. Leipzig: F. C. W. Vogel, 1883.

Heintel, Helmut. "Moritz Schiffs gescheiterter Habilitationsversuch an der Universität Göttingen im Jahre 1855." *Medizinhistorisches Journal* 15, no. 4 (1980): 378–84.

Heinzelmann, Ursula. *Beyond Bratwurst: A History of Food in Germany*. London: Reaktion Books, 2014.

Heischkel-Artelt, Edith, ed. *Ernährung und Ernährungslehre im 19. Jahrhundert*. Göttingen: Vandenhoeck and Ruprecht, 1976.

Henkelmann, Thomas. "Zur Geschichte der Psychosomatik in Heidelberg: V. v. Weizsäcker und A. Mitscherlich als Klinikgründer." *Psychotherapie, Psychosomatik, medizinische Psychologie* 42 (1992): 175–86.

Herrnstein, R. J. "Nature as Nurture: Behaviorism and the Instinct Doctrine." *Behaviorism* 1, no. 1 (Fall 1972): 23–52.

Hervey, Nick. "Gull, Sir William Withey, First Baronet (1816–1890)." *Oxford Dictionary of National Biography*. https://doi.org/10.1093/ref:odnb/11730.

Hintzsche, Erich. "Der Hunger in physiologischen Lehrbüchern von Haller bis Valentin." *Gesnerus* 20 (1963): 33–46.

Hippocrates. Loeb Classical Library. Vols. 1–11. Translated by W. H. S. Jones, Paul Potter, and Wesley D. Smith. Cambridge, MA: Harvard University Press, 1923–2018.

Hollander, Franklin. Introduction to Miner, *Regulation of Hunger and Appetite*, 3.

Holmes, Frederic L. *Between Biology and Medicine: The Formation of Intermediary Metabolism*. Berkeley, CA: Office for History of Science and Technology, 1992.

———. *Claude Bernard and Animal Chemistry: The Emergence of a Scientist*. Cambridge, MA: Harvard University Press, 1974.

———. "Claude Bernard and the *milieu intérieur*." *Archives internationales d'histoire des sciences* 16 (1963): 369–76.

———. "Claude Bernard, the *milieu intérieur*, and Regulatory Physiology." *History and Philosophy of the Life Sciences* 8, no. 1 (1986): 3–25.

———. "The Formation of the Munich School of Metabolism." In Coleman and Holmes, *The Investigative Enterprise*, 179–210.

———. "The Transformation of the Science of Nutrition." *Journal of the History of Biology* 8, no. 1 (Spring 1975): 135–44.

Hopkins, F. Gowland. "Cameron Prize Lectures on the Present Position of the Vitamin Problem." *British Medical Journal* 2, no. 3277 (October 20, 1923): 691–93; 2, no. 3278 (October 27, 1923): 748–50.

———. "The Study of Human Nutrition: The Outlook To-Day." *British Medical Journal* 1, no. 3872 (March 23, 1935): 571–77.

Houck, Lynne D., and Lee C. Drickamer, eds. *Foundations of Animal Behavior: Classic Papers with Commentaries*. Chicago: University of Chicago Press, 1996.

Hufeland, C. W. "Höchst merkwürdiges Beyspiel einer noch jezt fortdauernden, beynahe fünfzehn monatlichen Enthaltung von Speisen und Getränken bey einem sechzehenjährigen Mädgen." *Journal der practischen Arzneykunde und Wundarzneykunst* 8, pt. 2 (1799): 191–95.

Hunauld, Pierre. *Dissertation sur les vapeurs et les pertes de sang*. Paris: Jean-Noel Leloup, 1756.

Hunter, John. "On the Digestion of the Stomach after Death." *Philosophical Transactions* 72 (1772): 447–54.

———. "Some Observations on Digestion." In *Works of John Hunter, F.R.S.*, 4 vols., edited by James F. Palmer, 4:81–116. London: Longman, 1835–1837.

Hunter, W. S. "Summary Comments on the Heredity-Environment Symposium." *Psychological Review* 54, no. 6 (November 1947): 348–52.

Hyvert, Roger. *Précis de pathologie interne et de diagnostic*. 5th ed. Paris: Maloine, 1921.

"Introductory Statement." *Psychosomatic Medicine* 1, no. 1 (January 1939): 3–4.

Isaacson, Robert L. *The Limbic System*. New York: Plenum, 1974.

Ivey, Stokes P. "An Essay on Dyspepsia." Thesis, University of Pennsylvania, 1837.

["J.A.A."] "Food Habits of Birds." *The Auk* 10, no. 4 (October 1893): 347.

Janet, Pierre. *Les obsessions et la psychasthénie*. 2 vols. Paris: Félix Alcan, 1903.

Janowitz, Henry D., and Franklin Hollander. "The Time Factor in the Adjustment of Food Intake to Varied Caloric Requirement in the Dog: A Study of the Precision of Appetite Regulation." In Miner, *Regulation of Hunger and Appetite*, 56–67.

Jardine, N., J. A. Secord, and E. C. Spary. *Cultures of Natural History*. Cambridge: Cambridge University Press, 1996.

Jaucourt, Louis de. "Faim, f.f. (Physiol.)." In Diderot and d'Alembert, *Encyclopédie*, 6:373–76.

———. "Faim canine (Med.)." In Diderot and d'Alembert, *Encyclopédie*, 6:376–77.

Jensen, Randy L., James L. Stone, and Robert A. Hayne. "Introduction of the Human Horsley-Clarke Stereotactic Frame." *Neurosurgery* 38, no. 3 (1996): 563–67.

Jolliffe, Norman. *Reduce and Stay Reduced*. New York: Simon & Schuster, 1952.

Jones, Colin and Rebecca Spang. "*Sans-culottes, sans café, sans tabac*: Shifting Realms of Necessity and Luxury in Eighteenth-Century France." In Berg and Clifford, *Consumers and Luxury*, 37–62.

Jones, Ernest. "Anal-Erotic Character Traits." In *Papers on Psycho-Analysis*, 3rd ed., 680–704. London: Baillière, Tindall and Cox, 1923.

Jouanna, Jacques. *Hippocrates*. Translated by M. B. DeBevoise. Baltimore: Johns Hopkins University Press, 1999.

Jung, C. G. *Psychology of the Unconscious: A Study of the Transformations and Symbolisms of the Libido*. Translated by Beatrice M. Hinkle. Introduction by William McGuire. Princeton, NJ: Princeton University Press, 1992.

Jütte, Robert. *A History of the Senses from Antiquity to Cyberspace*. Translated by James Lynn. Malden, MA: Polity, 2005.

Kamminga, Harmke. "Nutrition for the People, or the Fate of Jacob Moleschott's Contest for a Humanist Science." In Kamminga and Cunningham, *Science and Culture of Nutrition*, 15–47.

Kamminga, Harmke, and Andrew Cunningham, eds. *The Science and Culture of Nutrition, 1840–1940*. Amsterdam: Rodopi, 1994.

Kass, Leon R. *The Hungry Soul: Eating and the Perfecting of Our Nature*. New York: Free Press, 1994.

Katz, David. "David Katz." In *A History of Psychology in Autobiography*, edited by Edwin G. Boring, Heinz Werner, Herbert S. Langfeld, and Robert M. Yerkes, 4:189–211. New York: Russell and Russell, 1952.

———. "La faim et l'appétit en psychologie générale et en biotypologie." *Biotypologie: Bulletin de la Société de Biotypologie* 6, no. 4 (December 1938): 233–51.

———. *Hunger und Appetit: Untersuchungen zur medizinischen Psychologie*. Leipzig: Johann Ambrosius Barth, 1932.

———. "Psychologische Probleme des Hungers und Appetits." *Der Nervenarzt* 1 (1928): 345–51.

———. "Psychologische Probleme des Hungers und Appetits, insbesondere beim Kinde." *Zeitschrift für Kinderforschung* 34 (1928): 158–97.

———. "Psychologische Untersuchungen über Hunger und Appetit." *Archiv für die gesamte Psychologie* 65 (1928): 269–320.

Kessler, David A. *The End of Overeating: Taking Control of the Insatiable American Appetite*. New York: Rodale, 2009.

Kevles, Daniel J. "'Into Hostile Political Camps': The Reorganization of International Science in World War I." *Isis* 62, no. 1 (Spring 1971): 47–60.

Keys, Ancel. *The Biology of Human Starvation*. 2 vols. Minneapolis: University of Minnesota Press, 1950.

King, Helen. "Green Sickness: Hippocrates, Galen and the Origins of the 'Disease of Virgins.'" *International Journal of the Classical Tradition* 2, no. 3 (1996): 372–87.

———. *Hippocrates' Woman: Reading the Female Body in Ancient Greece*. London: Routledge, 1998.

King-Hele, Desmond. *Erasmus Darwin*. New York: Charles Scribner's, 1963.

Kiple, Kenneth F., and Kriemhild Coneé Ornelas, eds. *The Cambridge World History of Food*. 2 vols. Cambridge: Cambridge University Press, 2000.

Kirkham, Tim C., and Steven J. Cooper, eds. *Appetite and Body Weight: Integrative Systems and the Development of Anti-Obesity Drugs*. Amsterdam: Elsevier, 2007.

———. "Introduction and Overview." In Kirkham and Cooper, *Appetite and Body Weight*, 1–4.

Kirkland, Anna. "The Environmental Account of Obesity: A Case for Feminist Skepticism." *Signs* 36, no. 2 (Winter 2011): 463–85.

Kisch, E. Heinrich. "Fettsucht." In Eulenberg, *Real-Encyclopädie der gesammten Heilkunde*, 7:551–70.

Klein, Edward C., Jr. "The Pernicious Type of Nervous Dyspepsia." *Annals of Internal Medicine* 7, no. 8 (1934): 960–65.

Klein, Marc. "Aspects biologiques de l'instinct reproducteur dans le comportement des mammifères." In Autuori et al., *L'instinct*, 287–339.

Knoeff, Rina. *Herman Boerhaave (1668–1738): Calvinist Chemist and Physician*. Amsterdam: Royal Netherlands Academy of Arts and Sciences, 2002.

Kohler, Robert E., Jr. "The Enzyme Theory and the Origin of Biochemistry." *Isis* 64, no. 2 (June 1973): 181–96.

———. "Walter Fletcher, F. G. Hopkins, and the Dunn Institute of Biochemistry: A Case Study in the Patronage of Science." *Isis* 69, no. 3 (1978): 331–55.

Korsmeyer, Carolyn. *Making Sense of Taste: Food and Philosophy*. Ithaca, NY: Cornell University Press, 1999.

Kremer, Richard L. "Physiology." In Bowler and Pickstone, *Modern Biological and Earth Sciences*, 342–66.

Kretschmer, Ernst. *Körperbau und Charakter: Untersuchungen zum Konstitutionsproblem und zur Lehre von den Temperamenten*. Berlin: Springer, 1925.

Kringelbach, Morten L. "Cortical Systems Involved in Appetite and Food Consumption." In Kirkham and Cooper, *Appetite and Body Weight*, 5–26.

Kruta, Vladislav. "Friedrich Tiedemann." In Gillispie, *Dictionary of Scientific Biography*, 13:402–4.

———. "Rudolf Peter Heinrich Heidenhain." In Gillispie, *Dictionary of Scientific Biography*, 6:224–27.

Kuczmarski, R. J., K. M. Flegal, S. M. Campbell, and C. L. Johnson. "Increasing Prevalence of Overweight among US Adults: The National Health and Nutrition Examination Surveys, 1960 to 1991." *Journal of the American Medical Association* 272, no. 3 (1994): 205–11.

Lafon, Max. "Quelques documents sur l'appétit et la consommation alimentaire chez les insectes." In *Le comportement alimentaire et l'appétit*, edited by H. Piéron, J.-J. Alloiteau, P. Aschkenasy-Lelu, A. Doumic, M. Lafon, J. Le Magnen, P. Royer, et al., 103–22. Paris: Éditions du Centre National de la Recherche Scientifique, 1952.

Lamotte, J.-A. "Propositions sur la faim, présentées et soutenues à l'École de Médecine de Paris, le 10 nivôse an XIII." Thesis, École de Médecine de Paris, 1804.

Landecker, Hannah. "Food as Exposure: Nutritional Epigenetics and the New Metabolism." *BioSocieties* 6, no. 2 (2011): 167–94.

Langenbach, Michael. "Phenomenology, Intentionality, and Mental Experiences: Edmund Husserl's *Logische Untersuchungen* and the First Edition of Karl Jaspers's *Allgemeine Psychopathologie*." *History of Psychiatry* 6 (1995): 209–24.

Laporte. "Appétit." In Vicq d'Azyr and Moreau de la Sarthe, *Encyclopédie méthodique,* 3:198.

Lasègue, Charles. "De l'anorexie hystérique." *Archives générales de médecine,* 6th ser., 21 (1873): 385–403.

———. "Des appétits en général et de l'appétit digestif en particulier." *Gazette des hôpitaux civils et militaires* 54, no. 2 (1881): 10–11.

Lattemann, Dianne Figlewicz, Nicole M. Sanders, Amy MacDonald Naleid, and Alfred J. Sipols. "Integration of Peripheral Adiposity Signals and Psychological Controls of Appetite." In Kirkham and Cooper, *Appetite and Body Weight,* 167–90.

Lawrence, Christopher. "Incommunicable Knowledge: Science, Technology, and the Clinical Art in Britain, 1850–1914." *Journal of Contemporary History* 20 (1985): 503–20.

———. "Still Incommunicable: Clinical Holists and Medical Knowledge in Interwar Britain." In *Greater Than the Parts: Holism in Biomedicine, 1920–1950,* edited by Christopher Lawrence and George Weisz, 94–111. New York: Oxford University Press, 1998.

Le Camus, Antoine. *La médecine de l'esprit.* 2 vols. New ed. Paris: Ganeau, 1769.

Legallois, J.-J.-C. *Expériences sur le principe de la vie.* Paris: d'Hautel, 1812.

Le Gendre, P. "L'opothérapie gastrique au moyen du suc gastrique de chien à estomac isolé (*gastérine de Frémont*)." *Bulletins et mémoires de la Société médicale des hôpitaux de Paris,* 3rd ser., 17 (1900): 61–70.

Lehrman, Daniel S. "A Critique of Konrad Lorenz's Theory of Instinctive Behavior." *Quarterly Review of Biology* 28 (1953): 337–63.

———. "On the Organization of Maternal Behavior and the Problem of Instinct." In Autuori et al., *L'instinct,* 475–520.

Le Magnen, J. *Hunger.* New York: Cambridge University Press, 1985.

Lenoir, Timothy. "Science for the Clinic: Science Policy and the Formation of Carl Ludwig's Institute in Leipzig." In Coleman and Holmes, *Investigative Enterprise,* 139–78.

———. *The Strategy of Life: Teleology and Mechanics in Nineteenth-Century German Biology.* Chicago: University of Chicago Press, 1982.

Lepovsky, Samuel. "The Action of Various Deleterious Compounds on Voluntary Food Intake." In Miner, *Regulation of Hunger and Appetite,* 132–36.

Lerner, Paul. *Hysterical Men: War, Psychiatry, and the Politics of Trauma in Germany, 1890–1930.* Ithaca, NY: Cornell University Press, 2003.

Leroux, J.-L. "Dissertation sur l'appétit et la faim, présentée et soutenue à l'École de Médecine de Paris, le 29 frimaire, an 13." Thesis, École de Médecine de Paris, 1804.

Lesch, John E. *Science and Medicine in France: The Emergence of Experimental Physiology, 1790–1855.* Cambridge, MA: Harvard University Press, 1984.

Lesses, Mark F., and Abraham Myerson. "Human Autonomic Pharmacology: XVI. Benzedrine Sulfate as an Aid in the Treatment of Obesity." *New England Journal of Medicine* 218, no. 3 (1938): 119–24.

Lester, Rebecca J. "Critical Therapeutics: Cultural Politics and Clinical Reality in Two Eating Disorder Treatment Centers." *Medical Anthropology Quarterly*, new ser., 21, no. 4 (December 2007): 369–87.

Leuret, François. "Note sur une nouvelle sonde destinée à l'alimentation des aliénés." *Gazette médicale de Paris*, 2nd ser., 13, nos. 34–35 (August 23 and 30, 1845): 540–41, 555–56.

Leuret, François, and J.-L. Lassaigne. *Recherches physiologiques et chimiques pour servir à l'histoire de la digestion*. Paris: Huzard, 1825.

Levenstein, Harvey. *Fear of Food: A History of Why We Worry about What We Eat*. Chicago: University of Chicago Press, 2012.

———. *Paradox of Plenty: A Social History of Eating in Modern America*. New York: Oxford University Press, 1993.

———. *Revolution at the Table: The Transformation of the American Diet*. New York: Oxford University Press, 1988.

Levine, A. S., and R. L. Atkinson. "Opioids in the Regulation of Food Intake and Energy Expenditure: Introduction." Symposium presented by the American Institute of Nutrition at the 70th Annual Meeting of the Federation of American Societies for Experimental Biology, St. Louis, MO, April 16, 1986. In *Federation Proceedings* 46, no. 1 (January 1987): 159–62.

Lewin, Kurt. *Vorsatz, Wille und Bedürfnis mit Vorbemerkungen über die psychischen Kräfte und Energien und die Struktur der Seele*. Berlin: Julius Springer, 1926. English translation in *Organization and Pathology of Thought: Selected Sources*. Edited by David Rapaport and Jerome Jaffe. New York: Columbia University Press, 1951.

Liebig, Justus. *Die organische Chemie in ihrer Anwendung auf Physiologie und Pathologie*. Braunschweig: Vieweg, 1842.

Limoges, Camille. "Edmond Perrier." In Gillispie, *Dictionary of Scientific Biography*, 10: 522–23.

Logan, Cheryl A. "The Altered Rationale for the Choice of a Standard Animal in Experimental Psychology: Henry H. Donaldson, Adolf Meyer, and 'the' Albino Rat." *History of Psychology* 2, no. 1 (1999): 3–24.

———. "'Are Norway Rats . . . Things?': Diversity versus Generality in the Use of Albino Rats in Experiments on Development and Sexuality." *Journal of the History of Biology* 34, no. 2 (2001): 287–314.

———. "Before There Were Standards: The Role of Test Animals in the Production of Empirical Generality in Physiology." *Journal of the History of Biology* 35, no. 2 (Summer 2002): 329–63.

Logue, Alexandra W. "The Origins of Behaviorism: Antecedents and Proclamation" and "The Growth of Behaviorism: Controversy and Diversity." In Buxton, *Points of View in the Modern History of Psychology*, 141–67, 169–96.

———. "Watson's Behaviorist Manifesto: Past Positive and Current Negative Consequences." In *Modern Perspectives on John B. Watson and Classical Behaviorism*, edited by James T. Todd and Edward K. Morris, 109–23. Westport, CT: Greenwood, 1994.

Longet, François-Achille. *Mémoire sur la véritable nature des nerfs pneumo-gastriques et les usages de leurs anastomoses*. Paris: Rignoux, 1849.

Longet, Francois-Achille, and Maurice Schiff. *Fragments sur les phénomènes chimiques de la digestion, extraits du Traité de physiologie de M. Longet*. Vol. 1. Paris: n.p., 1857.

Lôo, P. "Anorexie mentale, maladie de Simmonds, maladie de Sheehan." *Annales médico-psychologiques* 113, no. 1 (March 1955): 369–75.

López Piñero, José M. *Historical Origins of the Concept of Neurosis*. Translated by D. Berrios. Cambridge: Cambridge University Press, 1983.

Lorenz, Konrad. "Die angeborenen Formen möglicher Erfahrung." *Zeitschrift für Tierpsychologie* 5 (1943): 235–409.

Loudon, I. S. L. "Chlorosis, Anaemia, and Anorexia Nervosa." *British Medical Journal* 281 (1980): 1669–75.

Lowe, Charles Upton, Charles D. May, Herbert M. Stauffer, and E. D. B. Neuhauser. "Fibrosis of the Pancreas: Enterogastrone and the 'Duodenal Mechanism' in Relation to Increased Appetite." *American Journal of Diseases of Children* 79, no. 1 (1950): 91–98.

Lucas, Alexander R. "Toward the Understanding of Anorexia Nervosa as a Disease Entity." *Mayo Clinic Proceedings* 56, no. 4 (1981): 254–64.

Luciani, Luigi. *Das Hungern: Studien und Experimente am Menschen*. Foreword by Jacob Moleschott. Translated by M. O. Fraenkel. Hamburg: Leopold Voss, 1890.

Ludmerer, Kenneth M. *Learning to Heal: The Development of American Medical Education*. Baltimore: Johns Hopkins University Press, 1985.

Ludwig, Carl. *Lehrbuch der Physiologie des Menschen*. 2nd rev. ed. 2 vols. Leipzig: C. F. Winter, 1858–1861.

———. "Neue Versuche über die Beihilfe der Nerven zur Speichelabsonderung." In *Abhandlungen über den Speichel*, edited by Max von Frey, 3–25. Leipzig: Wilhelm Engelmann, 1890.

Lustig, A. J. "Ants and the Nature of Nature in Auguste Forel, Erich Wasmann, and William Morton Wheeler." In Daston and Vidal, *Moral Authority of Nature*, 282–307.

MacLeod, Robert B. "David Katz, 1884–1953." *Psychological Review* 61, no. 1 (January 1954): 1–4.

Magendantz, Heinz, and Samuel Proger. "Anorexia Nervosa or Hypopituitarism?" *Journal of the American Medical Association* 114, no. 20 (May 18, 1940): 1973–83.

Magendie, François. *Leçons sur les fonctions et les maladies du système nerveux*. 2 vols. Edited by Constantin James. Paris: Ebrard, 1839.

———. "Mémoire sur les propriétés nutritives des substances qui ne contiennent pas d'azote." *Annales de chimie et de physique* 3 (1816): 66–77.

———. *Mémoire sur le vomissement, lu à la première classe de l'Institut de France*. Paris: Crochard, 1813.

———. *Précis élémentaire de physiologie*. 2 vols. Paris: Méquignon-Marvis, 1816–1817.

———. "Some General Ideas on the Phenomena Peculiar to Living Bodies." Appendix to Albury, "Experiment and Explanation in the Physiology of Bichat and Magendie." In Coleman and Limoges, *Studies in History of Biology*, 1:107–15.

Mandler, George. *A History of Modern Experimental Psychology from James and Wundt to Cognitive Science*. Cambridge, MA: MIT Press, 2007.

Mani, Nikolaus. "Das Werk von Friedrich Tiedemann und Leopold Gmelin: 'Die Verdauung nach Versuchen' und seine Bedeutung für die Entwicklung der Ernährungslehre in der ersten Hälfte des 19. Jahrhunderts." *Gesnerus* 13 (1956): 190–214.

———. "Die wissenschaftliche Ernährungslehre im 19. Jahrhundert." In Heischkel-Artelt, *Ernährung und Ernährungslehre im 19. Jahrhundert*, 22–75.

Masserman, Jules H. "Psychodynamisms in Anorexia Nervosa and Neurotic Vomiting." *Psychoanalytic Quarterly* 10, no. 2 (1941): 211–42.

Maulitz, Russsell. "Rudolf Virchow, Julius Cohnheim and the Program of Pathology." *Bulletin of the History of Medicine* 52, no. 2 (1978): 162–82.

Mauss, Marcel. *The Gift: Forms and Functions of Exchange in Archaic Societies*. Translated by Ian Cunnison. Introduction by E. E. Evans-Pritchard. New York: W. W. Norton, 1967.

Mayer, Jean. "Regulation of Energy Intake and the Body Weight: The Glucostatic Theory and the Lipostatic Hypothesis." In Miner, *Regulation of Hunger and Appetite*, 15–43.

McClements, David Julian, César Vega, Anne E. McBride, and Eric Andrew Decker. "In Defense of Food Science." *Gastronomica* 11, no. 2 (Summer 2011): 76–84.

McCollum, E. V. *The Newer Knowledge of Nutrition: The Use of Food for the Preservation of Vitality and Health*. New York: Macmillan, 1918.

McCulloch, Marsha. "Appetite Hormones." *Today's Dietitian* 17, no. 7 (July 2015): 26–30.

McFarland, David, ed. *The Oxford Companion to Animal Behavior*. Foreword by Niko Tinbergen. Oxford: Oxford University Press, 1982.

McNeil, Maureen. *Under the Banner of Science: Erasmus Darwin and His Age*. Manchester: Manchester University Press, 1987.

Mead, Margaret. "The Factor of Food Habits." *Annals of the American Academy of Political and Social Science* 225, no. 1 (January 1943): 136–41.

Mendelson, Myer. "The Treatment of Obesity." *Psychosomatics* 1, no. 3 (May–June 1960): 158–60.

Mennell, Stephen. *All Manners of Food: Eating and Taste in England and France from the Middle Ages to the Present*. Oxford, UK: Basil Blackwell, 1985.

———. "On the Civilizing of Appetite." In *The Body: Social Process and Cultural Theory*, edited by Mike Featherstone, Mike Hepworth, and Bryan S. Turner, 126–56. London: Sage Publications, 1991.

Messer, Ellen. "Anthropological Perspectives on Diet." *Annual Review of Anthropology* 13 (1984): 205–49.

Métraux, Alexandre. "Kurt Lewin: Philosopher-Psychologist." *Science in Context* 5, no. 2 (1992): 373–84.

Meyer, Bernard C., and Leonard A. Weinroth. "Observations on Psychological Aspects of Anorexia Nervosa." *Psychosomatic Medicine* 19, no. 5 (September–October 1957): 389–98.

Micale, Mark S. *Approaching Hysteria: Disease and Its Interpretations*. Princeton, NJ: Princeton University Press, 1995.

Miller, Ian. *A Modern History of the Stomach*. London: Routledge, 2016.

Miller, J. P. "Phthisis—Its Successful Treatment." *Journal of the American Medical Association* 2 (April 1884): 657–60.

Mills, John A. *Control: A History of Behavioral Psychology*. New York: New York University Press, 1998.

Miner, Roy Waldo, ed. *The Regulation of Hunger and Appetite*. New York: New York Academy of Sciences, 1955.

Mintz, Sidney W. *Sweetness and Power: The Place of Sugar in Modern History*. New York: Viking, 1985.

Mintz, Sidney, W. Mintz, and Christine M. Du Bois. "The Anthropology of Food and Eating." *Annual Review of Anthropology* 31 (2002): 99–119.

Mitchell, S. Weir. *Fat and Blood: And How to Make Them*. Philadelphia: J. B. Lippincott, 1877.

Mohr, C. F. "Hypertrophie (markschwammige Entartung?) der *Hypophysis cerebri* und dadurch bedingter Druck auf die Hirngrundfläche, insbesondere auf die Sehnerven, das Chiasma derselben und den linkseitigen Hirnschenkel." *Wochenschrift für die gesammte Heilkunde* 6 (1840): 565–71.

Moleschott, Jacob. *Die Physiologie der Nahrungsmittel: Ein Handbuch der Diätetik*. Darmstadt: Carl Wilhelm Leske, 1850.

———. *Physiologie der Nahrungsmittel: Ein Handbuch der Diätetik*. 2nd rev. ed. Giessen: Ferber, 1859.

Monin, E., and Ph. Maréchal. *Stefano Merlatti: Histoire d'un jeûne célèbre. Précédée d'une Étude anecdotique, physiologique et médicale sur le jeûne et les jeûneurs*. Paris: C. Marpon et E. Flammarion, 1887.

Monroe, John Warne. *Laboratories of Faith: Mesmerism, Spiritism, and Occultism in Modern France*. Ithaca, NY: Cornell University Press, 2008.

Montègre, A.-F. Jenin de. *Expériences sur la digestion dans l'homme, présentées à la première class de l'Institut de France, le 8 septembre 1812. Suivies du Rapport des Commissaires nommés par l'Institut.* Paris: Le Normant, 1814.

Moreau de la Sarthe, Jacques-Louis. *Histoire naturelle de la femme, suivie d'un traité d'hygiène appliquée à son régime physique et moral aux différentes époques de la vie*. 3 vols. Paris: Duprat, 1803.

Morgan, C. Lloyd. *An Introduction to Comparative Psychology*. London: Walter Scott, 1894.

———. "The Limits of Animal Intelligence." *Fortnightly Review* 60 (1893): 223–39.

Morison, Alexander. *Outlines of Mental Diseases*. Edinburgh: MacLachlan and Stewart, 1824.

Morley, John E., Allen S. Levine, George K. Yim, and Martin T. Lowy. "Opioid Modulation of Appetite." *Neuroscience and Biobehavioral Reviews* 7 (1983): 281–305.

Mouton. "Appétit." In Adelon and Mérat, *Dictionnaire des sciences médicales*, 2: 258–60.

Müller, E. "Geschichte einer langen Enthaltsamkeit von allen Nahrungsmitteln." *Journal der practischen Arzneykunde und Wundarzneykunst* 24, pt. 2 (1806): 154–59.

Müller, Franz Carl, ed. *Handbuch der Neurasthenie*. Leipzig: F. C.W. Vogel, 1893.

Müller, Johannes. *Müller's Elements of Physiology* (1840–43). 4 vols. Translated by William Baly. Introduction by N. J. Wade. Bristol: Thoemmes Press, 2003.

Mundt, Christoph. "Impact of Karl Jaspers' *General Psychopathology*: The Range of Appraisal." In *One Century of Karl Jaspers'* General Psychopathology, edited by Giovanni Stanghellini and Thomas Fuchs, 42–54. Oxford: Oxford University Press, 2013.

Myers, Charles S. "Instinct and Intelligence." *British Journal of Psychology* 3 (1909–1910): 209–18.

Nahum, Louis H., and H. E. Himwich. "Insulin and Appetite. I. A Method for Increasing Weight in Thin Patients." *American Journal of the Medical Sciences* 183, no. 5 (May 1932): 608–13.

Nestle, Marion, and Malden Nesheim. *Why Calories Count: From Science to Politics*. Berkeley: University of California Press, 2012.

Nicolai, Wilhelm. "Ueber die Entstehung des Hungergefühls." Dissertation, Friedrich Wilhelms Universität Berlin, 1892.

Nicolas, Serge. "Henry Beaunis (1830–1921), directeur-fondateur du laboratoire de Psychologie physiologique de la Sorbonne." *L'année psychologique* 95, no. 2 (1995): 267–91.

Nicolle, Grace. "Pre-Psychotic Anorexia." *Proceedings of the Royal Society of Medicine* 32 (1938): 153–62.

Nieto-Galan, Agustí. "Mr Giovanni Succi Meets Dr. Luigi Luciani in Florence: Hunger Artists and Experimental Physiology in the Late 19th Century." *Social History of Medicine* 28, no. 1 (February 2015): 64–81.

Niewöhner, Jörg, Martin Döring, Michalis Kontopodis, Jeannette Madarász, and Christoph Heintze. "Cardiovascular Disease and Obesity Prevention in Germany: An Investigation into a Heterogeneous Engineering Project." *Science, Technology, & Human Values* 36, no. 5 (September 2011): 723–51.

Noguès, Georges. "L'anorexie mentale, ses rapports avec la psychophysiologie de la faim." Thesis, Université de Toulouse, 1913.

Noorden, Carl von. *Die Fettsucht*. 2nd. rev. ed. Vienna: Hölder, 1910.

Numbers, Ronald L., and William J. Orr, Jr. "William Beaumont's Reception at Home and Abroad." *Isis* 72, no. 4 (December 1981): 590–612.

Nutton, Vivian. *Ancient Medicine*. 2nd ed. London: Routledge, 2013.

Nye, Robert A. *Crime, Madness and Politics in Modern France: The Medical Concept of National Decline*. Princeton, NJ: Princeton University Press, 1984.

Oddy, Derek J., Peter J. Atkins, and Virginie Amilien, eds. *The Rise of Obesity in Europe: A Twentieth Century Food History*. Farnham, UK: Ashgate, 2009.

Oddy, Derek J., and Alain Drouard, eds. *The Food Industries of Europe in the Nineteenth and Twentieth Centuries*. Burlington, VT: Ashgate, 2013.

O'Donnell, John M. *The Origins of Behaviorism: American Psychology, 1870–1920*. New York: New York University Press, 1985.

Offen, Karen. *European Feminisms, 1700–1950: A Political History*. Stanford, CA: Stanford University Press, 2000.

Offer, Avner. "Body Weight and Self-Control in the United States and Britain since the 1950s." *Social History of Medicine* 14, no. 1 (2001): 79–106.

Ó Gráda, Cormac. *Famine: A Short History*. Princeton, NJ: Princeton University Press, 2009.

Ó Gráda, Cormac, and Jean-Michel Chevet. "Famine and Market in Ancien Régime France." *Journal of Economic History* 62, no. 3 (September 2002): 706–33.

O'Hara-May, Jane. "Measuring Man's Needs." *Journal of the History of Biology* 4, no. 2 (September 1971): 249–73.

Olmsted, J. M. D. *François Magendie, Pioneer in Experimental Physiology and Scientific Medicine in XIX Century France*. New York: Schuman's, 1944.

Olmsted, J. M. D., and E. Harris Olmsted. *Claude Bernard and the Experimental Method in Medicine*. New York: Henry Schuman, 1952.

Osler, William. "William Beaumont: A Pioneer American Physiologist." In Beaumont, *Experiments and Observations*, vii–xl.

Otis, Laura. *Müller's Lab*. Oxford: Oxford University Press, 2007.

Pagel, Walter. "Van Helmont's Ideas on Gastric Digestion and the Gastric Acid." *Bulletin of the History of Medicine* 30, no. 6 (1956): 524–36.

Paget, Stephen. "On Cases of Voracious Hunger and Thirst from Injury or Disease of the Brain." *Transactions of the Clinical Society of London* 30 (1897): 113–19.

Passmore, R. "William Cullen and Dietetics." In *William Cullen and the Eighteenth-Century Medical World*, edited by A. Doig, J. P. S. Ferguson, I. A. Milne, and R. Passmore, 167–85. Edinburgh: Edinburgh University Press, 1993.

Pavlov, I. P. *Conditioned Reflexes: An Investigation of the Physiological Activity of the Cerebral Cortex*. Translated by G. V. Anrep. London: Oxford University Press and Humphrey Milford, 1927.

———. "The Food Centre." In *Lectures on Conditioned Reflexes: Twenty-Five Years of Objective Study of the Higher Nervous Activity (Behaviour) of Animals*, edited and translated by W. Horsley Gantt, introduction by Walter B. Cannon, 1:147–55. New York: International, 1928.

———. *The Work of the Digestive Glands*. 2nd English ed. Translated by W. H. Thompson. London: Charles Griffin, 1910.

Pellevoisin, Martial. "De l'alimentation forcée chez les alienés." Thesis, Faculté de Médecine de Strasbourg, 1862.

Penfold, Steve. "Fast Food." In Pilcher, *Oxford Handbook of Food History*, 279–301.

Perrier, Edmond. *La femme dans la nature, dans les moeurs, dans la légende, dans la société: Tableau de son évolution physique et psychique*. Vol. 1. Paris: Bong, 1908–1910.

———. "L'instinct. Lecture faite devant les cinq Académies." *L'année psychologique* 8 (1901): 561–72.

Peter, J.-P. "Aux sources de la médicalisation, le regard et le mot: Le travail des topographies médicales." In *Populations et cultures: Études réunies en l'honneur de François Lebrun*, 103–11. Rennes: Amis de François Lebrun, 1989.

Petrick, Gabriella M. "Industrial Food." In Pilcher, *Oxford Handbook of Food History*, 258–78.

Pflüger, Eduard F. W. "Die teleologische Mechanik der lebendigen Natur." *Archiv für die gesammte Physiologie des Menschen und der Thiere* 15 (1877): 57–101.

———. *Ueber die Kunst der Verlängerung des menschlichen Lebens*. Bonn: Emil Strauss, 1890.

Pick, Alois. "Die Behandlung der Appetitlösigkeit mit besonderer Berücksichtigung ihrer nervösen Formen." *Wiener medizinische Wochenschrift* 63, no. 2 (January 1913): 106–15.

Pickstone, John V. "Physiology and Experimental Medicine." In *Companion to the History of Modern Science*, edited by R. C. Olby, 728–42. London: Routledge, 1990.

———. *Ways of Knowing: A New History of Science, Technology and Medicine*. Chicago: University of Chicago Press, 2000.

Piéron, H., J.-J. Alloiteau, P. Aschkenasy-Lelu, A. Doumic, M. Lafon, J. Le Magnen, P. Royer, et al., eds. *Le comportement alimentaire et l'appétit*. Paris: Éditions du Centre national de la recherche scientifique, 1952.

Pilcher, Jeffrey M., ed. *The Oxford Handbook of Food History*. New York: Oxford University Press, 2012.

Pinel, Philippe. "Mémoire sur la manie périodique, ou intermittente." *Mémoires de la Société Médicale d'Émulation* 1, year 5 (1797): 94–119.

———. *Nosographie philosophique ou méthode de l'analyse appliquée à la médecine*. 2 vols. Paris: Crapelet, 1798.

———. "Recherches et observations sur le traitement moral des aliénés." *Mémoires de la Société Médicale d'Émulation* 2, year 6 (1798): 215–55.

———. *Traité médico-philosophique sur l'aliénation mentale* (1809). 2nd ed. New York: Arno Press, 1976.

Pitcairn, Andrew. *The Philosophical and Mathematical Elements of Physick*. London: n.p., 1718.

Pollack, Andrew. "AMA Recognizes Obesity as a Disease." *New York Times*, June 18, 2013.

Porter, Roy. *Blood and Guts: A Short History of Medicine*. New York: W. W. Norton, 2003.

———. "Erasmus Darwin: Doctor of Evolution?" In *History, Humanity, and Evolution: Essays for John C. Greene*, edited by James R. Moore, 39–69. Cambridge: Cambridge University Press, 1989.

Postel, Jacques, and Claude Quétel, eds. *Nouvelle histoire de la psychiatrie*. Rev. ed. Paris: Dunod, 1994.

Poulain, J.-P. "Eléments pour une histoire de la médicalisation de l'obésité." *Obésité* 3 (2009): 7–16.

Quigley, J. P. "The Role of the Digestive Tract in Regulating the Ingestion of Food." In Miner, *Regulation of Hunger and Appetite*, 6–14.

Rabinbach, Anson. *The Human Motor: Energy, Fatigue, and the Origins of Modernity*. Berkeley: University of California Press, 1992.

Rabinow, Paul, and Nikolas Rose. "Biopower Today." *Biosocieties* 1 (2006): 195–217.

Rahman, Lincoln, Henry B. Richardson, and Herbert S. Ripley. "Anorexia Nervosa with Psychiatric Observations." *Psychosomatic Medicine* 1, no. 3 (July 1939): 335–65.

Raige-Delorme, J., and A. Dechambre, eds. *Dictionnaire encyclopédique des sciences médicales*. 100 vols. Paris: P. Asselin, S. de Labé, and V. Masson, 1865–1889.

Rather, L. J. "The 'Six Things Non-Natural': A Note on the Origin and Fate of a Doctrine and a Phrase." *Clio medica* 3 (1968): 335–47.

Réaumur, [René-Antoine Ferchault de]. *Art de faire éclorre, et d'élever en toute saison des oiseaux domestiques de toutes especes, soit par le moyen de la chaleur du fumier, soit par le moyen de celle du feu ordinaire*. 2nd ed. 2 vols. Paris: L'Imprimerie Royale, 1751.

———. "Sur la digestion des oiseaux." *Mémoires de l'Académie Royale des Sciences* (1752): 266–307, 461–95.

Reed, Edward S. *From Soul to Mind: The Emergence of Psychology from Erasmus Darwin to Williams James*. New Haven, CT: Yale University Press, 1997.

Reid, John. "An Experimental Investigation into the Functions of the Eighth Pair of Nerves, or the Glosso-Pharyngeal, Pneumogastric, and Spinal Accessory." *Edinburgh Medical and Surgical Journal* 49 (1838): 109–76; 51 (1839): 269–330.

Reill, Peter Hanns. *Vitalizing Nature in the Enlightenment*. Berkeley: University of California Press, 2005.

"Report of Diseases Treated at the New Town Dispensary, Edinburgh." *Edinburgh Medical and Surgical Journal* 12 (1816): 245–49; 13 (1817): 117–18.

Review of *Emotions and Bodily Changes: A Survey of Literature on Psychosomatic Interrelationships, 1910–1933*, by [Helen] Flanders Dunbar. *British Medical Journal* 1, no. 3925 (March 28, 1936): 642.

Review of *Essays on Insanity, Hypochondriasis, and Other Nervous Affections*, by John Reid. *Edinburgh Medical and Surgical Journal* 12 (1816): 467–73.

Review of *Observations on Mental Derangement*, by H. Bardsley. *Edinburgh Medical and Surgical Journal* 22 (1824): 279–86.

Review of *Practical Remarks upon Indigestion; particularly as connected with Bilious and Nervous Affections of the Head and other parts; including Observations upon the Disorders and Diseases of the Stomach and superior Parts of the Alimentary Canal, Illustrated by Cases*, by John Howship. *Edinburgh Medical and Surgical Journal* 25 (1826): 393–403.

Review of *Recherches physiologiques*, by F. Leuret and J. L. Lassaigne, and *Die Verdauung nach Versuchen*, by F. Tiedemann and L. Gmelin. *Edinburgh Medical and Surgical Journal* 27 (1827): 417–25; 28 (1827): 364–79.

Review of *Traité sur les gastralgies et les entéralgies ou maladies nerveuses de l'estomac et des intestins*, 4th ed., by J.-P.-T. Barras. *Gazette médicale de Paris*, 2nd ser., 13, no. 16 (April 19, 1845): 253–54.

Review of *A Treatise on Diet*, by J. A. Paris. *Edinburgh Medical and Surgical Journal* 27 (1827): 163–73.

Review of *A View of the Nervous Temperament*, by Thomas Trotter. *Edinburgh Medical and Surgical Journal* 3 (1807): 473–82.

Review of *The Work of the Digestive Glands*, 2nd ed., by Ivan Pavlov. *Lancet* 176, no. 4543 (September 24, 1910): 949.

Rey, Roselyne. *Naissance et développement du vitalisme en France de la deuxième moitié du 18e siècle à la fin du Premier Empire*. Oxford, UK: Voltaire Foundation, 2000.

Richardière, Henri-Alphonse. *Maladies de la nutrition: Goutte, obésité, diabète*. 6th ed. Paris: J. B. Baillière, 1920.

Richards, Audrey I. *Hunger and Work in a Savage Tribe: A Functional Study of Nutrition among the Southern Bantu*. Foreword by Bronislaw Malinowski. London: Routledge, 1932.

Richards, Robert J. *Darwin and the Emergence of Evolutionary Theories of Mind and Behavior*. Chicago: University of Chicago Press, 1987.

Richet, Charles. "Dégoût." In Richet, *Dictionnaire de Physiologie*, 4:754–60.

———, ed. *Dictionnaire de physiologie*. 10 vols. Paris: Félix Alcan, 1895–1928.

———. *Du suc gastrique chez l'homme et les animaux: Ses propriétés chimiques et physiologiques*. Paris: Baillière, 1878.

———. *L'homme et l'intelligence: Fragments de physiologie et de psychologie*. Paris: Félix Alcan, 1884.

Richter, Curt P. "Animal Behavior and Internal Drives." *Quarterly Review of Biology* 2, no. 3 (September 1927): 307–43.

———. "The Behavior of the Rat: A Study of General and Specific Activities." PhD diss., Johns Hopkins University, Baltimore, 1921.

———. "A Behavioristic Study of the Activity of the Rat." *Comparative Psychology Monographs* 1 (1922): 1–55.

———. *Biological Clocks in Medicine and Psychiatry*. Springfield, IL: Charles C. Thomas, 1965.

———. "Biology of Drives." *Psychosomatic Medicine* 3, no. 2 (1941): 105–10.

———. "Biology of Drives." *Journal of Comparative Physiology and Psychology* 40 (1947): 129–34.

———. "Increased Salt Appetite in Adrenalectomized Rats." *American Journal of Physiology* 115 (1936): 155–61.

———. "Salt Appetite of Mammals: Its Dependence on Instinct and Metabolism." In Autuori et al., *L'instinct*, 577–632.

Roberts, Mary Louise. *Civilization without Sexes: Reconstructing Gender in Postwar France, 1917–1927*. Chicago: University of Chicago Press, 1994.

Roche, Daniel. *A History of Everyday Things: The Birth of Consumption in France, 1600–1800*. Translated by Brian Pearce. New York: Cambridge University Press, 2000.

Rodin, Judith, Gordon Mogenson, and David Booth. "Editorial." *Appetite* 1, no. 1 (March 1980): n.p.

Roger, Jacques. *Buffon: A Life in Natural History*. Translated by Sarah Lucille Bonnefoi. Edited by L. Pearce Williams. Ithaca, NY: Cornell University Press, 1997.

———. *Les sciences de la vie dans la pensée française au XVIIIe siècle*. New ed. Paris: Albin Michel, 1993.

Rosen, George. "Carl Ludwig and his American Students." *Bulletin of the History of Medicine* 4 (1936): 609–50.

Rosenberg, Charles E. "Pathologies of Progress: The Idea of Civilization as Risk." *Bulletin of the History of Medicine* 72, no. 4 (1998): 714–30.

———. "The Place of George M. Beard in Nineteenth Century Psychiatry." *Bulletin of the History of Medicine* 36 (1962): 245–59.

Rosenheim, [Theodor]. "Magenneurosen." In Eulenberg, *Real-Encyclopädie der gesammten Heilkunde*, 14:392–426.

———. "Ueber nervöse Dyspepsie." *Berliner klinische Wochenschrift* 34 (1897): 908–10, 936–39, 965–67.

Rosenthal, Moriz. *Magenneurosen und Magencatarrh sowie deren Behandlung*. Vienna: Urban and Schwarzenberg, 1886.

Rosenzweig, Mark R. "The Mechanisms of Hunger and Thirst." In *Psychology in the Making: Histories of Selected Research Problems*, edited by Leo J. Postman, 73–143. New York: Knopf, 1962.

Rothschuh, Karl E. *History of Physiology*. Translated by Guenter B. Risse. Huntington, NY: Robert E. Krieger, 1973.

Roubaud, Émile. "Recherches biologiques sur les guêpes solitaires et sociales d'Afrique: La genèse de la vie sociale et l'évolution de l'instinct maternel chez les Vespides." *Annales des sciences naturelles: Zoologie et biologie animale*, 10th ser., 1 (1916): 1–160.

Rousseau, George. "Coleridge's Dreaming Gut: Digestion, Genius, Hypochondria." In Forth and Carden-Coyne, *Cultures of the Abdomen*, 105–26.

Roussel, Pierre. *Système physique et moral de la femme*. Paris: Vincent, 1775.

Roux, Joanny. "La faim, étude physio-psychologique." *Bulletin de la Société d'anthropologie de Lyon* 16 (1897): 409–55.

Rubner, Max. "Der Nahrungstrieb des Menschen." *Sitzungsberichte der Königlich-Preussischen Akademie der Wissenschaften zu Berlin* 11 (1920): 341–64.

Rullier, P. "Faim." In Adelon et al., *Dictionnaire de médecine*, 8:487–96.

Ryle, John A. "Anorexia Nervosa." *Lancet* 228, no. 5903 (October 17, 1936): 893–99.

Sauvages, François Boissier de. *Nosologie méthodique, dans laquelle les maladies sont rangées par classes, suivant le systême de Sydenham, & l'ordre des Botanistes*. 3 vols. Translated by Nicolas. Paris: Hérissant, 1771.

Scheindlin, Benjamin. "'Take One More Bite for Me': Clara Davis and the Feeding of Young Children." *Gastronomica* 5, no. 1 (Winter 2005): 65–69.

Schiff, Maurice [Moritz]. *Leçons sur la physiologie de la digestion, faites au Muséum d'histoire naturelle de Florence.* 2 vols., edited by Émile Levier. Florence: Hermann Loescher, 1867.

Schlöder, Christian, Maximilian Schochow, and Florian Steger. "Alter und Altern in der ärztlichen Praxis Friedrich Hoffmanns (1660–1742)." *Medizinhistorische Journal* 49, no. 3 (2014): 237–59.

Schmalz, E. "Zwei Geschichten von Frauen, welche angeblich sehr lange ohne Nahrung zu sich zu nehmen ihr Leben fortgesetzt haben." *Journal der practischen Heilkunde* 69, Supplement (1829): 317–29.

Schneirla, T. C. "Interrelationships of the 'Innate' and the 'Acquired' in Instinctive Behavior." In Autuori et al., *L'instinct*, 387–452.

Schofield, Robert E. *The Lunar Society of Birmingham: A Social History of Provincial Science and Industry in Eighteenth-Century England.* Oxford: Clarendon Press, 1963.

Schroeder-Gudehus, Brigitte. *Les scientifiques et la paix: La communauté scientifique internationale au cour des années 20.* Montreal: Presses de l'Université de Montréal, 1978.

Schulkin, Jay. *Curt Richter: A Life in the Laboratory.* Baltimore: Johns Hopkins University Press, 2005.

———. "Curt Richter: Psychobiology and the Concept of Instinct." *History of Psychology* 10, no. 4 (2007): 325–43.

Schwartz, Hillel. *Never Satisfied: A Cultural History of Diets, Fantasies, and Fat.* New York: Macmillan, 1986.

Scrinis, Gyorgy. "On the Ideology of Nutritionism." *Gastronomica* 8, no. 1 (Winter 2008): 39–48.

Sehnert, Keith W. "Development of Phentermine, an Appetite-Control Drug." *Clinical Medicine* 70 (February 1963): 400–403.

Semelaigne, René. *Les pionniers de la psychiatrie française avant et après Pinel.* Paris: J.-B. Baillière, 1930.

Senebier, Jean. "Considérations sur la méthode suivie par Monsieur l'Abbé Spallanzani dans ses expériences sur la digestion." In Spallanzani, *Expériences sur la digestion*, i–cxlix.

Sèze, [Victor] de. *Recherches phisiologiques et philosophiques sur la sensibilité ou la vie animale.* Paris: Prault, 1786.

Shapin, Steven. "How to Eat Like a Gentleman: Dietetics and Ethics in Early Modern England." In *Right Living: An Anglo-American Tradition of Self-Help Medicine and Hygiene*, edited by Charles E. Rosenberg, 21–58. Baltimore: Johns Hopkins University Press, 2003.

———. "Trusting George Cheyne: Scientific Expertise, Common Sense, and Moral Authority in Early Eighteenth-Century Dietetic Medicine." *Bulletin of the History of Medicine* 77, no. 2 (2003): 263–97.

Shorter, Edward. *From Paralysis to Fatigue: A History of Psychosomatic Illness in the Modern Era.* New York: Free Press, 1992.

———. *From the Mind into the Body: The Cultural Origins of Psychosomatic Symptoms*. New York: Free Press, 1994.

Sigerist, Henry E. *A History of Medicine*. Vol. 2, *Early Greek, Hindu, and Persian Medicine*. New York: Oxford University Press, 1961.

Simmonds, Morris. "Über embolische Prozesse in der Hypophysis." *Virchows Archiv für pathologische Anatomie* 217 (1914): 226–39.

———. "Ueber Hypophysisschwund mit tödlichem Ausgang." *Deutsche medicinische Wochenschrift* 40, no. 7 (February 1914): 322–23.

Skandalakis, Lee J., and John E. Skandalakis. "The History and Surgical Anatomy of the Vagus Nerve." *Surgery, Gynecology & Obstetrics* 162, no. 1 (1986): 75–85.

Skinner, B. F. *The Behavior of Organisms: An Experimental Analysis*. New York: D. Appleton-Century, 1938.

———. "On the Conditions of Elicitation of Certain Eating Reflexes." *Proceedings of the National Academy of Sciences* 16, no. 6 (1930): 433–38.

Skultans, Vieda. *Madness and Morals: Ideas on Insanity in the Nineteenth Century*. London: Routledge and Kegan Paul, 1975.

Sleigh, Charlotte. *Six Legs Better: A Cultural History of Myrmecology*. Baltimore: Johns Hopkins University Press, 2007.

Smith, David, and Malcolm Nicolson. "The 'Glasgow School' of Paton, Findlay and Cathcart: Conservative Thought in Chemical Physiology, Nutrition and Public Health." *Social Studies of Science* 19, no. 2 (May 1989): 195–238.

Smith, Roger. "The Background of Physiological Psychology in Natural Philosophy." *History of Science* 11 (1973): 75–123.

———. "The Human Significance of Biology: Carpenter, Darwin and the *Vera causa*." In *Nature and the Victorian Imagination*, edited by U. C. Knoepflmacher and G. B. Tennyson, 216–30. Berkeley: University of California Press, 1977.

———. *Inhibition: History and Meaning in the Sciences of Mind and Brain*. Berkeley: University of California Press, 1992.

Smith, Wesley D. *The Hippocratic Tradition*. Ithaca, NY: Cornell University Press, 1979.

Snapper, Isidore. "Food Preferences in Man: Special Cravings and Aversions." In Miner, *Regulation of Hunger and Appetite*, 92–106.

Sollier, Paul. "Anorexie hystérique (sitieirgie hystérique): Formes pathogéniques—Traitement moral." *Revue de médecine* 11 (1891): 625–50.

Soukup, Rudolf W. "Arnold Durig (1872–1961): Einer der letzten großen Physiologen, die das gesamte Fach noch souverän überblickten." In *Die wissenschaftliche Welt von gestern*, edited by Rudolf W. Soukup and W. Gerhard Pohl, 128–32. Vienna: Böhlau, 2004.

Spallanzani, Lazzaro. *Expériences sur la digestion de l'homme et de différentes espèces d'animaux*. New ed. Translated and introduced by J. Senebier. Lausanne: Mourer, Cadet, 1785.

Spang, Rebecca L. *The Invention of the Restaurant: Paris and Modern Gastronomic Culture*. Cambridge, MA: Harvard University Press, 2000.

Spary, E. C. *Eating the Enlightenment: Food and the Sciences in Paris*. Chicago: University of Chicago, 2012.

———. *Feeding France: New Sciences of Food, 1760–1815*. Cambridge: Cambridge University Press, 2014.

———. "The 'Nature' of Enlightenment." In *The Sciences in Enlightened Europe*, edited by William Clark, Jan Golinski, and Simon Schaffer, 272–304. Chicago: University of Chicago Press, 1999.

Spary, E. C., and Anya Zilberstein, eds. "Food Matters: Critical Histories of Food and the Sciences." *Osiris* 35 (forthcoming).

Speakman, John R. "Gene Environment Interactions and the Origin of the Modern Obesity Epidemic: A Novel 'Nonadaptive Drift' Scenario." In Kirkham and Cooper, *Appetite and Body Weight*, 301–22.

"Special Section." *Perspectives on Psychological Science* 5, no. 6 (November 2010): 714–75.

Spink, John S. "'Sentiment,' 'sensible,' 'sensibilité': Les Mots, les idées, d'après les 'moralistes' français et britanniques du début du dix-huitième siècle." *Zagadnienia Rodzajów Literackich* 20 (1977): 33–47.

Spurzheim, J. G. *Outlines of Phrenology*. 3rd ed. Boston: Marsh, Capen and Lyon, 1834.

Stahnisch, Frank. "Den Hunger standardisieren: François Magendies Fütterungsversuche zur Gelatinekost 1831–1841." *Medizinhistorisches Journal* 39 (2004): 103–34.

Stainbrook, Edward. "Psychosomatic Medicine in the Nineteenth Century." *Psychosomatic Medicine* 14, no. 3 (1952): 211–27.

Stanghellini, Giovanni. "The Ethics of Incomprehensibility." In Stanghellini and Fuchs, *One Century*, 166–81.

Stanghellini, Giovanni, and Thomas Fuchs, eds. *One Century of Karl Jaspers' "General Psychopathology."* Oxford: Oxford University Press, 2013.

Stark, William. *The Works of the late William Stark, M.D.: Consisting of Clinical and Anatomical Observations, with Experiments, Dietetical and Statical, Revised and Published from his Original MSS by J. Carmichael Smyth, M.D., F.R.S.* London: J. Johnson, 1788.

Starling, Ernest H. "The Croonian Lectures on the Chemical Correlation of the Functions of the Body." *Lancet* 2, no. 4275 (August 5, 1905): 339–41.

Starobinski, Jean. "Le concept de cénesthésie et les idées neuropsychologiques de Moritz Schiff." *Gesnerus* 34 (1977): 2–20.

———. "Sur la chlorose." *Romantisme* 31 (1981): 113–30.

Staum, Martin S. *Cabanis: Enlightenment and Medical Philosophy in the French Revolution*. Princeton, NJ: Princeton University Press, 1980.

———. *Nature and Nurture in French Social Sciences, 1859–1914 and Beyond*. Montreal: McGill-Queen's University Press, 2011.

Stearns, Peter. *Fat History: Bodies and Beauty in the Modern West*. New York: New York University Press, 1997.

Steckerl, Fritz, ed. *The Fragments of Praxagoras of Cos and His School*. Leiden: E. J. Brill, 1958.

Steinbrugge, Lieselotte. *The Moral Sex: Woman's Nature in the French Enlightenment.* New York: Oxford University Press, 1995.

Steinke, Hubert. *Irritating Experiments: Haller's Concept and the European Controversy on Irritability and Sensibility, 1750–90.* Amsterdam: Rodopi, 2005.

Stellar, Eliot. "The Biological Nature of Appetite." *Proceedings of the American Philosophical Society* 136, no. 4 (December 1992): 507–20.

Sternberg, Wilhelm. "Der Appetit in der exakten Medizin." *Zeitschrift für Sinnesphysiologie* 45 (1911): 433–59.

———. *Das Nahrungs-Bedürfnis, der Appetit und der Hunger: Eine diätetische Studie.* Leipzig: Johann Ambrosius Barth, 1913.

Stettler, Antoinette. "Zeichen lesen und Zeichen deuten: Zur Geschichte der Medizinischen Semiotik." *Gesnerus* 44 (1987): 33–54.

Steudel, Johannes. "Johannes Peter Müller." In Gillispie, *Dictionary of Scientific Biography,* 9:567–74.

Stiller, Berthold. "Die Pathologie des Appetits." *Archiv für Verdauungskrankheiten* 21 (1915): 23–34.

Stolberg, Michael. "'Abhorreas pinguedinem': Fat and Obesity in Early Modern Medicine, ca. 1500–1750." *Studies in History and Philosophy of Biological and Biomedical Sciences* 43 (2012): 370–78.

———. "A Woman's Hell? Medical Perceptions of Menopause in Preindustrial Europe." *Bulletin of the History of Medicine* 73, no. 3 (Fall 1999): 404–28.

Story, M., and J. E. Brown. "Do Young Children Instinctively Know What to Eat? The Studies of Clara Davis Revisited." *New England Journal of Medicine* 316, no. 2 (1987): 103–6.

Strominger, Jack L., and John R. Brobeck. "A Mechanism of Regulation of Food Intake." *Yale Journal of Biology and Medicine* 25, no. 5 (1953): 383–90.

Swazey, Judith P. "*Action propre* and *Action commune*: The Localization of Cerebral Function." *Journal of the History of Biology* 3, no. 2 (Autumn 1970): 213–34.

———. "Charles Scott Sherrington." In Gillispie, *Dictionary of Scientific Biography,* 12:395–403.

———. *Reflexes and Motor Integration: Sherrington's Concept of Integrative Action.* Cambridge, MA: Harvard University Press, 1969.

Talley, Nicholas J. "Dyspepsia and Non-Ulcer Dyspepsia: An Historical Perspective." *Medical Journal of Australia* 145 (December 1 and 15, 1986): 614–18.

Teich, Mikuláš. "Science and Food during the Great War: Britain and Germany." In Kamminga and Cunningham, *Science and Culture of Nutrition,* 213–34.

Teitelbaum, Philip. "Appetite." *Proceedings of the American Philosophical Society: Symposium on "Psychology: A Behavioral Reinterpretation"* 108, no. 6 (December 15, 1964): 464–72.

Temkin, Owsei. *Galenism: Rise and Decline of a Medical Philosophy.* Ithaca, NY: Cornell University Press, 1973.

———. *Hippocrates in a World of Pagans and Christians.* Baltimore: Johns Hopkins University Press, 1991.

Teuteberg, Hans Jürgen, and Günter Wiegelmann. *Der Wandel der Nahrungsgewohnheiten unter dem Einfluß der Industrialisierung*. Göttingen: Vandenhoeck and Ruprecht, 1972.

Thomas, Andrew J. "An Essay on Dyspepsia." Thesis, University of Pennsylvania, 1853.

Thomas, Joseph. *Le cancer*. Paris: A. Maloine, 1906.

Thomas, Keith. *Man and the Natural World: A History of the Modern Sensibility*. New York: Pantheon, 1983.

Thomas, Marion. "Rethinking the History of Ethology: French Animal Behavior Studies in the Third Republic (1870–1940)." PhD diss., University of Manchester, 2003.

Thompson, Claire, Steven Cummins, Tim Brown, and Rosemary Kyle. "What Does It Mean to Be a 'Picky Eater'?: A Qualitative Study of Food-Related Identities and Practices." *Appetite* 84 (2015): 235–39.

Thoms, Ulrike. *Anstaltskost im Rationalisierungsprozeß: Die Ernährung in Krankenhäusern and Gefängnissen im 18. und 19. Jahrhundert*. Stuttgart: Franz Steiner, 2005.

Thorndike, Edward L. *Animal Intelligence: Experimental Studies*. 1911. Introduction by Darryl Bruce. New Brunswick, NJ: Transaction, 2000.

Tiedemann, Friedrich. *Physiologie des Menschen*. Vol. 3, *Nahrungs-Bedürfniss, Nahrungs-Trieb und Nahrungs-Mittel des Menschen*. Darmstadt: Carl Wilhelm Leske, 1836.

Tiedemann, Friedrich, and Leopold Gmelin. *Recherches expérimentales, physiologiques et chimiques sur la digestion, considérée dans les quatre classes d'animaux vertébrés*. 2 vols. Translated by A.-J.-L. Jourdan. Paris: J.-B. Baillière, 1826–1827.

———. *Die Verdauung, nach Versuchen*. 2 vols. Heidelberg: Karl Groos, 1826–1827.

Tinbergen, Niko. "Ueber die Ernährung einer Waldohreulenbrut (Asio otus otus (L.))." *Beiträge zur Fortpflanzungsbiologie der Vögel* 8 (1932): 54–55.

Todes, Daniel P. "From the Machine to the Ghost Within: Pavlov's Transition from Digestive Physiology to Conditional Reflexes." *American Psychologist* 52 (1997): 947–55.

———. "Pavlov's Physiology Factory." *Isis* 88 (1997): 205–46.

———. *Pavlov's Physiology Factory: Experiment, Interpretation, Laboratory Enterprise*. Baltimore: Johns Hopkins University Press, 2002.

Torlais, Jean. *Un esprit encyclopédique en dehors de 'L'Encyclopédie': Réaumur*. 2nd ed. Paris: Blanchard, 1961.

Treitel, Corinna. "Max Rubner and the Biopolitics of Rational Nutrition." *Central European History* 41, no. 1 (2008): 1–25.

Trémolières, J., and J. Claudian. "La diététique des profondeurs (Essai historique)." *Rein et foie, maladies de la nutrition* 13, no. 13 (1971): 9–84.

Trifiletti, Arezzo di, P. Misino, P. Giannantoni, B. Giannantoni, A. Cascino, L. Fazi, F. Rossi Fanelli, et al. "Comparison of the Performance of Four Different Tools in Diagnosing Disease-Associated Anorexia and Their Relationship with Nutritional, Functional and Clinical Outcome Measures in Hospitalized Patients." *Clinical Nutrition* 32, no. 4 (August 2013): 527–32.

Trubek, Amy. "The Revolt against Homogeneity." In *Food in Time and Place: The American Historical Association Companion to Food History*, edited by Paul Freedman, Joyce E. Chaplin, and Ken Albala. Oakland: University of California Press, 2014.

Turner, Bryan S. "The Discourse of Diet." *Theory, Culture & Society* 1, no. 1 (1982): 23–32.

———. "The Government of the Body: Medical Regimens and the Rationalization of Diet." *British Journal of Sociology* 33, no. 2 (1982): 254–69.

Ulijaszek, Stanley J., and Hayley Lofink. "Obesity in Biocultural Perspective." *Annual Review of Anthropology* 35 (2006): 337–60.

Van der Eijk, Philip J. "Aristotle's Psycho-physiological Account of the Soul-Body Relationship." In *Psyche and Soma: Physicians and Metaphysicians on the Mind-Body Problem from Antiquity to the Enlightenment*, edited by John P. Wright and Paul Potter, 57–77. Oxford: Clarendon Press, 2000.

Vandereycken, Walter. "Can Eating Disorders Become 'Contagious' in Group Therapy and Specialized Inpatient Care?" *European Eating Disorders Review* 19 (2011): 289–95.

Vandereycken, Walter, Tilmann Habermas, Ron van Deth, and Rolf Meermann. "German Publications on Anorexia Nervosa in the Nineteenth Century." *International Journal of Eating Disorders* 10, no. 4 (1991): 473–90.

Vandereycken, Walter, and Ron van Deth. *From Fasting Saints to Anorexic Girls: The History of Self-Starvation*. New York: New York University Press, 1994.

Vanderlinden, Johan, and Walter Vandereycken. "Guidelines for the Family Therapeutic Approach to Eating Disorders." *Psychotherapy and Psychosomatics* 56, nos. 1–2 (1991): 36–42.

Van Deth, Ron, and Walter Vandereycken. "Eating Disorders: Clinical Section." In *A History of Clinical Psychiatry: The Origin and History of Psychiatric Disorders*, edited by G. E. Berrios and Roy Porter, 593–601. New York: New York University Press, 1995.

Venel, G.-F. "Digestion." In Diderot and d'Alembert, *Encyclopédie*, 4:999–1003.

Vernon, James. *Hunger: A Modern History*. Cambridge, MA: Harvard University Press, 2007.

Viaud, Gaston. "Taxies et tropismes dans le comportement instinctif." In Autuori et al., *L'instinct*, 5–49.

Vicq d'Azyr, Félix, and Jacques-Louis Moreau de la Sarthe, eds. *Encyclopédie méthodique: Médecine*. 13 vols. Paris: Panckoucke, 1787–1830. Available at the website of the Bibliothèque interuniversitaire de santé, at http://www.biusante.parisdescartes.fr.

Vidal, Fernando. *The Sciences of the Soul: The Early Modern Origins of Psychology*. Translated by Saskia Brown. Chicago: University of Chicago Press, 2011.

Vigarello, Georges. *Les métamorphoses du gras: Histoire de l'obésité du Moyen-Age au XXe siècle*. Paris: Seuil, 2010.

Vila, Anne C. *Enlightenment and Pathology: Sensibility in the Literature and Medicine of Eighteenth-Century France*. Baltimore: Johns Hopkins University Press, 1998.

———. "The Philosophe's Stomach: Hedonism, Hypochondria, and the Intellectual in Enlightenment France." In Forth and Carden-Coyne, *Cultures of the Abdomen*, 89–104.

———. *Suffering Scholars: Pathologies of the Intellectual in Enlightenment France*. Philadelphia: University of Pennsylvania Press, 2018.

Vincent, Swale. "The Arris and Gale Lecture on a Critical Examination of Current Views on Internal Secretion." *Lancet* 200, no. 5163 (August 12, 1922): 313–20.

Viner, Russell. "Putting Stress in Life: Hans Selye and the Making of Stress Theory." *Social Studies of Science* 29, no. 3 (June 1999): 391–410.

Wallace, J. Sim. "Some Principles of Dietetics." *Lancet* 185, no. 4789 (June 12, 1915): 1229–32.

Waller, John V., M. Ralph Kaufman, and Felix Deutsch. "Anorexia Nervosa: A Psychosomatic Entity." *Psychosomatic Medicine* 11, no. 1 (January 1940): 3–16.

Warner, John Harley. "Ideals of Science and their Discontents in Late Nineteenth-Century American Medicine." *Isis* 82, no. 3 (1991): 454–78.

Watson, John B. *Behaviorism*. New York: W. W. Norton, 1925.

———. "Psychology as the Behaviorist Views It." *Psychological Review* 20, no. 2 (1913): 158–77.

Weatherall, Mark. "Bread and Newspapers: The Making of 'A Revolution in the Science of Food.'" In Kamminga and Cunningham, *Science and Culture of Nutrition*, 179–212.

Weidman, Nadine M. *Constructing Scientific Psychology: Karl Lashley's Mind-Brain Debates*. New York: Cambridge University Press, 1999.

Weiner, Dora B. *The Citizen-Patient in Revolutionary and Imperial Paris*. Baltimore: Johns Hopkins University Press, 1993.

———. *Comprendre et soigner: Philippe Pinel (1745–1826), La médecine de l'esprit*. Paris: Fayard, 1999.

Weizsäcker, Viktor von. "Über Traume bei sogenannter endogener Magersucht." *Deutsche medizinische Wochenschrift* 63, nos. 7–8 (February 1937): 253–57, 294–97.

Whorton, James C. *Inner Hygiene: Constipation and the Pursuit of Health in Modern Society*. New York: Oxford University Press, 2000.

Williams, Elizabeth A. *A Cultural History of Medical Vitalism in Enlightenment Montpellier*. Aldershot, UK: Ashgate, 2003.

———. "Food and Feeling: 'Digestive Force' and the Nature of Morbidity in Vitalist Medicine." In *Vital Matters: Eighteenth-Century Views of Conception, Life, and Death*, edited by Helen Deutsch and Mary Terrall, 203–21. Toronto: University of Toronto Press, 2012.

———. "Gags, Funnels and Tubes: Forced Feeding of the Insane and of Suffragettes." *Endeavour* 32, no. 4 (2008): 134–40.

———. "Hysteria and the Court Physician in Enlightenment France." *Eighteenth-Century Studies* 35, no. 2 (2002): 247–55.

———. "Neuroses of the Stomach: Eating, Gender, and Psychopathology in French Medicine, 1800–1870." *Isis* 98 (2007): 54–79.

———. *The Physical and the Moral: Anthropology, Physiology, and Philosophical Medicine in France, 1750–1850*. New York: Cambridge University Press, 1994.

———. "Stomach and Psyche: Eating, Digestion, and Mental Illness in the Medicine of Philippe Pinel." *Bulletin of the History of Medicine* 84 (2010): 358–86.

———. "Of Two Lives One? Jean-Charles-Marguerite-Guillaume Grimaud and the Question of Holism in Vitalist Medicine." *Science in Context* 21, no. 4 (2008): 593–613.

Wilson, A. M., Jr. "Sensibility in France in the Eighteenth Century: A Study in Word History." *French Quarterly* 13 (1931): 35–46.

Wilson, Bee. *First Bite: How We Learn to Eat*. London: Fourth Estate, 2015.

Winick, Myron, ed. *Hunger Disease: Studies by the Jewish Physicians in the Warsaw Ghetto*. New York: John Wiley, 1979.

Wolf, Stewart. *Brain, Mind, and Medicine: Charles Richet and the Origins of Physiological Psychology*. New Brunswick, NJ: Transaction, 1993.

———. "The Psyche and the Stomach: A Historical Vignette." *Gastroenterology* 80, no. 3 (1981): 605–14.

Wolfe, Charles T., ed. "Vitalism without Metaphysics? Medical Vitalism in the Enlightenment." Special issue, *Science in Context* 21, no. 4 (December 2008).

Wolfe, Elin L., A. Clifford Barger, and Saul Benison. *Walter B. Cannon, Science and Society*. Cambridge, MA: Boston Medical Library in the Francis A. Countway Library of Medicine, distributed by Harvard University Press, 2000.

Wood, Cyrus B. "On Indigestion or Dyspepsia." Thesis, University of Pennsylvania, 1818.

Wood, T. B., and F. G. Hopkins. *Food Economy in War Time*. Cambridge: Cambridge University Press, 1916.

Woodward, William R. "Johannes Müller, Hermann Lotze, Jakob Henle und die Konstruktion des vegetativen Nervensystems." In *Johannes Müller und die Philosophie*, edited by Michael Hagner and Bettina Wahrig-Schmidt, 155–71. Berlin: Akademie Verlag, 1992.

Wundt, Wilhelm. *Grundzüge der physiologischen Psychologie*. Leipzig: Wilhelm Engelmann, 1874.

Young, Paul Thomas. "The Experimental Analysis of Appetite." *Psychological Bulletin* 38, no. 3 (March 1941): 129–64.

Young, Robert M. *Mind, Brain, and Adaptation in the Nineteenth Century: Cerebral Localization and its Biological Context from Gall to Ferrier*. Oxford, UK: Clarendon Press, 1970.

Zanobio, Bruno, and Gianluigi Porta. "Luigi Luciani." In Gillispie, *Dictionary of Scientific Biography*, 8:535–36.

Zedler, Johann Heinrich, ed. *Grosses vollständiges Universal-Lexicon aller Wissenschafften und Künste*. 64 vols. and 4 supplements. Leipzig: Zedler, 1731–1754. http://www.zedler-lexikon.de/index.html?c.

Zweiniger-Bargielowska, Ina. "The Culture of the Abdomen: Obesity and Reducing in Britain, circa 1900–1939." *Journal of British Studies* 44, no. 2 (April 2005): 239–73.

Index

Page numbers in italics refer to figures.